PRINCIPLES OF OPTICAL ENGINEERING

PRINCIPLES OF OPTICAL ENGINEERING

FRANCIS T. S. YU
IAM-CHOON KHOO
Department of Electrical Engineering
The Pennsylvania State University

JOHN WILEY & SONS
NEW YORK CHICHESTER BRISBANE TORONTO SINGAPORE

Library of Congress Cataloging-in-Publication Data

Yu, Francis T. S., 1934–
Principles of optical engineering/Francis T. S. Yu, Iam-Choon Khoo.
Includes bibliographical references.
ISBN 0-471-60567-0
1. Optical instruments. 2. Optics. I. Khoo, Iam-Choon.
II. Title.
TS513.Y8 1990
621.36—dc20

90-11916
CIP

Printed in the United States of America

10 9 8 7 6 5 4 3 2

To Our Parents

PREFACE

Optics, in conjunction with electronics, has played an ever-increasing role in a wide range of applications encompassing uses in the fundamental sciences as well as in practical optical systems. This trend has been due mostly to the discovery and rapid development of versatile lasers covering the ultraviolet to far-infrared regions, and a parallel and perhaps more expansive development of solid-state and optical materials. The introduction of optical fibers and semiconductor lasers in communication has also stimulated a profound relationship between optics and communication engineering. There are more electrical engineers working on the research and development of modern optical systems than ever before, and the number will continue to increase in the years to come. It is our purpose in writing this book to introduce undergraduates to the basic concepts in modern optics and to fill the gap for the interested students advancing to higher-level electrooptics courses. The contents of this book were designed mainly for junior- and senior-level engineering students.

This book, when in manuscript form, was used as undergraduate course notes, and the materials were chosen to fit the needs of first-term junior electrical engineering students. However, the book may also be used in other disciplines. It contains ten chapters, the first of which deals with linear system transforms. The discussion of these transforms will serve as a background to the electrooptics system approach. Chapters 2 and 3 review the important fundamental concepts of classical optics. Chapter 4 discusses some of the existing photodetectors and electrooptic devices. In Chapter 5 we present the electromagnetic nature of light, in Chapter 6 the principles of diffraction optics, and in Chapter 7 basic principles of lasers. Chapter 8 introduces the phenomenon of wavefront reconstruction, and Chapter 9 reviews the basic systems for optical signal processing. Chapter 10 serves as an introduction to the fundamentals of optical fibers. We have found that it is possible to teach the whole book without significant omission in a full-semester course. The contents, however, are not intended to cover the vast domain of modern engineering optics. Rather, the book comprises fundamental materials that we believe are important and useful to undergraduate electrical engineering students.

Optical engineering is at the threshold of being applied on a widely extending, unprecedented scale. We believe that this book will serve as an accurate, up-to-date introductory text to more advanced texts or courses in the exciting field of electro-optics. In view of the vast number of contributions in this field, we apologize in advance for the possible (and inevitable) omission of some appropriate references in various chapters of this book.

We are indebted to our undergraduate students for their active participation, feedback, and criticism over the years. We are also indebted to Mrs. Debby Pruger for her patient typing of the manuscript. Most of all, we are grateful to the unwavering support of our families.

FRANCIS T. S. YU
IAM-CHOON KHOO

January 1990
University Park, Pennsylvania

CONTENTS

PRINCIPLES OF OPTICAL ENGINEERING

1 LINEAR SYSTEM TRANSFORMS

Generally speaking, an optical system can be represented by an input–output block diagram. In most instances the concept of the system theory can be easily applied to optical imaging and processing systems. There are, however, no general techniques to solve nonlinear optical systems, except for a few special cases. Nonlinear problems are usually solved by means of graphical and approximation procedures. Although no optical system is strictly linear, a great number of optical systems can be treated as linear, and these can be approached by linear system analysis.

The analysis of optical systems usually involves linear spatially invariant concepts. It is the aim of this chapter to introduce some of the fundamentals. Since most of the optical systems can be described by two-dimensional spatial variables, we shall, in various parts of this book, use two-dimensional notation.

1.1 LINEAR SPATIALLY INVARIANT SYSTEM

It is well known that the behavior of a physical system may be described by the relationship of the output response to the input excitation, as shown in Figure 1.1. We assume that the output response and the input excitation are measurable quantities. Let us say that the output response $g_1(x, y)$ is caused by an input excitation $f_1(x, y)$, and a second response $g_2(x, y)$ is produced by $f_2(x, y)$. The input–

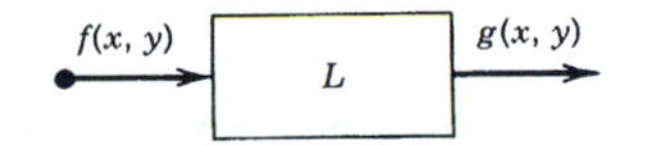

FIGURE 1.1 Block diagram representation of a physical system.

output relationships are written in the following forms,

$$f_1(x, y) \rightarrow g_1(x, y) \tag{1.1}$$

and

$$f_2(x, y) \rightarrow g_2(x, y), \tag{1.2}$$

where (x, y) represent a two-dimensional spatial coordinate system.

If the sum of these two input excitations are applied to the input of the physical system, the corresponding output response would be a linear combination of $g_1(x, y)$ and $g_2(x, y)$, that is,

$$f_1(x, y) + f_2(x, y) \rightarrow g_1(x, y) + g_2(x, y). \tag{1.3}$$

Equation 1.3, in conjunction with Eqs. 1.1 and 1.2, represents the well-known *additivity* property of a linear system, which is also known as the *superposition* property. In other words, a necessary condition for a physical system to be linear is that the *principle of superposition* be held. The principle of superposition implies that the presence of one excitation in the system does not affect the response caused by other excitations.

If we multiply the input excitation by an arbitrary constant K, that is, $Kf_1(x, y)$, and the output response is equal to $Kg_1(x, y)$,

$$Kf_1(x, y) \rightarrow Kg_1(x, y), \tag{1.4}$$

then the system is said to posses the *homogeneity* property. In other words, a linear system is capable of preserving the scale factor of an input excitation. Thus, if a physical system is to be a linear system, the system must possess the additivity and homogeneity properties given in Eqs. 1.3 and 1.4.

There is, however, another important physical aspect that characterizes a linear system with constant parameters. This physical property is *spatial invariance*, which is analogous to the *time invariance* of an electrical network. If the output response of a physical system remains unaltered (except with appropriate translation) with respect to the input excitation, the physical system is said to possess the spatially invariant property. The qualification of a spatially invariant system is that if

$$f(x, y) \rightarrow g(x, y),$$

then

$$f(x - x_0, y - y_0) \rightarrow g(x - x_0, y - y_0), \tag{1.5}$$

where $f(x, y)$ and $g(x, y)$ are the corresponding input excitation and output response, respectively, and x_0 and y_0 are some arbitrary constants.

Thus, if a linear system possesses the spatially invariant property of Eq. 1.5, the system is known as a *linear spatially invariant system.*

We note that the concept of linear spatial invariance is rather important in the analysis of the electrooptic systems. This property would simplify otherwise rather complicated formulations.

Example 1.1

Given an ideal linear-phase low-pass filter, the amplitude and phase distributions are

$$A(\omega) = \begin{cases} A, & |\omega| \leq |\omega_c| \\ 0, & |\omega| > |\omega_c| \end{cases}$$

and

$$\phi(\omega) = -t_0\omega,$$

where ω is in radians per second and t_0 is an arbitrary positive constant. To show the linearity of this filter, we let the input signals $f_1(t)$ and $f_2(t)$ be

$$f_1(t) = A \sin \omega_1 t, \qquad |\omega_1| \leq |\omega_c|$$

and

$$f_2(t) = B \cos \omega_2 t, \qquad |\omega_2| \leq |\omega_c|.$$

The corresponding output responses can be shown as

$$g_1(t) = A^2 \sin \omega_1 (t - t_0)$$

and

$$g_2(t) = AB \cos \omega_2 (t - t_0).$$

To prove that the filter is a linear filter, we would let the input signal be the sum of $K_1 f_1(t)$ and $K_2 f_2(t)$,

$$\begin{aligned} f(t) &= K_1 f_1(t) + K_2 f_2(t) \\ &= K_1 A \sin \omega_1 t + K_2 B \cos \omega_2 t, \end{aligned}$$

where K_1 and K_2 are arbitrary constants. The corresponding output response is therefore

$$\begin{aligned} g(t) &= K_1 A^2 \sin \omega_1 (t - t_0) + K_2 AB \cos \omega_2 (t - t_0) \\ &= K_1 g_1(t) + K_2 g_2(t). \end{aligned}$$

From this equation we see that the filter possesses the properties of homogeneity and additivity. It is therefore a linear filter. ■

Example 1.2

If the input excitation of Example 1.1 is delayed by a time factor of τ,

$$f_1'(t) = A \sin \omega_1(t - \tau) = f_1(t - \tau),$$

the output response can be shown as

$$g_1'(t) = A^2 \sin \omega_1(t - \tau - t_0) = g_1(t - \tau),$$

which is delayed by the same τ. Thus, the ideal low-pass filter is a linear time-invariant filter.

Notice that the same linearity concept can be applied in the spatial domain as in the time domain. ■

One of the apparent distinctions of a linear spatially invariant system is that the system's *transfer function* can be uniquely described by the *spatial impulse response.* In contrast with the temporal (i.e., time) impulse response $h(t)$, the spatial impulse response $h(x, y)$ describes a function of two spatial variables x and y. In other words, if we assume that an impulse excitation $\delta(x, y)$ is applied to a linear spatially invariant system, as shown in Figure 1.2, then the output response must be the spatial impulse response of the system. The important aspect of a spatial impulse response of a linear spatially invariant system is that the system response can be uniquely described by an impulse response. The Fourier transform of a spatial impulse response describes the *system transfer function* $H(p, q)$,

$$H(p, q) = \iint_{-\infty}^{\infty} h(x, y) \exp[-i(px + qy)]\, dx\, dy, \tag{1.6}$$

where p and q are the angular spatial-frequency variables, in radians per unit distance, as opposed to the temporal frequency ω, in radians per unit time. The concept of Fourier transformation will be discussed in the following sections.

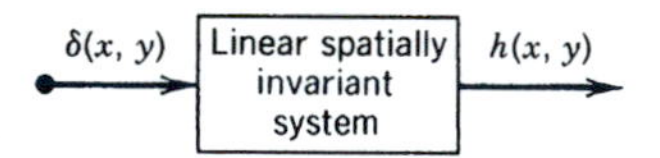

FIGURE 1.2 Description of a spatial impulse response.

Example 1.3

If the input excitation of the ideal low-pass filter of Example 1.1 is $\delta(t)$, with

$$\delta(t) = \begin{cases} \infty, & t = 0 \\ 0, & \text{otherwise,} \end{cases}$$

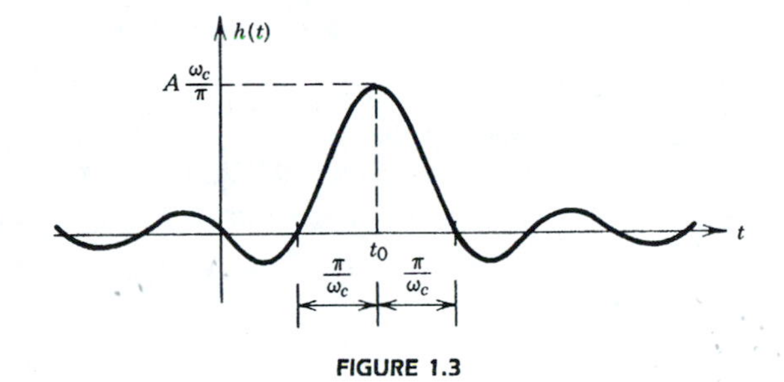

FIGURE 1.3

then the impulse response of the filter can be shown as

$$h(t) = \frac{A\omega_c}{\pi} \frac{\sin \omega_c(t - t_0)}{\omega_c(t - t_0)},$$

as sketched in Figure 1.3. ■

■ Example 1.4

Let the spatial impulse response of a linear optical system be

$$h(x, y) = \begin{cases} 1, & |x| \leq |a|, \quad |y| \leq |a| \\ 0, & |x| > |a|, \quad |y| > |a|, \end{cases}$$

where a is an arbitrary constant. Determine the system transfer function of the optical system.

To evaluate the system transfer function, we would substitute the spatial impulse response of $h(x, y)$ into Eq. 1.6,

$$H(p, q) = \int_{-a}^{a} \int_{-a}^{a} h(x, y) \exp[-i(px + qy)]\, dx\, dy,$$

which can be shown as

$$H(p, q) = 4a^2 \frac{\sin ap}{ap} \frac{\sin aq}{aq}.$$ ■

1.2 FOURIER TRANSFORMATION AND FOURIER SPECTRUM

In the analysis of a linear spatially invariant optical system, the use of the spatial frequency domain and Fourier transformation are critically important. In other words, the analysis can conveniently be carried out in the frequency domain with Fourier transformation.

Let us consider a complex function $f(x, y)$ that satisfied the following sufficient conditions.

1. $f(x, y)$ must be sectionally continuous, but with a finite number of discontinuities in every finite region of the (x, y) domain,
2. $f(x, y)$ must be absolutely integrable over the spatial domain (x, y), that is,

$$\iint_{-\infty}^{\infty} |f(x, y)|\, dx\, dy < \infty. \tag{1.7}$$

Then $f(x, y)$ can be written as

$$f(x, y) = \frac{1}{4\pi^2} \iint_{-\infty}^{\infty} F(p, q) \exp[+i(px + qy)]\, dp\, dq, \tag{1.8}$$

where

$$F(p, q) = \iint_{-\infty}^{\infty} f(x, y) \exp[-i(px + qy)]\, dx\, dy, \tag{1.9}$$

and (p, q) is the angular spatial-frequency coordinate system, in radians per unit distance.

Equations 1.8 and 1.9 are the well-known *Fourier transform pair*; Eq. 1.9 is called the *direct Fourier transform* and Eq. 1.8 the *inverse Fourier transform*. Equations 1.7 and 1.8 can be written in the briefer forms

$$f(x, y) = \mathscr{F}^{-1}[F(p, q)] \tag{1.10}$$

$$F(p, q) = \mathscr{F}[f(x, y)], \tag{1.11}$$

where $\mathscr{F}^{-1}$ and $\mathscr{F}$ denote the inverse and direct Fourier transformations, respectively.

We note that $F(p, q)$ is, in general, a complex quantity. It can be represented by an amplitude and a phase factor, that is,

$$F(p, q) = |F(p, q)| \exp[-i\phi(p, q)], \tag{1.12}$$

where $|F(p, q)|$ is generally referred to as the *amplitude spectrum*, $\phi(p, q)$ is known as the *phase spectrum*, and $F(p, q)$ is called the *Fourier spectrum* or *spatial-frequency spectrum* of $f(x, y)$.

It is, however, interesting to show that a Fourier transform is a *linear transformation*. The linearity of a mathematical transformation is essentially that of a linear system; namely it has the same additivity and homogeneity properties. For example, given that C_1 and C_2 are two arbitrary complex constants and that

$f_1(x, y)$ and $f_2(x, y)$ are Fourier-transformable functions, that is,

$$\mathscr{F}[f_1(x, y)] = F_1(p, q) \tag{1.13}$$

and

$$\mathscr{F}[f_2(x, y)] = F_2(p, q) \tag{1.14}$$

we can show that

$$\begin{aligned}\mathscr{F}[C_1 f_1(x, y) + C_2 f_2(x, y)] &= \iint_{-\infty}^{\infty} [C_1 f_1(x, y) + C_2 f_2(x, y)] \exp[-i(px + qy)]\, dx\, dy \\ &= C_1 \iint f_1(x, y) \exp[-i(px + qy)]\, dx\, dy \\ &\quad + C_2 \iint f_2(x, y) \exp[-i(px + qy)]\, dx\, dy \\ &= C_1 F_1(p, q) + C_2 F_2(p, q),\end{aligned} \tag{1.15}$$

which has the properties of additivity and homogeneity. Thus, the Fourier transformation is, in fact, a linear transformation.

Engineering students are made familiar with the concept of time frequency, but spatial frequency is rarely mentioned in most engineering courses. It is our aim to elaborate the space and spatial-frequency concepts.

In optical signal processing in particular, image functions are generally represented by a two-dimensional spatial variable. For example, the complex quantity of a wave field can be described by an (x, y) orthogonal coordinate system, that is, $f(x, y)$. The corresponding Fourier transform $F(p, q)$ represents another complex wave field that describes a spatial frequency domain (p, q), rather than a temporal frequency domain, in cycles per second. Spatial frequency is, in fact, a concept taken directly from temporal frequency.

Example 1.5

Consider a two-dimensional spatial Fourier transformer, as shown in Figure 1.4. If the input spatial function is given as

$$f(x, y) = \sin(20\pi x) \cos(30\pi y),$$

determine the corresponding output Fourier transform.

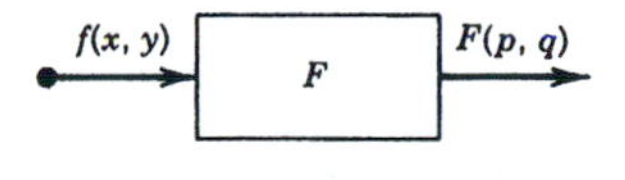

FIGURE 1.4

By substituting $f(x, y)$ in Eq. 1.9, we have

$$F(p, q) = \iint_{-\infty}^{\infty} \sin(20\pi x)\cos(30\pi y)\exp[-i(px + qy)]\,dx\,dy.$$

The preceding equation can be written as

$$F(p, q) = \int_{-\infty}^{\infty} \sin(20\pi x)e^{-ipx}\,dx \int_{-\infty}^{\infty} \cos(30\pi y)e^{-iqy}\,dy.$$

Since

$$\sin\theta \triangleq \frac{e^{i\theta} - e^{-i\theta}}{2i} \qquad \text{and} \qquad \cos\theta \triangleq \frac{e^{i\theta} + e^{-i\theta}}{2}$$

the equation just given becomes

$$F(p, q) = \left\{\frac{1}{2i}\int_{-\infty}^{\infty} \exp[-i(p - 20\pi)x]\,dx - \frac{1}{2i}\int_{-\infty}^{\infty} \exp[-i(p + 20\pi)x]\,dx\right\}$$
$$\times \left\{\frac{1}{2}\int_{-\infty}^{\infty} \exp[-i(q - 30\pi)y]\,dy + \frac{1}{2}\int_{-\infty}^{\infty} \exp[-i(q + 30\pi)y]\,dy\right\}.$$

This equation can be further reduced to

$$F(p, q) = \frac{1}{4i}[\delta(P - 20\pi) - \delta(P + 20\pi)][\delta(q - 30\pi) + \delta(q + 30\pi)]$$
$$= \frac{1}{4i}[\delta(p - 20\pi, q - 30\pi) - \delta(q + 20\pi, q - 30\pi)$$
$$+ \delta(p - 20\pi, q + 30\pi) - \delta(q + 20\pi, q + 30\pi)],$$

where δ denotes the Dirac delta function, as will be defined in Section 1.3. ■

■ Example 1.6

Show that the inverse Fourier transformation of Eq. 1.8 is also a linear transformation. For simplicity of notation, we let

$$f(x, y) = \mathscr{F}^{-1}[F(p, q)].$$

We assume that

$$f_1(x, y) = \mathscr{F}^{-1}[F_1(p, q)]$$

and

$$f_2(x, y) = \mathscr{F}^{-1}[F_2(p, q)].$$

To show that $\mathscr{F}^{-1}$ is a linear transformation, we have

$$\mathscr{F}^{-1}[C_1F_1(p,q) + C_2F_2(p,q)] = C_1\mathscr{F}^{-1}[F_1(p,q)] + C_2\mathscr{F}^{-1}[F_2(p,q)]$$
$$= C_1f_1(x,y) + C_2f_2(x,y).$$

Since the inverse Fourier transformation possesses the homogeneity and the additivity properties, the inverse Fourier transformation is a linear transformation. ■

■ Example 1.7

Calculate the spatial-frequency content of $f(x,y)$, which is referred to in Example 1.5.

By inspecting the corresponding Fourier spectrum of $F(p,q)$, we can identify the spatial-frequency content of $f(x,y)$ as

$$f_x = 10, \quad f_y = 15;$$
$$f_x = -10, \quad f_y = 15;$$
$$f_x = 10, \quad f_y = -15;$$
$$f_x = -10, \quad f_y = -15,$$

which represent four discrete spectra points; variables f_x and f_y are the spatial frequencies in cycles per unit distance. ■

1.3 DIRAC DELTA FUNCTION

We shall now define a two-dimensional Dirac delta function $\delta(x,y)$, as described in the (x,y) spatial domain. Similar to the temporal Dirac delta function $\delta(t)$, $\delta(x,y)$ is defined as

$$\delta(x,y) \triangleq \begin{cases} \infty, & \text{for } x = y = 0 \\ 0, & \text{otherwise} \end{cases} \tag{1.16}$$

and

$$\iint_{-\infty}^{\infty} \delta(x,y)\,dx\,dy = 1. \tag{1.17}$$

Thus, for $\delta(x - x_0, y - y_0)$, it is defined as

$$\delta(x - x_0, y - y_0) \triangleq \begin{cases} \infty, & \text{for } x = x_0, \quad y = y_0 \\ 0, & \text{otherwise} \end{cases} \tag{1.18}$$

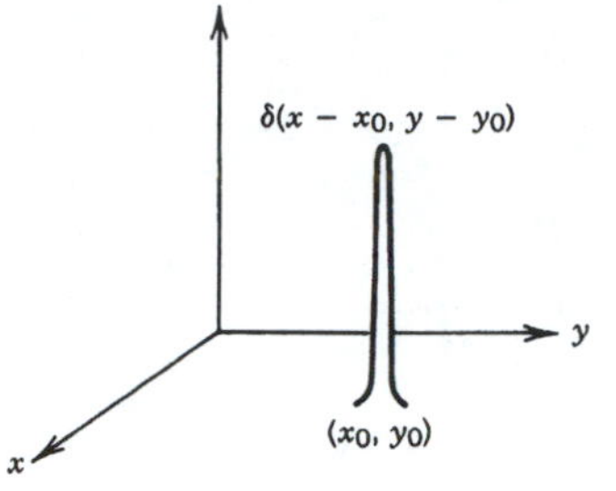

FIGURE 1.5 Sketch of $\delta(x - x_0, y - y_0)$.

and

$$\iint_{-\infty}^{\infty} \delta(x - x_0, y - y_0)\, dx\, dy = 1, \tag{1.19}$$

where x_0 and y_0 are arbitrary constants. In other words, $\delta(x - x_0, y - y_0)$ exists only at (x_0, y_0), and has zero value over the x, y plane elsewhere, as sketched in Figure 1.5.

With reference to Eq. 1.9, the Fourier transformation of the delta function can be obtained as follows:

$$F(p, q) = \mathscr{F}[\delta(x - x_0, y - y_0)] = \exp[-i(px_0 + qy_0)]. \tag{1.20}$$

The amplitude and phase spectra of a delta function can be identified as

$$|F(p, q)| = 1 \tag{1.21}$$

and

$$\phi(p, q) = -(px_0 + qy_0). \tag{1.22}$$

Equation 1.21 shows the amplitude spectral distribution, which is extended uniformly with a unit height over the spatial-frequency plane, as can be seen in Figure 1.6*a*. And Eq. 1.22 shows a linear-phase spectral distribution over the spatial frequency domain, as shown in Figure 1.6*b*.

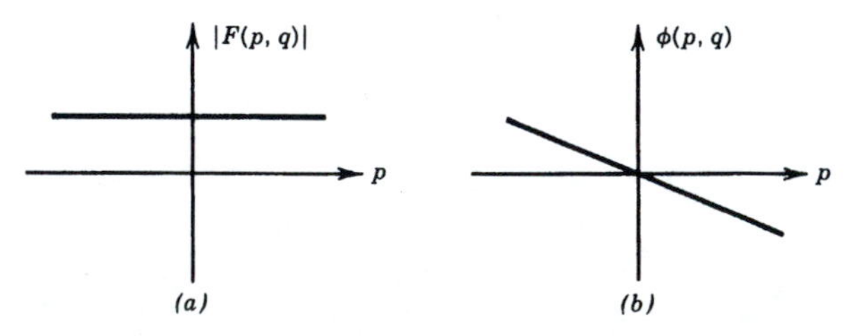

FIGURE 1.6 One-dimensional representation of a Fourier spectrum.

Example 1.8

Prove that the following relationship is true:

$$\iint_{-\infty}^{\infty} \delta(x, y) f(x, y)\, dx\, dy = f(0, 0).$$

Equation 1.16 indicates that $\delta(x, y)$ exists only at $x = 0$ and $y = 0$, that is,

$$\delta(x, y) = \begin{cases} \infty, & \text{for } x = y = 0 \\ 0, & \text{otherwise.} \end{cases}$$

Thus, the integral equation can be written as

$$\iint_{-\infty}^{\infty} \delta(x, y) f(x, y)\, dx\, dy = f(x, y)\Big|_{\substack{x=0\\y=0}} \iint_{-\varepsilon}^{\varepsilon} \delta(x, y)\, dx\, dy.$$

Using Eq. 1.17, we have proved that

$$\iint_{-\infty}^{\infty} \delta(x, y) f(x, y)\, dx\, dy = f(0, 0)$$

is true. ■

Example 1.9

Show that the relationship

$$\iint_{-\infty}^{\infty} \delta(x - x_0, y - y_0) f(x, y)\, dx\, dy = f(x_0, y_0)$$

holds, where $f(x, y)$ is assumed continuous at (x_0, y_0). By replacing the following independent variables, $x' = x - x_0$ and $y' = y - y_0$, we can write the integration equation as

$$\iint_{-\infty}^{\infty} \delta(x - x_0, y - y_0) f(x, y)\, dx\, dy = \iint_{-\infty}^{\infty} \delta(x', y') f(x' + x_0, y' + y_0)\, dx'\, dy'.$$

Since

$$\iint_{-\infty}^{\infty} \delta(x, y) f(x, y)\, dx\, dy = f(0, 0),$$

the preceding equation is reduced to

$$\iint_{-\infty}^{\infty} \delta(x - x_0, y - y_0) f(x, y)\, dx\, dy = f(x + x_0, y + y_0)\Big|_{\substack{x=0\\y=0}}$$

$$= f(x_0, y_0).$$ ■

1.4 CONVOLUTION AND CORRELATION

1.4.1 Convolution

The concept of convolution can be easily illustrated by a block diagram of an input–output linear system, like that shown in Figure 1.7. Let us assume that a band-limited input excitation $f(x)$ is applied at the input end of a linear system. Since $f(x)$ is band-limited, $f(x)$ can be approximated by a sampling function, such as

$$f(x) = \sum_{n=-N/2}^{N/2} f(x)\, \delta(x - n\, \Delta x), \tag{1.23}$$

where N is the total sampling points, $\delta(x)$ denotes the Dirac delta function, and Δx is the sampling distance known as the *Nyquist sampling interval*. The Nyquist sampling interval can be obtained from the sampling frequency, that is,

$$\Delta x = \frac{1}{f_s}, \tag{1.24}$$

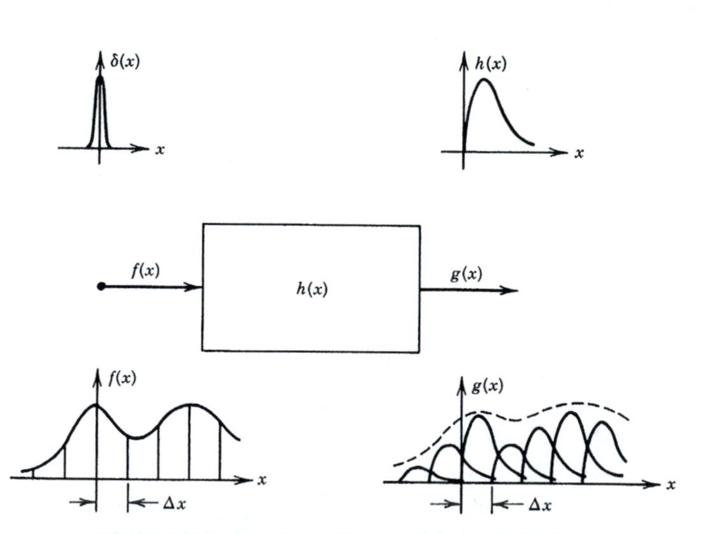

FIGURE 1.7 Response from a linear spatially invariant system.

where f_s is *Shannon's sampling frequency.* The sampling frequency satisfies the following inequality,

$$f_s \geq 2f_m, \tag{1.25}$$

where f_m is the highest frequency limit of input excitation $f(x)$.

Referring to Eq. 1.23, we see that as the sampling interval Δx approaches zero, the total sampling point N approaches infinity. Equation 1.23 then converges on the well-known *convolution integral*, which is,

$$f(x) = \lim_{\substack{\Delta x \to 0 \\ N \to \infty}} \sum_{-N/2}^{N/2} f(x)\,\delta(x - n\,\Delta x) = \int_{-\infty}^{\infty} f(x')\,\delta(x - x')\,dx'. \tag{1.26}$$

Thus, we see that convolution of a function $f(x)$ with a delta function yields the function itself.

Since the block diagram system is linear and spatially invariant, the output excitation is

$$g(x) = \sum_{-N/2}^{N/2} f(x)h(x - n\,\Delta x), \tag{1.27}$$

as sketched in Figure 1.7, where $h(x)$ is the spatial impulse response of the system. Again, we show that, as $\Delta x \to 0$, $N \to \infty$, Eq. 1.27 will converge on the following convolution integral:

$$g(x) = \lim_{\substack{\Delta x \to 0 \\ N \to \infty}} \sum_{-N/2}^{N/2} f(x)h(x - n\,\Delta x) = \int_{-\infty}^{\infty} f(x')h(x - x')\,dx'. \tag{1.28}$$

To simplify this equation, we write

$$g(x) = f(x) * h(x),$$

where the asterisk denotes the convolution operation. Thus, for a linear, spatially invariant system, the output response is the convolution of the input excitation with respect to the spatial impulse response of the system.

A simple example of the convolution integral is shown in the sketches in Figure 1.8. Figures 1.8*a* and *b* show the functions $f(x')$ and $h(x')$ as defined in the x' spatial domain. The function $h(-x')$, defined in Figure 1.8*c*, is the image of $h(x')$ with respect to the vertical axis at $x' = 0$. The function $h(x - x')$ represents the translation of $h(-x')$ on the x' axis for a given x, as shown in Figure 1.8*d*. We further note that since x is a spatial variable, then $h(x - x')$ would translate over the x' domain. The product of $f(x')h(x - x')$, for a given x, is plotted in Figure 1.8*e*, and the shaded area, bounded by $f(x')h(x - x')$, represents the convolution integral,

$$\int_{-\infty}^{\infty} f(x')h(x - x')\,dx',$$

for a given x.

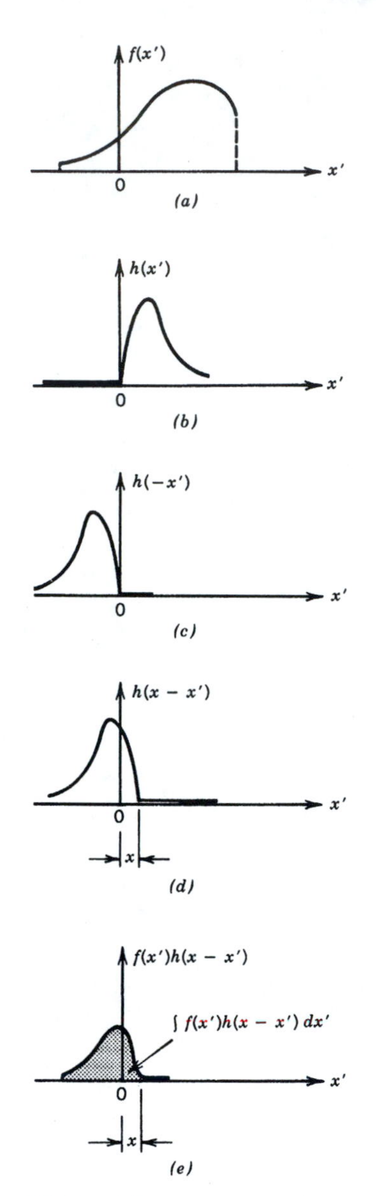

FIGURE 1.8 The concept of convolution.

Example 1.10

Given two arbitrary functions as depicted in Figures 1.9*a* and *b*, evaluate the convolution integral of $f_1(x)$ and $f_2(x)$. Let us demonstrate the usefulness of the graphical approach to the convolution integral. Figures 1.9*c* and *d* show the appropriate function $f_2(x)$ convoluting against $f_1(x)$, and the shaded areas represent the con-

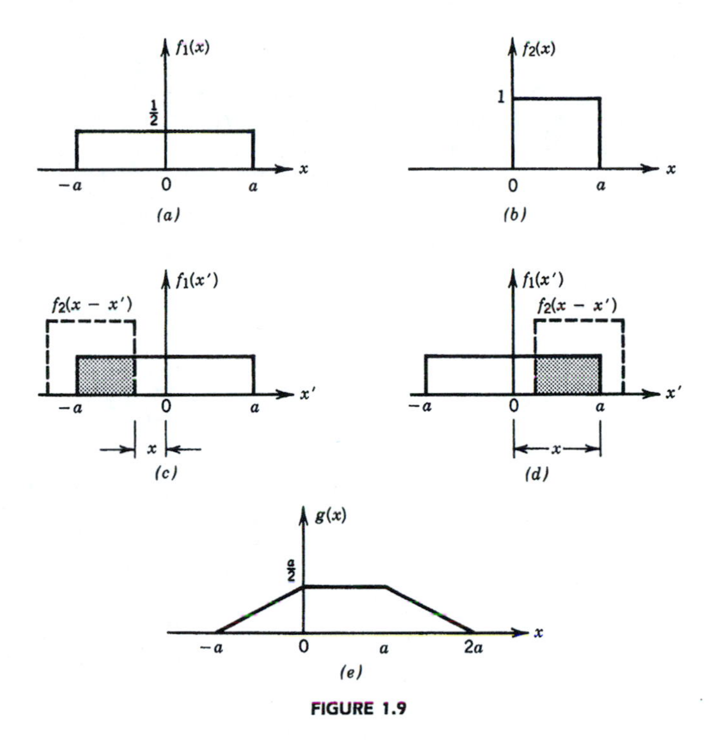

FIGURE 1.9

volution of these two functions for a given x. Function $g(x)$ represents the result of the convolution integral as depicted in Figure 1.9*e*. ■

■ Example 1.11

Show that the convolution of a function $f(x)$ against $\delta(x - x_0)$ is equal to $f(x - x_0)$, that is,

$$\int_{-\infty}^{\infty} f(x')\,\delta(x' - x + x_0)\,dx' = f(x - x_0).$$

If we replace $x - x_0$ by τ, the convolution integral becomes

$$\int_{-\infty}^{\infty} f(x')\,\delta(x' - \tau)\,dx'.$$

Since the convolution of any function with a delta function yields the function itself, we conclude that

$$\int_{-\infty}^{\infty} f(x')\,\delta(x' - \tau)\,dx' = f(\tau) = f(x - x_0).$$ ■

1.4.2 Correlation

The concept of correlation can also be interpreted by an input–output linear system, as illustrated in Figure 1.10, where $h'(x) = h(-x)$ represents the spatial impulse response of the system.

Since the output response can be written as

$$g(x) = \sum_{-N/2}^{N/2} f(x)h'(x + n\,\Delta x) = \sum_{-N/2}^{N/2} f(x)h(-x - n\,\Delta x) \tag{1.29}$$

and then, because $\Delta x \to 0$ and $N \to \infty$, we show that

$$g(x) = \lim_{\substack{\Delta x \to 0 \\ N \to \infty}} \sum_{-N/2}^{N/2} f(x)h(-x - n\,\Delta x) = \int_{-\infty}^{\infty} f(x')h(x + x')\,dx'. \tag{1.30}$$

The preceding integral equation is known as the *correlation integral* of functions $f(x)$ and $h(x)$. The operation given in Eq. 1.30 can be represented by the sketches in Figure 1.11. Figures 1.11*a* and *b* show the functional distribution of $f(x')$ and $h(x')$, and Figure 1.11*c* illustrates the translation of $h(x')$ over the x' axis. Figure 1.11*d* shows the functional distribution of the product of $f(x')$ and $h(x + x')$, and the shaded area in this figure represents the correlation integral for a given x. It is beneficial for the reader to comprehend the concept of autocorrelation, because its operation plays a key role in the extraction of the signal from random noise (e.g., radar detection). To illustrate the autocorrelation operation, we assume that the impulse response in Figure 1.11 is

$$h'(x) = f(-x).$$

Then the convolution integral of Eq. 1.30 becomes

$$R_{11}(x) = \int_{-\infty}^{\infty} f(x')f(x' + x)\,dx', \tag{1.31}$$

where $R_{11}(x)$ denotes the *autocorrelation function*, and the integral is known as the *autocorrelation integral*. To simplify Eq. 1.31, we write it as

$$R_{11}(x) = f(x) \circledast f(x),$$

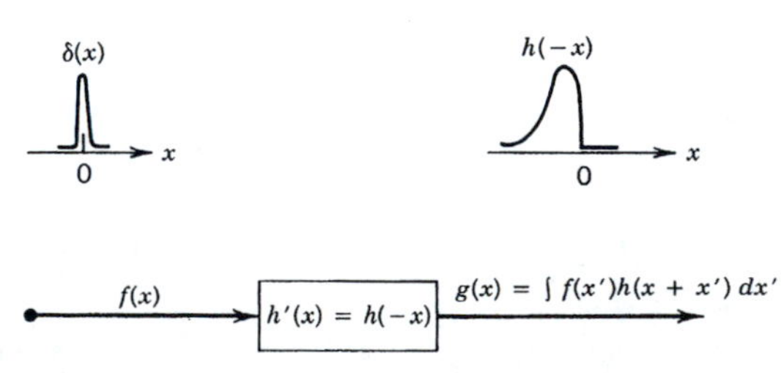

FIGURE 1.10 Convolution operation $h'(x)$ represents the spatial impulse response.

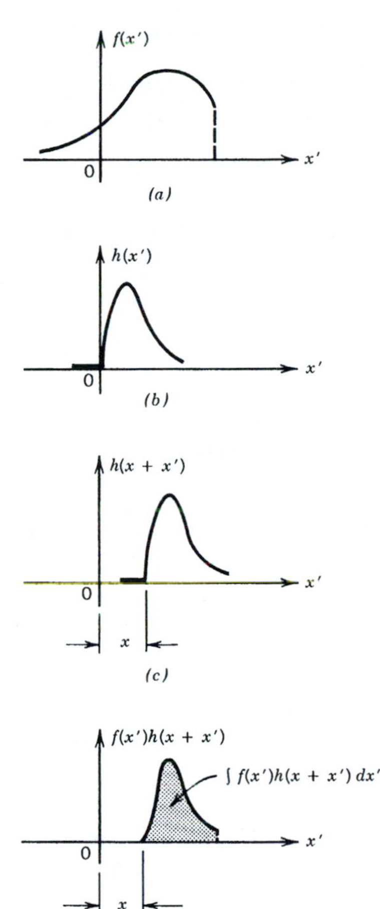

FIGURE 1.11 The concept of correlation.

with ⊛ denoting the correlation operation. Similarly, if $h'(x) = u(-x)$, where $u(x)$ is any arbitrary function other than $f(x)$, Eq. 1.30 becomes

$$R_{12}(x) = \int_{-\infty}^{\infty} f(x')u(x' + x)\,dx', \tag{1.32}$$

which is the *cross-correlation function* of $f(x)$ and $u(x)$; the subscripts 1 and 2 denote the cross-correlation operation. Equation 1.32 is known as the *cross-correlation integral* and can be simplified to read

$$R_{12}(x) = f(x) \circledast u(x). \tag{1.33}$$

PROPERTIES OF AN AUTOCORRELATION FUNCTION

It is worthwhile to state some of the basic properties of an autocorrelation function:

$$R_{11}(x) = R_{11}(-x) \tag{1.34}$$

$$R_{11}(0) > 0 \tag{1.35}$$

$$R_{11}(0) \geq R_{11}(x). \tag{1.36}$$

To show that the autocorrelation function is an even function, we utilize the definition of Eq. 1.31,

$$R_{11}(-x) = \int_{-\infty}^{\infty} f(x')f(x' - x)\,dx'.$$

By letting $x' - x = \alpha$, we have

$$R_{11}(-x) = \int_{-\infty}^{\infty} f(\alpha + x)f(\alpha)\,d\alpha = R_{11}(x).$$

To show that $R_{11}(0)$ is a positive quantity, we use the definition of $R_{11}(x)$,

$$R_{11}(x) = \int_{-\infty}^{\infty} f(x')f(x' + x)\,dx'.$$

Then, with $x = 0$, we have

$$R_{11}(0) = \int_{-\infty}^{\infty} f^2(x')\,dx'.$$

Since $f^2(x)$ is a nonnegative quantity, we conclude that

$$R_{11}(0) \geq 0.$$

To show that $R_{11}(0) \geq R_{11}(x)$ is true, we note that

$$\int_{-\infty}^{\infty} [f(x') - f(x' + x)]^2\,dx' \geq 0.$$

Expanding this expression, we have

$$\int_{-\infty}^{\infty} f^2(x')\,dx' + \int_{-\infty}^{\infty} f^2(x' + x)\,dx' - 2\int_{-\infty}^{\infty} f(x')f(x' + x)\,dx' \geq 0.$$

Since the first two terms have the same value, each being equal to $R_{11}(0)$, it is therefore trivial that

$$R_{11}(0) \geq R_{11}(x).$$

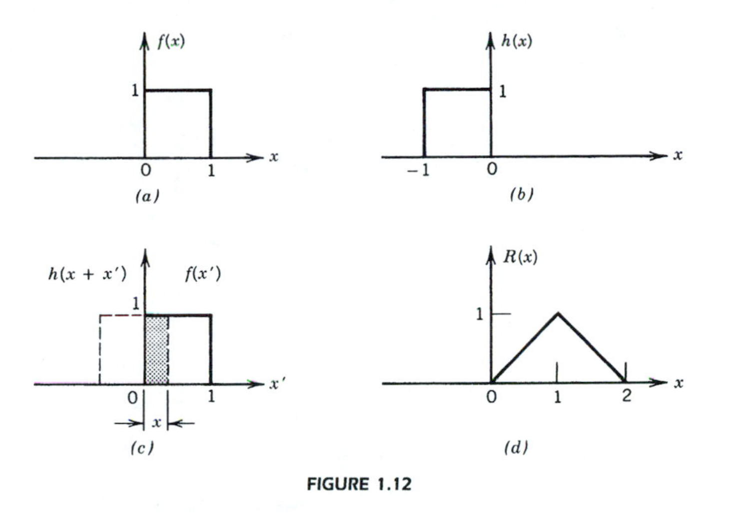

FIGURE 1.12

Example 1.12

Given two rectangular functions as defined in the Figures 1.12*a* and *b*, evaluate the correlation function between them. We again use the graphical approach to obtain the correlation function.

Figure 1.12*c* shows that the appropriate function $h(x)$ correlates against $f(x)$, with the shaded area representing the correlation of these two functions for a given x. The term $R(x)$ is the correlation function obtained with this graphical approach, as shown in Figure 1.12*d*. ■

Example 1.13

Given a periodic function, such as

$$f(x) = \cos(px + \phi),$$

where ϕ is an arbitrary phase shift, calculate the corresponding autocorrelation function.

By substituting $f(x)$ into Eq. 1.31, we can obtain the autocorrelation function:

$$R_{11}(x) = \int_{-\infty}^{\infty} \cos(px' + \phi)\cos(px' + \phi + px)\,dx'.$$

But

$$\cos A \cos B = \tfrac{1}{2}[\cos(A + B) + \cos(A - B)].$$

The autocorrelation can therefore be shown as

$$R_{11}(x) = \frac{1}{2}\int_{-\infty}^{\infty} [\cos(2px' + 2\phi + px) + \cos(px)]\,dx'$$
$$= \tfrac{1}{2}\cos px,$$

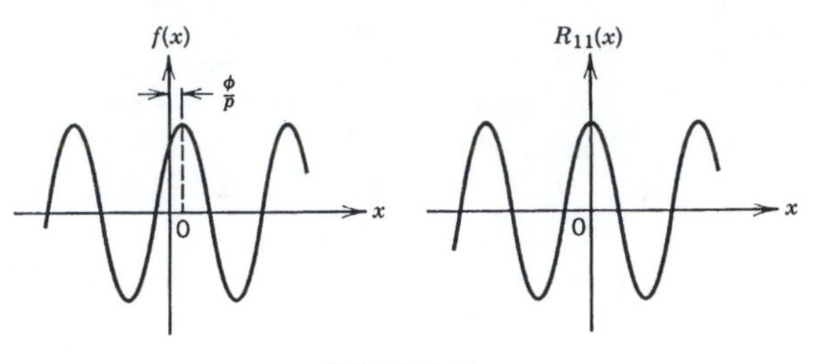

FIGURE 1.13

which is again a periodic function of the same period, as sketched in Figure 1.13. Notice that $R_{11}(x)$ is independent on the phase factor ϕ of $f(x)$. ■

1.5 PROPERTIES OF FOURIER TRANSFORMATION

Fourier transformation offers myriad ways for analyzing linear, spatially invariant systems. In this section we describe the basic properties of the Fourier transform that are frequently used in optical engineering.

1.5.1 Fourier Translation Property

If $f(x, y)$ is Fourier-transformable, that is,

$$\mathscr{F}[f(x, y)] = F(p, q),$$

then

$$\mathscr{F}[f(x - x_0, y - y_0)] = F(p, q) \exp[-i(px_0 + qy_0)],$$

where x_0 and y_0 are arbitrary real constants. To show that this property holds, we write

$$\mathscr{F}[f(x - x_0, y - y_0)] = \iint_{-\infty}^{\infty} f(x - x_0, y - y_0) \exp[-i(px + qy)]\, dx\, dy.$$

By substituting the variables $x' = x - x_0$ and $y' = y - y_0$ in the preceding equation, we show that

$$\mathscr{F}[f(x - x_0, y - y_0)] = \exp[-i(px_0 + qy_0)] \iint_{-\infty}^{\infty} f(x', y') \exp[-i(px' + qy)]\, dx'\, dy'$$

$$= \exp[-i(px_0 + qy_0)] F(p, q).$$

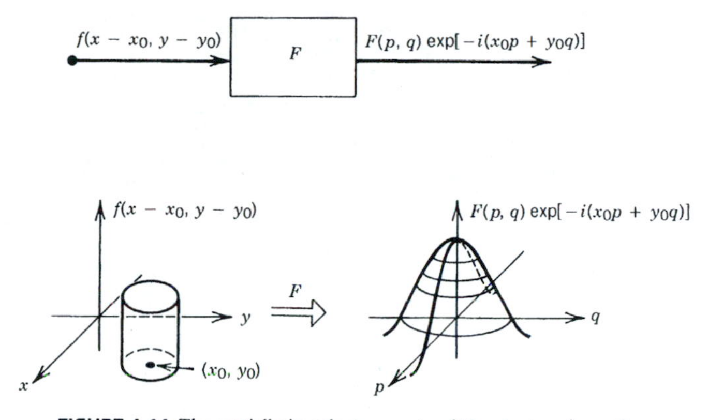

FIGURE 1.14 The spatially invariant property of Fourier transformation.

From this, we see that the translation of an object function, $f(x, y)$, in the spatial domain causes a linear phase shift in the Fourier or spatial frequency domain. In other words, for a real-object function, $f(x, y)$, the corresponding Fourier transform is spatially invariant in the spatial-frequency plane, as illustrated in Figure 1.14.

Example 1.14

Find the Fourier transform of a rectangular pulse of duration d, as shown in Figure 1.15*a*. If the rectangular function is shifted from the origin to $x = d/2$, as depicted in Figure 1.15*b*, find the corresponding Fourier transform.

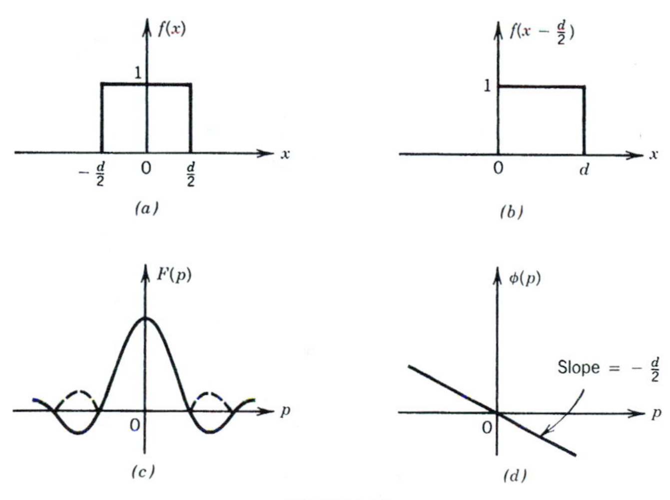

FIGURE 1.15

From the definition of Eq. 1.9, we get the following Fourier transform of Figure 1.15*a*:

$$F(p) = \int_{-\infty}^{\infty} \operatorname{rect}\left(\frac{x}{d}\right) e^{-ipx}\,dx$$

$$= \int_{-d/2}^{d/2} e^{-ipx}\,dx$$

$$= \frac{1}{ip}\left[\exp\left(i\frac{pd}{2}\right) - \exp\left(-i\frac{pd}{2}\right)\right]$$

$$= d\,\frac{\sin(pd/2)}{pd/2}.$$

Figure 1.15*c* shows the distribution of $F(p)$, and the dotted line indicates the magnitude $|F(p)|$.

For the shifted rectangular function of Figure 1.15*b*, the Fourier transform can be shown as

$$F'(p) = \int_{-\infty}^{\infty} \operatorname{rect}\left(\frac{x - d/2}{d}\right) e^{-ipx}\,dx$$

$$= \int_{0}^{d} e^{-ipx}\,dx$$

$$= \frac{1}{ip}(1 - e^{-ipd})$$

$$= d\,\frac{\sin(pd/2)}{pd/2}\exp\left(-i\frac{pd}{2}\right) = F(p)\exp\left(-i\frac{pd}{2}\right).$$

From this equation we see that the amplitude variation is essentially identical to that of Figure 1.15*c*. There is, however, a linear-phase factor added to this result, as sketched in Figure 1.15*d*. ■

1.5.2 Reciprocal Translation Property

In view of the definition of the Fourier transform given in Eqs. 1.8 and 1.9, we note that the inverse Fourier transform is essentially the same as the direct Fourier transform, except with a positive kernel. In other words, if we reverse the positive and negative coordinate axes of (x, y), the inverse Fourier transform is the same as the direct Fourier transform. By virtue of Fourier translation, the reciprocal translation theorem can therefore be stated as follows. If $f(x, y)$ is Fourier-transformable, that is,

$$\mathscr{F}[f(x, y)] = F(p, q),$$

then

$$\mathscr{F}^{-1}[F(p - p_0, q - q_0)] = f(x, y)\exp[i(xp_0 + yq_0)].$$

By substituting $p' = p - p_0$ and $q' = q - q_0$ in the preceding equation, we have

$$\mathscr{F}^{-1}[F(p - p_0, q - q_0)]$$

$$= \exp[i(p_0 x + q_0 y)] \frac{1}{4\pi^2} \iint_{-\infty}^{\infty} F(p', q') \exp[i(p'x + q'y)]\, dp'\, dq'$$

$$= \exp[i(p_0 x + q_0 y)] f(x, y).$$

Thus, we see that the translation of the Fourier spectrum, $F(p, q)$, in the spatial-frequency plane would also introduce a linear-phase factor in the object function.

Example 1.15

Using Example 1.14, let us shift the Fourier spectrum of Figure 1.15*c* to $p = p_0$, as shown in Figure 1.16*a*, and then evaluate the corresponding inverse Fourier transform.

The shifted Fourier spectrum can be written as

$$F(p - p_0) = d \frac{\sin[(p - p_0)d/2]}{(p - p_0)d/2}.$$

The inverse Fourier transform can be obtained by

$$f(x) = \frac{d}{2\pi} \int_{-\infty}^{\infty} \frac{\sin[(p - p_0)d/2]}{(p - p_0)d/2} e^{ipx}\, dp.$$

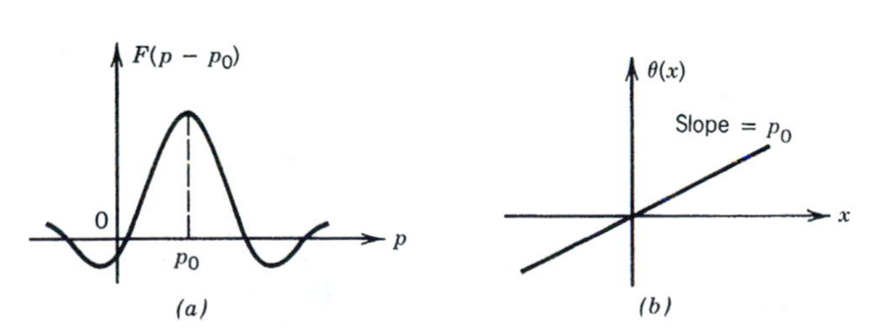

FIGURE 1.16

By letting $p - p_0 = p'$, we have

$$f(x) = \frac{d}{2\pi} e^{ip_0x} \int_{-\infty}^{\infty} \frac{\sin(p'd/2)}{p'd/2} e^{ip'x} \, dp'$$
$$= e^{ip_0x} \operatorname{rect}\left(\frac{x}{d}\right),$$

which is a rectangular function of finite duration d multiplexed with a linear-phase factor $\theta = p_0 x$. Sketches of $f(x)$ are shown in Figure 1.16*b*. ■

1.5.3 Scale Changes of Fourier Transforms

If $f(x, y)$ is Fourier-transformable, that is,

$$\mathscr{F}[f(x, y)] = F(p, q),$$

then

$$\mathscr{F}[f(ax, by)] = \frac{1}{ab} F\left(\frac{p}{a}, \frac{q}{b}\right),$$

where a and b are arbitrary positive constants. By direct substitution of the direct Fourier transformation of Eq. 1.9, we have

$$\mathscr{F}[f(ax, by)] = \iint_{-\infty}^{\infty} f(ax, by) \exp[-i(px + qy)] \, dx \, dy$$
$$= \frac{1}{ab} \iint_{-\infty}^{\infty} f(ax, by) \exp\left[-i\left(\frac{p}{a} ax + \frac{q}{b} bx\right)\right] d(ax) \, d(by)$$
$$= \frac{1}{ab} F\left(\frac{p}{a}, \frac{q}{b}\right).$$

From this result we see that a scale reduction of $f(x, y)$ in the spatial domain will enlarge the Fourier transform $F(p, q)$ in the spatial frequency domain. The overall amplitude spectrum of $F(p, q)$ is also proportionally reduced by a factor of $1/ab$.

■ Example 1.16

Given a set of rectangular functions with various durations, as depicted in the left-hand column of Figure 1.17, show that alternating the duration will also affect the height and width of the Fourier spectra.

The Fourier transform of Figure 1.17*a* can be shown as

$$F(p) = \int_{-\infty}^{\infty} \operatorname{rect}\left(\frac{x}{\Delta x}\right) e^{-ipx} \, dx$$
$$= \Delta x \frac{\sin[(p/2)\Delta x]}{(p/2)\Delta x},$$

which is sketched in Figure 1.17*b*.

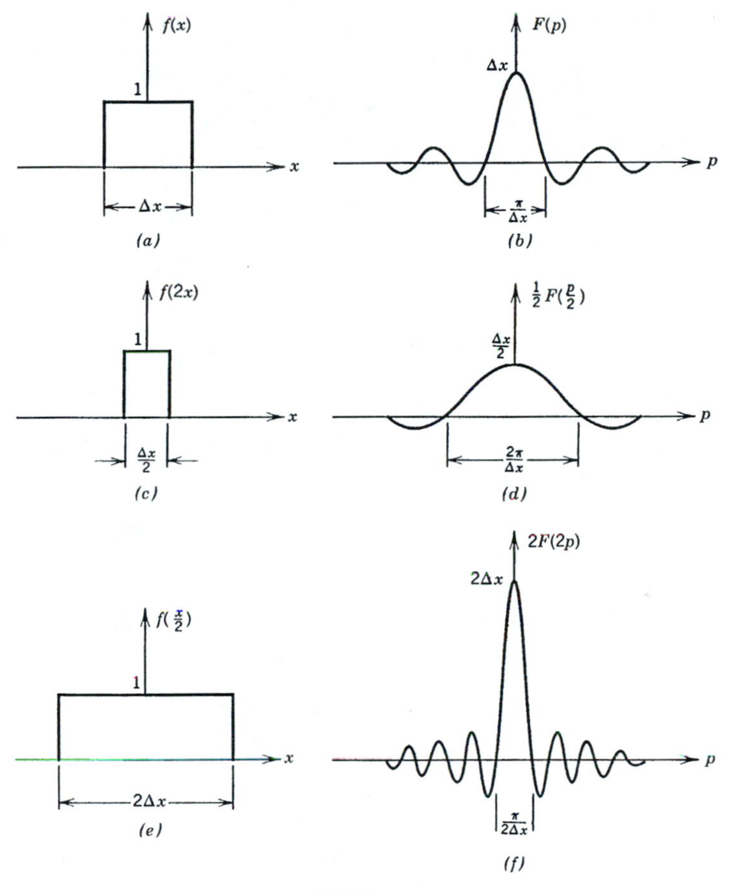

FIGURE 1.17

Similarly, the Fourier transforms of Figures 1.17*c* and *e* can be shown, respectively, as

$$F_1(p) = \int_{-\infty}^{\infty} \operatorname{rect}\left(\frac{x}{\Delta x/2}\right) e^{-ipx}\, dx$$

$$= \frac{\Delta x}{2} \frac{\sin[(p/4)\,\Delta x]}{(p/4)\,\Delta x} = \frac{1}{2} F\left(\frac{p}{2}\right)$$

and

$$F_2(p) = \int_{-\infty}^{\infty} \operatorname{rect}\left(\frac{x}{2\,\Delta x}\right) e^{-ipx}\, dx$$

$$= 2\,\Delta x \frac{\sin(p\,\Delta x)}{(p\,\Delta x)} = 2F(2p).$$

The sketches of $F_1(p)$ and $F_2(p)$ are shown in Figures 1.17*d* and *f*. By comparing these figures with those in the right-hand column, we see that the spectral width is inversely proportional to the duration, but that the height is proportional to the width of the rectangular pulse. ■

1.5.4 Convolution Property

If $f_1(x, y)$ and $f_2(x, y)$ are Fourier-transformable, that is,

$$\mathscr{F}[f_1(x, y)] = F_1(p, q)$$

and

$$\mathscr{F}[f_2(x, y)] = F_2(p, q),$$

then

$$\mathscr{F}\left[\iint_{-\infty}^{\infty} f_1(x, y) f_2(\alpha - x, \beta - y)\, dx\, dy\right] = F_1(p, q) F_2(p, q).$$

Since the convolution integral is a function of α and β variables,

$$C(\alpha, \beta) = \iint_{-\infty}^{\infty} f_1(x, y) f_2(\alpha - x, \beta - y)\, dx\, dy,$$

the Fourier transform of $C(\alpha, \beta)$ can be written as

$$\mathscr{F}\left[\iint_{-\infty}^{\infty} f_1(x, y) f_2(\alpha - x, \beta - y)\, dx\, dy\right] = \iint_{-\infty}^{\infty} f_1(x, y) \mathscr{F}[f_2(\alpha - x, \beta - y)]\, dx\, dy.$$

By substituting the definition of the Fourier transform given in Eq. 1.9 and letting $\alpha - x = \alpha'$ and $\beta - y = \beta'$, we can write this equation as

$$\begin{aligned}
&\iint_{-\infty}^{\infty} f_1(x, y) \left\{ \iint_{-\infty}^{\infty} f_2(\alpha - x, \beta - y) \exp[-i(\alpha p + \beta q)]\, d\alpha\, d\beta \right\} dx\, dy \\
&= \iint_{-\infty}^{\infty} f_1(x, y) \left\{ \exp[-i(px + qy)] \iint_{-\infty}^{\infty} f_2(\alpha', \beta') \exp[-i(\alpha' p + \beta' q)]\, d\alpha'\, d\beta' \right\} dx\, dy \\
&= \iint_{-\infty}^{\infty} f_1(x, y) \exp[-i(px + qy)]\, dx\, dy\, F_2(p, q) \\
&= F_1(p, q) F_2(p, q).
\end{aligned}$$

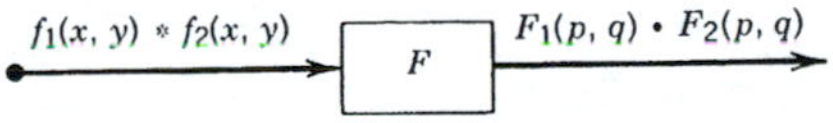

FIGURE 1.18 The convolution property of Fourier transformation.

In other words, the Fourier transform of $f_1(x, y)$ convolved with $f_2(x, y)$ equals the product of their Fourier transformations. For convenience of notation, we write the convolution integral as

$$\iint_{-\infty}^{\infty} f_1(x, y) f_2(\alpha - x, \beta - y)\, dx\, dy = f_1(x, y) * f_2(x, y),$$

where the asterisk represents the convolution operation. According to the convolution property, the preceding equation can be written in the following simplified form,

$$\mathcal{F}[f_1(x, y) * f_2(x, y)] = F_1(p, q) F_2(p, q),$$

which is illustrated by the block diagram of Figure 1.18.

Example 1.17

Given the following Fourier-transformable functions,

$$f_1(x) = \cos(p_0 x)$$

and

$$f_2(x) = \frac{p_1}{\pi} \frac{\sin(p_1 x)}{p_1 x}, \qquad p_1 \gg p_0,$$

find the Fourier transform of $f_1(x)$ convolved with $f_2(x)$.

The Fourier transform of $f_1(x)$ can be obtained by the definition of Eq. 1.9,

$$F_1(p) = \int_{-\infty}^{\infty} \cos(p_0 x) e^{-ipx}\, dx.$$

Since

$$\cos p_0 x = \tfrac{1}{2}(e^{ip_0 x} + e^{-ip_0 x}),$$

then

$$F_1(p) = \frac{1}{2} \int_{-\infty}^{\infty} (e^{ip_0 x} + e^{-ip_0 x}) e^{-ipx}\, dx$$
$$= \tfrac{1}{2}\delta(p - p_0) + \tfrac{1}{2}\delta(p + p_0).$$

Similarly,

$$F_2(p) = \frac{p_1}{\pi} \int_{-\infty}^{\infty} \frac{\sin(p_1 x)}{p_1 x} e^{-ipx}\, dx$$
$$= \mathrm{rect}\left(\frac{p}{2p_1}\right).$$

Thus we have

$$\mathscr{F}[f_1(x) * f_2(x)] = \tfrac{1}{2}[\delta(p - p_0) + \delta(p + p_0)]\, \mathrm{rect}\left(\frac{p}{2p_1}\right). \qquad \blacksquare$$

1.5.5 Cross-Correlation Property

If $f_1(x, y)$ and $f_2(x, y)$ are Fourier-transformable, that is,

$$\mathscr{F}[f_1(x, y)] = F_1(p, q)$$

and

$$\mathscr{F}[f_2(x, y)] = F_2(p, q),$$

then

$$\mathscr{F}\left[\iint_{-\infty}^{\infty} f_1^*(x, y) f_2(x + \alpha, y + \beta)\, dx\, dy\right] = F_1^*(p, q) F_2(p, q),$$

where the asterisk represents the complex conjugate.

Again, we note that the cross-correlation integral is a function of the α and β variables,

$$R_{12}(\alpha, \beta) = \iint_{-\infty}^{\infty} f_1^*(x, y) f_2(x + \alpha, y + \beta)\, dx\, dy,$$

and that the Fourier transform of $R_{12}(\alpha, \beta)$ is operating with the α and β variables. It is therefore trivial that

$$\mathscr{F}\left[\iint_{-\infty}^{\infty} f_1^*(x, y) f_2(x + \alpha, y + \beta)\, dx\, dy\right]$$
$$= \iint_{-\infty}^{\infty} f_1^*(x, y) \mathscr{F}[f_2(x + \alpha, y + \beta)]\, dx\, dy.$$

If we let $\alpha' = x + \alpha$ and $\beta' = y + \beta$, then by the definition of Fourier transformation, we have

$$\iint_{-\infty}^{\infty} f_1^*(x, y)\left\{\iint_{-\infty}^{\infty} f_2(x + \alpha, y + \beta)\exp[-i(\alpha p + \beta q)]\, d\alpha\, d\beta\right\} dx\, dy$$

$$= \iint_{-\infty}^{\infty} f_1^*(x, y)\exp[i(px + qy)]\left\{\iint_{-\infty}^{\infty} f_2(\alpha', \beta')\exp[-i(\alpha' p + \beta' q)]\, d\alpha'\, d\beta'\right\} dx\, dy$$

$$= \iint_{-\infty}^{\infty} f_1^*(x, y)\exp[i(px + qy)]\, dx\, dy F_2(p, q)$$

$$= F_1^*(p, q)F_2(p, q).$$

Example 1.18

Given the Fourier-transformable functions

$$f_1(x) = \sin(p_0 x)$$

and

$$f_2(x) = \mathrm{rect}\left(\frac{x - \Delta x}{\Delta x}\right),$$

find the Fourier transform of $f_1(x)$ cross-correlated with $f_2(x)$.

The Fourier transformations of $f_1(x)$ and $f_2(x)$ can be shown, respectively, as

$$F_1(p) = \int_{-\infty}^{\infty} \sin(p_0 x)\, e^{-ipx}\, dx$$

$$= \frac{1}{2i}\int_{-\infty}^{\infty} (e^{ip_0 x} - e^{-ip_0 x})e^{-ipx}\, dx$$

$$= \frac{1}{2i}[\delta(p - p_0) - \delta(p + p_0)]$$

and

$$F_2(p) = \Delta x\, \frac{\sin[(p/2)\Delta x]}{(p/2)\Delta x}\, e^{-ip\Delta x}.$$

The Fourier transform of their cross-correlation is therefore

$$\mathscr{F}[f_1(x) \circledast f_2(x)] = F_1^*(p)F_2(p)$$

$$= \tfrac{1}{2}[\delta(p - p_0) - \delta(p + p_0)]\, \Delta x\, \frac{\sin[(p/2)\Delta x]}{(p/2)\Delta x}\exp[-i(p\,\Delta x + \pi/2)].$$

■

1.5.6 Autocorrelation Property

If $f(x, y)$ is Fourier-transformable, that is,

$$\mathscr{F}[f(x, y)] = F(p, q),$$

then

$$\mathscr{F}\left[\iint_{-\infty}^{\infty} f^*(x, y) f(x + \alpha, y + \beta)\, dx\, dy\right] = |F(p, q)|^2,$$

where the asterisk represents the complex conjugate. This equation is written more simply as

$$\mathscr{F}[R_{11}(\alpha, \beta)] = \mathscr{F}[f^*(x, y) \circledast f(x, y)] = |F(p, q)|^2,$$

where $R_{11}(\alpha, \beta) = f^*(x, y) \circledast f(x, y)$ is known as the *autocorrelation function*, and $\circledast$ denotes the correlation operation. Conversely, we have

$$\mathscr{F}^{-1}[|F(p, q)|^2] = R_{11}(\alpha, \beta).$$

In other words, the autocorrelation function and the power spectral density are the Fourier transforms of each other. This result is also known as the *Wiener–Khinchine theorem*. The proof of this theorem is similar to that of the cross-correlation property, which can be shown as

$$\begin{aligned}
\mathscr{F}&\left[\iint_{-\infty}^{\infty} f^*(x, y) f(x + \alpha, y + \beta)\, dx\, dy\right] \\
&= \iint_{-\infty}^{\infty} f^*(x, y) \mathscr{F}[f(x + \alpha, y + \beta)]\, dx\, dy \\
&= \iint_{-\infty}^{\infty} f^*(x, y) F(p, q) \exp[i(px + qy)]\, dx\, dy \\
&= F^*(p, q) F(p, q) = |F(p, q)|^2.
\end{aligned}$$

Example 1.19

Find the power spectral density and the autocorrelation function of a shifted rectangular pulse, as shown in Figure 1.19. The Fourier transform of $f(x)$ is

$$\begin{aligned}
F(p) &= \int_{-\infty}^{\infty} \operatorname{rect}\left(\frac{x - x_0}{\Delta x}\right) e^{-ipx}\, dx \\
&= \Delta x \frac{\sin[(p/2)\Delta x]}{(p/2)\Delta x} e^{-ipx_0}.
\end{aligned}$$

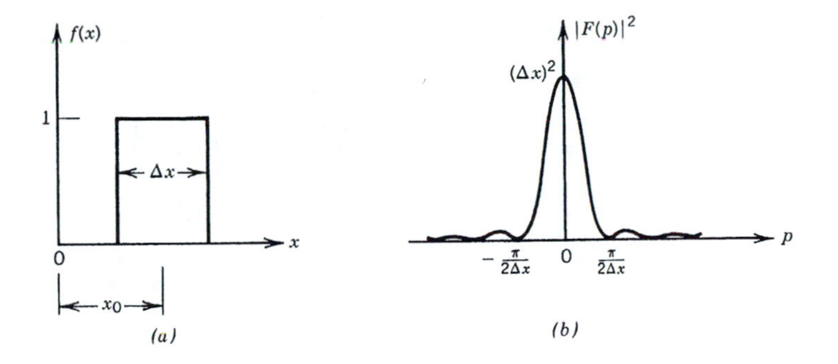

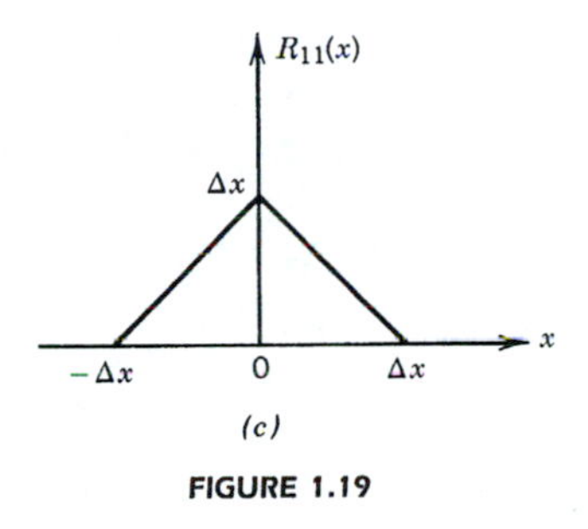

FIGURE 1.19

The power spectral density can be written as

$$
\begin{aligned}
|F(p)|^2 &= F^*(p)F(p) \\
&= \left\{\Delta x \frac{\sin[(p/2)\,\Delta x]}{(p/2)\,\Delta x}\right\}^2,
\end{aligned}
$$

which is sketched in Figure 1.19*b*.

The autocorrelation function can be found by applying the Wiener–Khinchine theorem,

$$
\begin{aligned}
R_{11}(x) &= \mathscr{F}^{-1}[|F(p,q)|^2] \\
&= \frac{(\Delta x)^2}{2\pi}\int_{-\infty}^{\infty}\left\{\frac{\sin[(p/2)\,\Delta x]}{(p/2)\,\Delta x}\right\}^2 e^{-ipx}\,dp \\
&= \begin{cases} x + \Delta x, & -\Delta x \le x < 0 \\ -x + \Delta x, & 0 < x \le \Delta x \\ 0, & \text{otherwise} \end{cases}
\end{aligned}
$$

which is sketched in Figure 1.19*c*. ■

1.5.7 Conservation Property

If $f(x, y)$ is Fourier-transformable, that is,

$$\mathscr{F}[f(x, y)] = F(p, q),$$

then

$$\iint_{-\infty}^{\infty} |f(x, y)|^2 \, dx \, dy = \frac{1}{4\pi^2} \iint_{-\infty}^{\infty} |F(p, q)|^2 \, dp \, dq.$$

This result is also known as *Parseval's theorem.* In other words, Parseval's theorem implies the conservation of energy. To show that this property holds, we write

$$\begin{aligned}
\iint_{-\infty}^{\infty} |f(x, y)|^2 \, dx \, dy &= \iint_{-\infty}^{\infty} f(x, y) f^*(x, y) \, dx \, dy \\
&= \iint_{-\infty}^{\infty} dx \, dy \left\{ \frac{1}{4\pi^2} \iint_{-\infty}^{\infty} F(p', q') \exp[i(xp' + yq')] \, dp' \, dq' \right\} \\
&\quad \times \left\{ \frac{1}{4\pi^2} \iint_{-\infty}^{\infty} F^*(p'', q'') \exp[-i(xp'' + yq'')] \, dp'' \, dq'' \right\} \\
&= \frac{1}{4\pi^2} \iint_{-\infty}^{\infty} F(p', q') \, dp' \, dq' \iint_{-\infty}^{\infty} F^*(p'', q'') \, dp'' \, dq'' \\
&\quad \times \left\{ \frac{1}{4\pi^2} \iint_{-\infty}^{\infty} \exp\{i[x(p' - p'') + y(q' - q'')]\} \, dx \, dy \right\} \\
&= \frac{1}{4\pi^2} \iint_{-\infty}^{\infty} F(p', q') \, dp' \, dq' \iint_{-\infty}^{\infty} F^*(p'', q'') \, dp'' \, dq'' \delta(p' - p'', q' - q'') \\
&= \frac{1}{4\pi^2} \iint_{-\infty}^{\infty} |F(p, q)|^2 \, dp \, dq.
\end{aligned}$$

Example 1.20

By referring to function $f(x)$ and its Fourier transform $F(p)$ in Example 1.19, we can show that

$$\begin{aligned}
\int_{-\infty}^{\infty} |f(x)|^2 \, dx &= \int_{-\infty}^{\infty} \left| \operatorname{rect}\left(\frac{x - x_0}{\Delta x} \right) \right|^2 dx \\
&= (\Delta x)^2
\end{aligned}$$

and

$$\frac{1}{2\pi}\int_{-\infty}^{\infty}|F(p)|^2\,dp = \frac{(\Delta x)^2}{2\pi}\int_{-\infty}^{\infty}\frac{\sin^2[(p/2)\,\Delta x]}{[(p/2)\,\Delta x]^2}\,dp$$

$$= (\Delta x)^2. \qquad \blacksquare$$

1.5.8 Symmetric Properties

Assume that $f(x, y)$ is Fourier-transformable, that is,

$$\mathscr{F}[f(x, y)] = F(p, q).$$

1. If $f(x, y)$ is a real function over (x, y) domain,

$$f^*(x, y) = f(x, y),$$

then

$$F^*(-p, -q) = F(p, q).$$

To show that this property holds, we take the conjugate of the Fourier transform,

$$F^*(p', q') = \frac{1}{4\pi^2}\iint_{-\infty}^{\infty} f^*(x, y)\exp[i(p'x + q'y)]\,dx\,dy.$$

If we let $p' = -p$ and $q' = -q$, we have

$$F^*(-p, -q) = \frac{1}{4\pi^2}\iint_{-\infty}^{\infty} f^*(x, y)\exp[-i(px + qy)]\,dx\,dy.$$

Since $f(x, y)$ is real, that is, $f^*(x, y) = f(x, y)$, we see that

$$F^*(-p, -q) = F(p, q).$$

2. If $f(x, y)$ is a real and even function,

$$f^*(x, y) = f(x, y) = f(-x, -y),$$

then $F(p, q)$ is also a real and even function,

$$F(p, q) = F^*(p, q) = F^*(-p, -q).$$

To prove that this property holds, we show that

$$F(p, q) = \frac{1}{4\pi^2}\iint_{-\infty}^{\infty} f(x, y)\exp[-i(px + qy)]\,dx\,dy$$

$$= \frac{1}{4\pi^2}\iint_{-\infty}^{\infty} f(-x, -y)\exp[-i(px + qy)]\,dx\,dy.$$

By letting $x = -x'$ and $y = -y'$, we can write the preceding equation as

$$F(p,q) = \frac{1}{4\pi^2} \iint_{-\infty}^{\infty} f(x', y') \exp[i(px' + qy')] \, dx' \, dy'.$$

Since $f(x', y')$ is real, that is, $f(x', y') = f^*(x', y')$, we see that

$$F(p,q) = F^*(p,q) = F^*(-p,-q).$$

3. If $f(x, y)$ is a real and odd function,

$$f(x,y) = f^*(x,y) = -f(-x,-y),$$

then

$$F(p,q) = -F(-p,-q) = F^*(-p,-q).$$

The proof of this property can be obtained in

$$F(p,q) = \frac{1}{4\pi^2} \iint_{-\infty}^{\infty} f(x,y) \exp[-i(px + qy)] \, dx \, dy$$

$$= \frac{1}{4\pi^2} \iint -f(-x,-y) \exp[-i(px + qy)] \, dx \, dy.$$

By letting $x' = -x$ and $y' = -y$, we have

$$F(p,q) = -\frac{1}{4\pi^2} \iint_{-\infty}^{\infty} f(x', y') \exp[+i(px' + qy')] \, dx \, dy$$

$$= -F(-p,-q).$$

And assuming that $f(x, y)$ is real, we can see that

$$F(p,q) = -F(-p,-q) = F^*(-p,-q).$$

REFERENCES

1. A. Papoulis, *The Fourier Integral and Its Applications*, McGraw-Hill, New York, 1962.
2. D. K. Cheng, *Analysis of Linear System*, Addison-Wesley, Reading, Mass. 1959.
3. F. T. S. Yu, *Optical Information Processing*, Wiley-Interscience, New York, 1983, Chapter 1.

PROBLEMS

1.1 If the transfer function of a linear spatially invariant system is

$$H(p) = \frac{ip}{1 + ip},$$

determine the spatial impulse response.

1.2 Show that the additivity and homogeneity properties of Problem 1.1. hold.

1.3 If the arbitrary input excitation of Problem 1.1 is translated to $x = x_0$, that is, $f(x - x_0)$, show that the output excitation is $g(x - x_0)$.

1.4 If for a second-order nonlinear phase transfer function of a system

$$H(p) = Ae^{-ip^2}$$

an input excitation is given by

$$f(x) = \cos(p_1 x) + \cos(p_2 x),$$

calculated the output response.

1.5 Assume an ideal linear-phase system for which the transfer function is

$$H(p) = \begin{cases} e^{-i\alpha p}, & |p| \leq |p_c| \\ 0, & |p| > |p_c|, \end{cases}$$

where α is an arbitrary constant and p_c is the angular cutoff spatial frequency. If the input excitation is

$$f(x) = \delta(x) + \delta(x - x_0),$$

calculate the output response $g(x)$.

1.6 We assume that the input excitation of Problem 1.5 is a rectangular function of finite duration Δx. Sketch the output responses for such cases as

$$\Delta x \ll \frac{\pi}{p_c}, \qquad \Delta x = \frac{\pi}{p_c}, \qquad \text{and} \qquad \Delta x \gg \frac{\pi}{p_c}.$$

1.7 Determine the Fourier transforms of the following functions.

(a) $f(x) = \sin(p_1 x) + \cos(p_2 x)$,
where p_1 and p_2 are arbitrary angular spatial frequencies.

(b) $f(x) = e^{ip_0 x} + e^{-ip_0 x}$.

(c) $f(x) = \sum_{n=-\infty}^{\infty} \delta(x - n\alpha)$,
where α is an arbitrary constant.

1.8 Determine the Fourier transforms of

$$f_1(x) = \cos(p_1 x)$$

and

$$f_2(x) = \sin(p_1 x).$$

Using the results, show that the Fourier transformation is a linear transformation.

1.9 Find the Fourier transform of $f(x)$ defined by

$$f(x) = \begin{cases} e^{-\alpha x}, & x > 0 \\ 0, & x < 0, \end{cases}$$

where $\alpha > 0$.

1.10 Using Problem 1.9, show that the following Fourier transform is shift-invariant.

$$|\mathscr{F}[f(x - x_0)]| = |\mathscr{F}[f(x)]|.$$

1.11 Show that multiplying $f(x)$ by $\cos(p_0 x)$ translates its spectrum to $\pm p_0$ in the angular spatial frequency domain p.

1.12 Given a band-limited signal $f(x)$, in which the upper angular spatial-frequency limit is p_m. If $f(x)$ is sampled by a periodic impulse function of period d, show that a periodic sequence of the spectra will be generated in the spatial frequency domain of p. Then determine what sampling interval d is required so that the spectra do not overlap.

1.13 We assume that the spatial impulse response of a linear spatially invariant system is given by

$$h(x) = \frac{p_c}{\pi} \frac{\sin(p_c x)}{p_c x}.$$

Find the output response, with an implicit integral form produced by rectangular excitation, such as

$$f(x) = \operatorname{rect}\left(\frac{x}{\Delta x}\right) = \begin{cases} 1, & |x| \leq \dfrac{\Delta x}{2} \\ 0, & \text{otherwise,} \end{cases}$$

where Δx is the pulse duration.

If the pulse duration of the input excitation is very large compared to the pulse width of the impulse response, that is, if

$$\Delta x \gg \frac{2\pi}{p_c},$$

determine the asymptotic output response.

1.14 Show that the inverse Fourier transform of $F^*(p)$ is $f^*(-x)$, or

$$\mathscr{F}^{-1}[F^*(p)] = f^*(-x).$$

1.15 Let us assume that the transfer function of a linear spatially invariant system is given by

$$H(p) = F^*(p),$$

which is equal to the complex conjugate of the input Fourier transform.

(a) Find the corresponding output response $g(x)$.

(b) If the input excitation is an arbitrary function $u(x)$, where $u(x) \neq f(x)$, evaluate the output response.

(c) State the functional operations of parts a and b and the significance of the Wiener–Khinchine theorem.

1.16 Find the Fourier transform of a rectangular pulse train of duration Δx and period T.

2 PRINCIPLES OF REFLECTION AND REFRACTION

In this chapter, we discuss the basic phenomena of *reflection* and *refraction* at a boundary surface that has been formed by the meeting of two different media. Although light is electromagnetic in nature, its wavelength is much shorter than that of a radio wave. The velocity of light wave propagation in a dielectric medium is (as will be seen in Chapter 5)

$$v = \frac{1}{\sqrt{\mu\varepsilon}}, \tag{2.1}$$

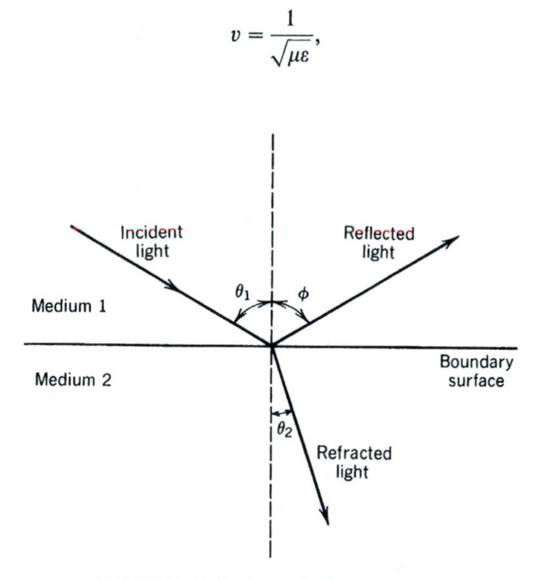

FIGURE 2.1 Reflection and refraction of light.

where μ and ε are the permeability and the permittivity (i.e., the dielectric constant) of the medium.

Let us consider a monochromatic (i.e., a single wavelength) light wave that impinges on a plane boundary surface between two media as depicted in Figure 2.1. If we assume that the two media have different permeabilities and permittivities, the velocities of the light waves within these two media will be different. For simplicity, we assume that these two media are transparent; hence, the incident light wave would penetrate the second medium continuously. From past experience, we know that penetrating light rays will bend away from the direction of incidence. This bending causes optical illusions such as the broken appearance of a tablespoon when it is immersed in clear water. Thus, we see that when a reflected light ray and a refracted light ray originate at the boundary surface, only a part of the incident ray is transmitted (i.e., refracted), while the rest is reflected.

2.1 SNELL'S LAW OF REFRACTION

Let us now discuss one of the most important theories of refraction, Snell's law of refraction. The angle of the incident light ray θ_1 is called the *incident angle*; the angle of the refracted light ray θ_2 is called the *refraction angle*; and the angle of the reflected light ray ϕ, is called the *reflection angle*.

Since the boundary surface between the two media is assumed to be a planar surface, it can be shown, by the law of refraction, that the reflection angle is equivalent to the incident angle. It can also be shown that the ratio of the sine of the incident angle to the sine of the refraction angle is equal to the ratio of their velocities of wave propagation in the two media; that is,

$$\frac{\sin\theta_1}{\sin\theta_2} = \frac{v_1}{v_2}, \tag{2.2}$$

where v_1 and v_2 are the velocities of the wave propagation in medium 1 and medium 2, respectively.

To show that Eq. 2.2 holds, we assume a monochromatic plane wave illumination, as depicted in Figure 2.2; here the dark lines represent the constant propagation of the phase train. Since the two media are assumed to be different, the velocities of wave propagation are therefore not the same. Thus the reflected and refracted wavefronts can be represented as

$$\overline{AP_2} = v_1 t, \qquad \overline{P_1B} = v_1 t, \qquad \text{and} \qquad \overline{P_1C} = v_2 t, \tag{2.3}$$

where t denotes the time variable.

From the triangles of P_1AP_2, P_1BP_2, and P_1CP_2 and the law of sines, we have

$$\overline{P_1P_2} = \frac{\overline{AP_2}}{\sin\theta_1} = \frac{\overline{P_1B}}{\sin\phi} = \frac{\overline{P_1C}}{\sin\theta_2}. \tag{2.4}$$

With substitutions from Eq. 2.2, we reduce Eq. 2.4 to

$$\frac{v_1}{\sin\theta_1} = \frac{v_1}{\sin\phi} = \frac{v_2}{\sin\theta_2}, \tag{2.5}$$

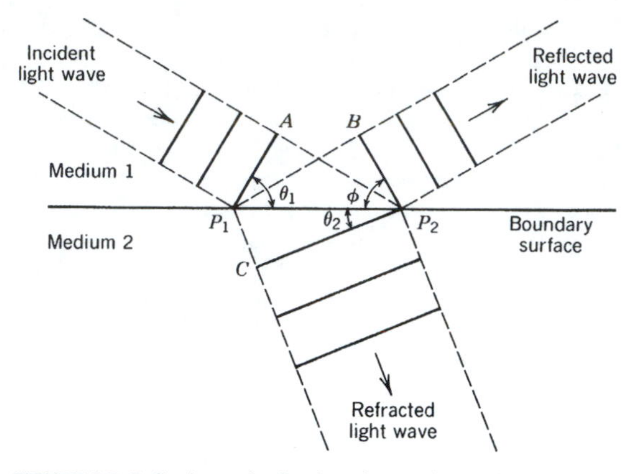

FIGURE 2.2 Reflection and refraction of monochromatic wave trains.

from which we conclude that

$$\phi = \theta_1, \tag{2.6}$$

that is, the angle of reflection is equal to the angle of incident.

We now define the *indices of refraction* of the two media by the following equations,

$$\eta_1 \triangleq \frac{c}{v_1}, \tag{2.7}$$

$$\eta_2 \triangleq \frac{c}{v_2}, \tag{2.8}$$

where c is the velocity of light propagation in a vacuum ($c = 3 \times 10^8$ m/sec). Thus, Eq. 2.2 can be written in terms of the refractive index ratio

$$\frac{\sin\theta_2}{\sin\theta_1} = \frac{\eta_1}{\eta_2}, \tag{2.9}$$

or alternatively,

$$\eta_1 \sin\theta_1 = \eta_2 \sin\theta_2, \tag{2.10}$$

where Eq. 2.9 or Eq. 2.10 is best known as *Snell's law of refraction.*

Strictly speaking, to find out how an incident light wave is reflected and refracted from a boundary surface, we should start with the boundary conditions from the electromagnetic standpoint. It can be shown that the reflected and refracted light waves are affected by the polarization of the incident light wave. In Figure 2.3, however, we illustrate a reflected light wave coming from an unpolarized incident light source. Thus, we see that in order to obtain a total reflection, the angle of incidence should be smaller for $\eta_1 > \eta_2$ than for $\eta_1 < \eta_2$.

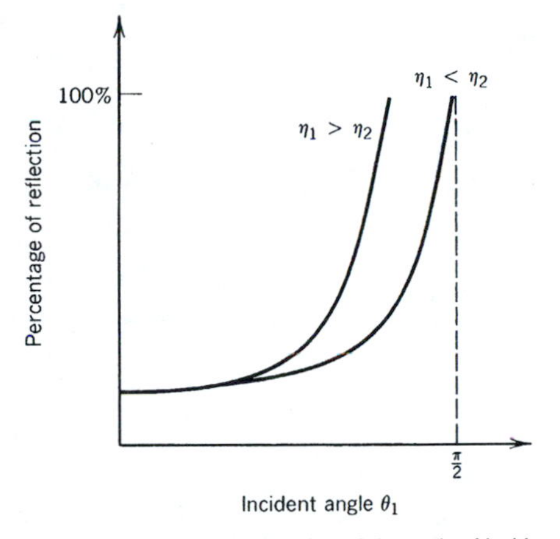

FIGURE 2.3 Reflection as a function of the angle of incidence.

Example 2.1

In Figure 2.1 we assume that the refractive index for medium 1 is $\eta_1 = 1.3$, and that for medium 2 it is $\eta_2 = 1.5$. If a light ray is traveling in the upper medium and impinging on the boundary surface at an angle of 45° to the normal axis, determine the corresponding angles of reflection and refraction.

We note that

$$\eta_1 = 1.3, \qquad \theta_1 = 45°, \qquad \eta_2 = 1.5.$$

From Eq. 2.6 the angle of reflection is

$$\phi = \theta_1 = 45°.$$

By Snell's law of refraction, as expressed in Eq. 2.10, the angle of refraction can be evaluated as follows:

$$1.3 \sin 45° = 1.5 \sin \theta_2$$
$$\sin \theta_2 = 0.613$$
$$\theta_2 = 37.8°.$$

■

Example 2.2

Given the two media described in Example 2.1 and the light ray impinging from the lower medium on the boundary surface, at an angle of incidence equal to 35°, calculate the angle of refraction in the upper medium.

We note that

$$\eta_1 = 1.5, \qquad \eta_2 = 1.3, \qquad \text{and} \qquad \theta_1 = 35°.$$

By applying Snell's law, we can calculate the angle of refraction in the upper medium:

$$1.5 \sin 35^\circ = 1.3 \sin \theta_2$$
$$\sin \theta_2 = 0.66$$
$$\theta_2 = 41.4^\circ.$$

Again we see that the angle of reflection is equal to the angle of incidence:

$$\phi = \theta_1 = 35^\circ. \qquad \blacksquare$$

2.2 HUYGENS' PRINCIPLE

We now discuss *Huygens' principle*, one of the most significant principles in diffraction optics. Through Huygens' principle, it is possible to obtain, by using graphs, the shape of a wavefront at any instant when given a wavefront of an earlier instant. The principle essentially states that every point of a wavefront can be regarded as a secondary point source from which a small spherical wavelet is generated and spreads in all directions at a velocity of wave propagation. Thus, a new wavefront can be constructed graphically by drawing a surface tangent to all the secondary spherical wavelets. If the velocity of propagation is not constant at all parts of the wavefront, each spherical wavelet should be drawn to an appropriate wave velocity.

Figure 2.4 gives an illustrated example for the use of Huygens' principle. The known wavefront is shown as the surface Σ, called the *primary wavefront*, and the directions of wave propagation are indicated by small arrows. To determine the wavefront after an interval of time Δt with a wave velocity v, we construct a series of spheres (i.e., the secondary spherical wavelets) of radius $r = v\,\Delta t$ from each point of the surface wavefront Σ. These spheres represent the secondary wavelets. Now if we draw a surface tangent to all the surfaces of the spheres, we get the

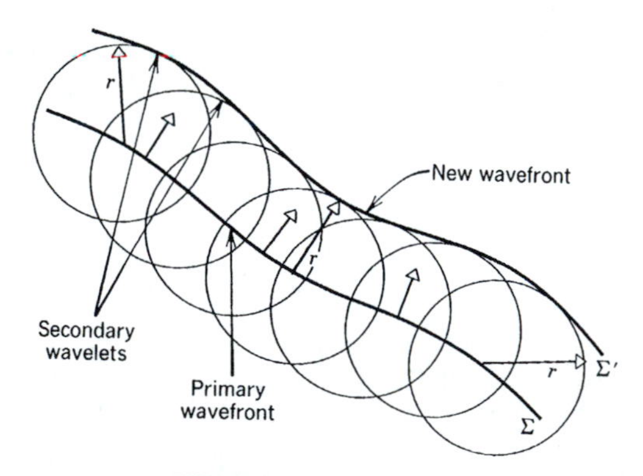

FIGURE 2.4 Huygens' principle.

shape of a new wavefront, called the *secondary wavefront*. We note that the wave velocities at every point on the surface are assumed to be equal. This is a typical example of the light wave propagation in a homogeneous isotropic medium.

Furthermore, according to Huygens' principle, a backward wavefront should form from the secondary spherical wavelets. This phenomenon has never been observed, however. The discrepancy can be explained by examining the wave propagation within the surrounding closed surface. Nonetheless, the application of Huygens' principle lies in predicting a new wavefront. During the period immediately after Huygens' principle was developed, little physical significance was attached to the secondary wavelets, but as the wave nature of light propagation became more fully understood, Huygens' principle took on a deeper physical meaning.

Huygens' principle is a very useful technique for demonstrating the *diffraction phenomenon*, which is discussed in chapter 6.

Example 2.3

Let Σ be the primary wavefront at time t. At each point on Σ draw a spherical wavelet of radius $r = v\,\Delta t$, where $v = c/\eta$, and c is the velocity of light. Although η varies continuously in the medium, the radii of the wavelets vary according to η, as shown in Figure 2.5. As a result, the surface Σ', which is tangent to all the secondary wavelets, is the new wavefront. By similar procedures, we can obtain another new wavefront Σ'' from Σ', and so on, as shown in the figure. Accuracy in predicting the new wavefronts varies inversely with time Δt. Thus, smaller values of Δt correspond to more accurate predictions of the new wavefront.

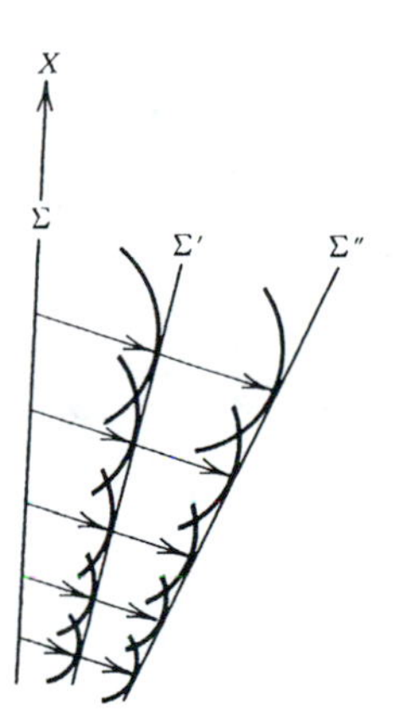

FIGURE 2.5 Wavefront propagation in a nonhomogeneous medium.

This phenomenon in wave propagation in a nonhomogeneous medium may explain several interesting optical illusions in the earth's atmosphere, namely the mirage and looming effects of distant objects. ■

2.3 REFRACTION AND REFLECTION

In Section 2.1 we derived Snell's law of refraction, which is very useful in geometrical optics and in physical optics. In Section 2.2 we discussed the applications of Huygens' principle for predicting new wavefronts in either homogeneous or nonhonogeneous media. We now show that the laws of refraction and reflection can also be derived from Huygens' principle.

Let us consider a plane monochromatic light wave that impinges on a plane boundary surface Σ between two homogeneous media. The velocities of light wave propagation in each of the media are different. When applying Huygens' principle, we see that all the points on surface Σ act as centers of propagation for the secondary spherical wavelets. The secondary wavelets, originating at the boundary surface Σ, give rise to both forward and backward propagations. This results in both refracted and reflected wavefronts, as shown in Figure 2.6.

We note that the wavelets propagating backward have a velocity equal to that of the incident light wave, since these wavelets are in the same medium as the incident wave. The radii of these reflected wavelets can be determined by the following equation,

$$r_n = \frac{c\,\Delta t_n}{\eta_1}, \qquad n = 1, 2, 3, \ldots, \tag{2.11}$$

where c is the velocity of light, η_1 is the refractive index of the upper medium, and Δt_n is the time interval after the incident wavefront touches the boundary surface. The oblique light ray of line $\overline{P_1B}$ is shown in Figure 2.6. The radius of

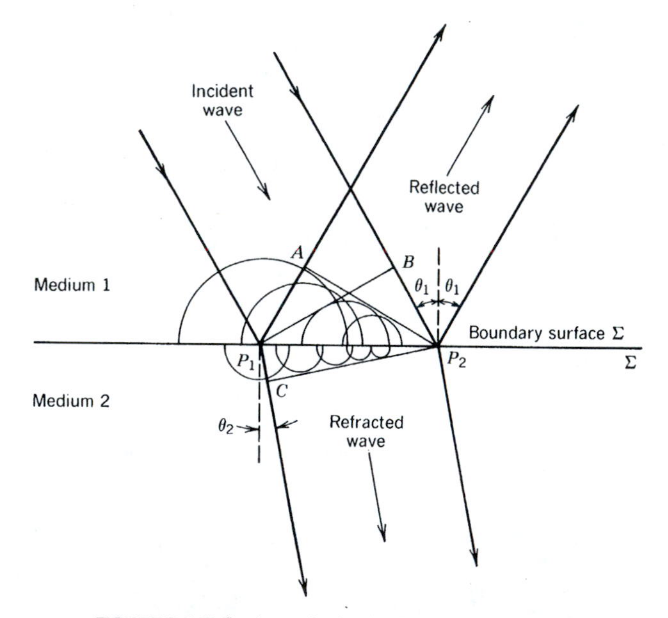

FIGURE 2.6 Reflection and refraction by Huygens' principle.

the secondary wavelet, with its center at point P_1, is

$$r_1 = \overline{P_1 A} = \frac{c\,\Delta t_1}{\eta_1}. \tag{2.12}$$

The radii of the secondary wavelets with centers at other points on the boundary surface in line $\overline{P_1 P_2}$ can be determined by

$$r_n = \overline{OP_2} \sin\theta_1, \tag{2.13}$$

where $\overline{OP_2}$ is the distance between the center of a secondary wavelet and point P_2, as shown in the figure, and θ_1 is the angle of reflection.

Similarly, we can also determine the radii of the refracted secondary wavelets by

$$r'_n = \frac{c\,\Delta t_n}{\eta_2}, \qquad n = 1, 2, 3, \ldots, \tag{2.14}$$

where η_2 is the refractive index of the lower medium. Again if the radius of the refracted wavelet at point P_1 is

$$r'_1 = \overline{P_1 C} = \frac{c\,\Delta t_1}{\eta_2}, \tag{2.15}$$

the radii of the other refracted wavelets, whose centers are on line $\overline{P_1 P_2}$, must be

$$r'_n = \overline{OP_2} \sin\theta_2, \tag{2.16}$$

where $\overline{OP_2}$ is the distance between the center of a refracted wavelet and the point P_2, and θ_2 is the angle of refraction.

To maintain plane-reflected and plane-refracted wavefronts (from Figure 2.6), we see that

$$\overline{BP_2} = \overline{P_1 A} = \frac{c}{\eta_1} \Delta t_1 = v_1\, \Delta t_1, \tag{2.17}$$

where v_1 is the velocity of wave propagation in the upper medium. By the law of sines, we have

$$\overline{P_1 P_2} = \frac{\overline{BP_2}}{\sin\theta_1} = \frac{\overline{P_2 A}}{\sin\theta_1} = \frac{\overline{PC}}{\sin\theta_2}. \tag{2.18}$$

Since the ratio of the wave propagations in the two media is

$$\frac{\sin\theta_1}{\sin\theta_2} = \frac{v_1}{v_2}, \tag{2.19}$$

where v_2 is the velocity of wave propagation in the lower medium, and

$$v_1 = \frac{c}{\eta_1}, \qquad v_2 = \frac{c}{\eta_2}, \tag{2.20}$$

we can combine Eqs. 2.19 and 2.20 to obtain the following,

$$\frac{\sin\theta_1}{\sin\theta_2} = \frac{\eta_2}{\eta_1}. \tag{2.21}$$

This equation gives Snell's law of refraction as derived from Huygens' principle.

2.4 SPHERICAL WAVE AND IMAGE FORMATION

In order to gain a deeper appreciation of Snell's law of refraction, we begin with the following situation. A spherical wavefront is reflected and refracted at a plane boundary surface.

Let us consider a monochromatic point source S located in the upper medium of a boundary surface, as shown in Figure 2.7. By applying Huygens' principle, we can show that a divergent spherical wavefront, with the same radius of curvature, is reflected from the boundary surface, and a refracted spherical wavefront, with a different radius of curvature, penetrates the lower medium.

If we use a light ray representation, Figure 2.7 can be replaced by Figure 2.8. In this figure we see that when an incident light ray hits the boundary surface, it is reflected with the same angle of incidence.

If we extend the reflected light rays below the boundary surface and through the lower medium, they converge at point S', which lies opposite S at an equal distance from the boundary surface. Similarly, if we extend the refracted light rays into the upper medium, the extended light rays converge at point S'', which lies on line SS'.

We have shown that although the wavefronts originate from point S, they appear to diverge from point S'. This phenomenon is caused by reflection from a plane boundary surface. Standing in the upper medium, we see that image point S' lies below the boundary surface, directly opposite object point S. Point S' is

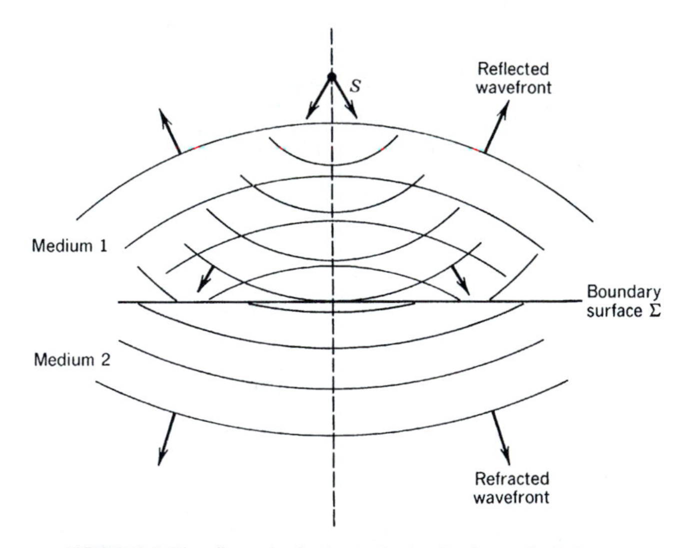

FIGURE 2.7 The effects of reflection and refraction by a spherical wave.

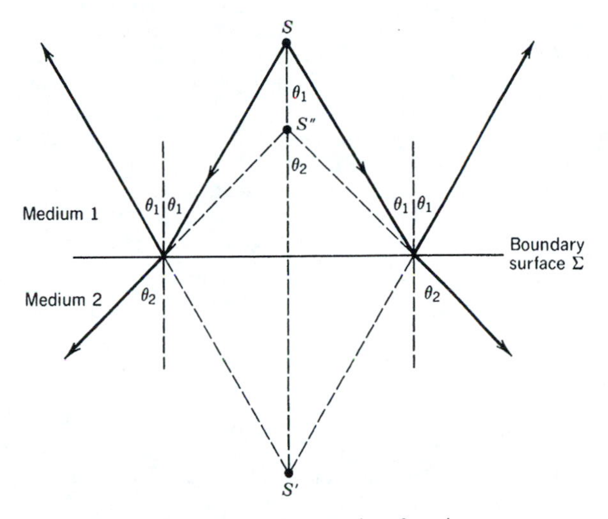

FIGURE 2.8 Light ray representation of a point source.

called the *virtual image* because the light rays appear to converge below the boundary surface. Images formed above the boundary surface, on the other hand, are called *real images*.

Since we can think of an extended, diffuse (i.e., scattered) object as composed of many infinitesimal points, the image of the object can therefore be found by the ray-tracing technique. For example, a finite extended object lies above a plane mirror. Every point of the object results in an image point behind the mirror, as illustrated in Figure 2.9. Looking from above the mirror, an observer sees a virtual

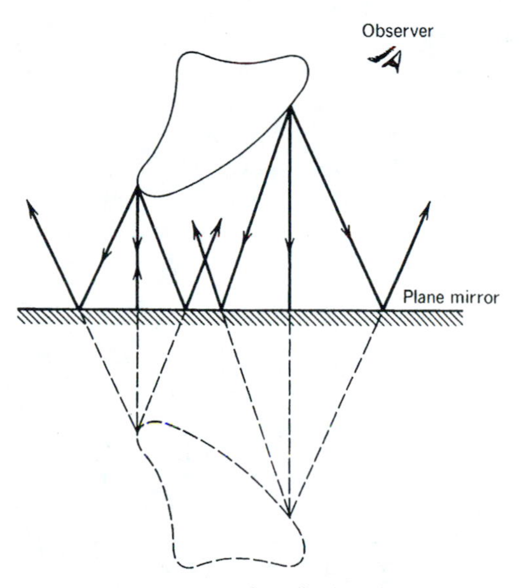

FIGURE 2.9 Image formation by reflection.

image of the object. Notice that a small cone of reflected light rays carries the entire virtual image. Since these reflected light rays are identical to the light rays that emanate from the object, the observer perceives them as they actually originated from the object.

If the observation takes place in the lower medium of Figure 2.8, the virtual image S'' is seen above the boundary surface. It is located on the same normal line as the object but lies at a different distance from the boundary surface. The distance between the image point and the boundary surface depends on the relative refractive index between the two media. For example, if $\eta_1 > \eta_2$, the virtual image is closer to the boundary surface, but if $\eta_1 < \eta_2$, the image is farther away from the boundary surface. The second phenomenon can be easily demonstrated by placing an object in clear water. The "lifting" effect of the object is evidently due to the refractive phenomenon of the light waves.

Example 2.4

Using the ray-tracing method, locate the virtual images formed by the two plane mirrors M_1 and M_2 of Figure 2.10.

Let us extend mirror M_1 by line m_1 and mirror M_2 by line m_2, as shown in the figure. Draw perpendicular lines from object points A and B with respect to the mirror plane of M_1. If we continue the ray tracings behind mirror plane M_1 at distances equal to the distances from object points A and B, we can find the position of the virtual image $A'B'$ created by M_1. Similarly, by using image $A'B'$ and the mirror plane of M_2, we can find another virtual image $A''B''$.

We stress that the locations of these virtual images can be easily verified by employing the ray-tracing method. For example, two light rays originating from point A of the object strike mirror plane M_1. Using the law of reflection, we can show that the light rays from the object are reflected from mirror M_1 toward mirror M_2, and are then reflected from mirror M_2, as shown by the arrowheads in the figure. If the reflected rays from mirrors M_1 and M_2 are projected backward behind the mirrors, two virtual images can be found. The resulting images coincide

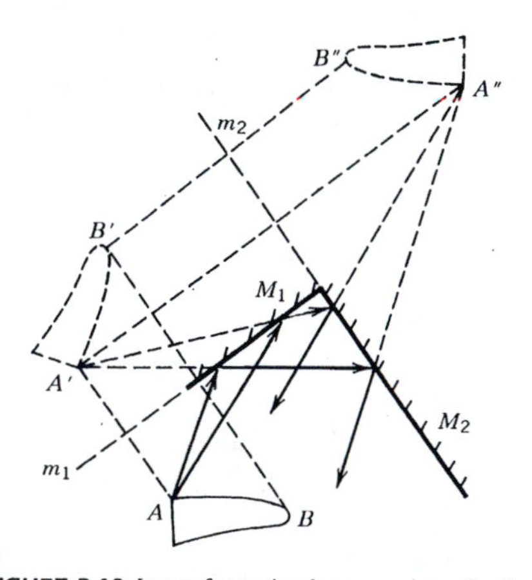

FIGURE 2.10 Image formation by successive reflections.

perfectly with the images formed by the extended-mirrors method that we described first. ■

Example 2.5

Calculate the distance from point S'' in Figure 2.8 to the boundary surface.

Let us call this distance y', and the distance between the object point S and the boundary surface y. From Snell's law of refraction (Eq. 2.10), we have

$$\eta_1 \sin\theta_1 = \eta_2 \sin\theta_2.$$

From the right triangle in Figure 2.8, we see that

$$y_1 \tan\theta_1 = y_2' \tan\theta_2. \tag{2.22}$$

It is apparent that

$$\frac{y\tan\theta_1}{\eta_1 \sin\theta_1} = \frac{y'\tan\theta_2}{\eta_2 \sin\theta_2}. \tag{2.23}$$

Furthermore, since

$$\tan\theta_1 = \frac{\sin\theta_1}{\cos\theta_1}$$

and

$$\tan\theta_2 = \frac{\sin\theta_2}{\cos\theta_2},$$

Eq. 2.23 can be written as

$$y' = y\frac{\eta_2}{\eta_1}\frac{\cos\theta_2}{\cos\theta_1}. \tag{2.24}$$

We note further that the ratio $\cos\theta_2/\cos\theta_1$ varies with respect to the angle of incidence, θ_1, and that y' is not the same for the light rays that diverge from S. Thus, if the refracted wavefront is not spherical, it will not appear to diverge from a point. However, if the observer is looking vertically upward from a point near the normal line below the boundary surface, the angles of refraction and incidence are very small:

$$\cos\theta_2 \simeq \cos\theta_1. \tag{2.25}$$

Equation 2.24 can be reduced to

$$y' = y\frac{\eta_2}{\eta_1}. \tag{2.26}$$

Thus, if $\eta_1 > \eta_2$, the image appears to be closer to the boundary surface, and if $\eta_1 < \eta_2$, the image appears to be farther away from the boundary surface. ■

2.5 TOTAL REFLECTION AND DISPERSION

Total reflection is a phenomenon that we frequently encounter in everyday life. To illustrate the phenomenon, we sketch a number of light rays emanating from a point source S, located in the lower medium. These light rays strike the plane boundary surface Σ of the upper medium shown in Figure 2.11. From Snell's law of refraction we can obtain the sine of the refraction angle,

$$\sin\theta_r = \frac{\eta_i}{\eta_r}\sin\theta_i, \tag{2.27}$$

where θ_i is the angle of incidence.

If the ratio of η_i and η_r is greater than one (i.e., $\eta_i > \eta_r$), then $\sin\theta_r$ is greater than $\sin\theta_i$. Thus, when θ_r is 90°, θ_i is still smaller than 90° (i.e., $\theta_i < 90°$). But when $\theta_r = 90°$, the refracted light ray will be propagated on the boundary surface in the upper medium, as shown in the figure. This angle of incidence, which produces a refracted light ray tangent to the boundary surface, is called the *critical angle of incidence* and is represented by θ_c in the figure. If, however, the angle of incidence is greater than the critical angle of incidence (i.e., $\theta_i > \theta_c$), all the light rays will be reflected back into the lower medium. This second phenomenon is known as *total internal reflection*. By using Snell's law of refraction, we see that total internal reflection can occur only when a light ray from one medium strikes the boundary surface of another medium that has a *smaller* index of refraction. Thus, the critical angle of incidence is

$$\theta_c = \arcsin\left(\frac{\eta_r}{\eta_i}\right), \tag{2.28}$$

where $\theta_r = 90°$.

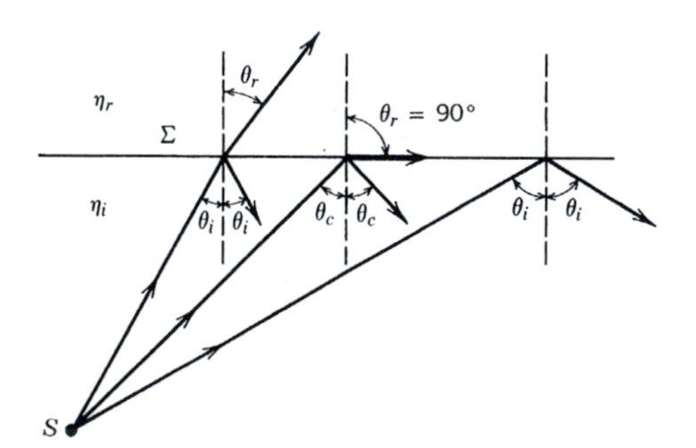

FIGURE 2.11 Effects on refraction with different incident angles, $\eta_i > \eta_r$.

Example 2.6

A light ray travels under the surface of a clear-water lake. The index of refraction of the water is $\eta = 1.33$. Compute the critical angle of incidence.

Since it is a water-to-air surface boundary, we have

$$\theta_c = \arcsin\left(\frac{1}{1.33}\right) = 48.75°.$$

Thus, if the angle of incidence is greater than the critical angle (i.e., $\theta_i > \theta_c$), the light ray will be totally reflected back under the surface. In practice, however, the boundary surface is usually not uniform. Therefore, when the incident angle exceeds the critical angle, we expect only a small portion of light rays to be refracted. ■

Example 2.7

Suppose a light ray is incident at an angle θ_i on the upper surface of a transparent glass plate, as shown in Figure 2.12. Assuming that both surfaces of the glass plate are optically smooth and parallel to one another, compute the corresponding angles of refraction below both the first and second surfaces of the glass plate.

Let η be the index of refraction of the glass plate and η_a be the index of refraction of air. Applying Snell's law to the refraction on the first boundary surface we have

$$\eta_a \sin \theta_{i1} = \eta \sin \theta_{r1}.$$

And on the second boundary surface, we have

$$\eta \sin \theta_{i2} = \eta_a \sin \theta_{r2}.$$

Since

$$\theta_{i2} = \theta_{r1},$$

we obtain

$$\theta_{r2} = \theta_{i1}.$$

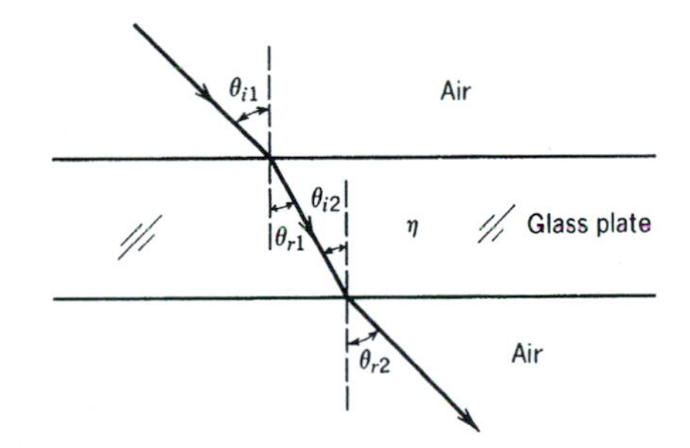

FIGURE 2.12 Refraction by the two parallel surfaces of a glass plate.

Thus, we see that the first angle of incidence (where the light is traveling from air to glass) is equal to the second angle of refraction (where the light is traveling from glass to air). The final emergent light is parallel to but not in line with the incident ray. In fact, after passing through any number of transparent parallel plates, a ray of light is displaced only from its original path. The refracted ray will therefore still be parallel to the incident light. ■

We emphasize that total internal reflection inside a glass prism is of considerable importance when dealing with optical instruments. The critical angle for light passing from air to a glass surface can be shown to be

$$\sin \theta_c = \frac{1}{1.5} = 0.67,$$

$$\theta_c = 42^\circ,$$

where we assume that the refractive index for glass is $\eta = 1.5$.

Total reflection is one of the major advantages of the glass prism over the standard metallic-coated reflector, since no metallic surface coating offers 100 percent reflection. Furthermore, the reflective property of glass prisms is permanent and not subject to tarnishing or any deterioration. When the glass prism has a surface coating of nonreflective film, however, a small amount of light is lost entering and leaving the surfaces of the glass prism.

A commonly used reflecting prism, a 45°–45°–90° prism, is shown in Figure 2.13. We assume that the light is normally incident on one of the vertical faces and that it strikes the inclined surface of the prism at an incident angle of 45°. Since the incident angle is larger than the critical angle, the light will be totally reflected at an angle equal to the incident. Thus, the light is reflected 90° by the prism.

We will, however, obtain different results when we project the light on the same prism, but at a different angle of incidence, as shown in Figure 2.14. After entering

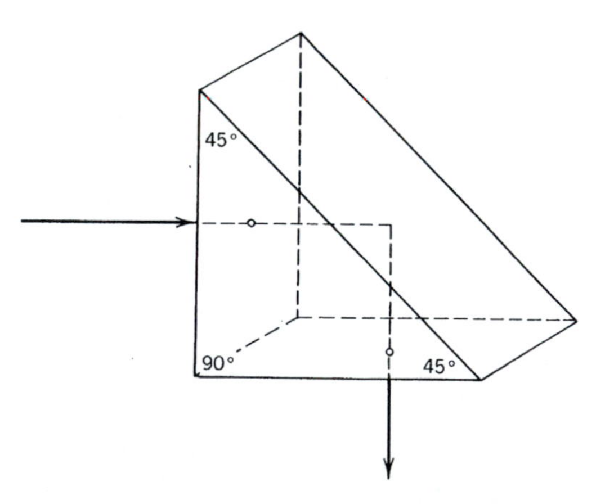

FIGURE 2.13 A 45°–45°–90° totally reflecting prism.

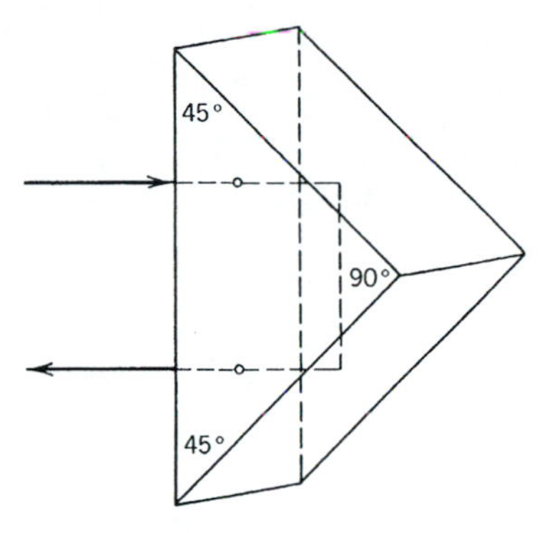

FIGURE 2.14 A totally reflecting Porro prism.

the prism, the light will be totally reflected back opposite to the incident light. In geometric optics, a 45°–45°–90° prism arranged in this manner is often called a *Porro* prism. Two or more *Porro* prisms can also be used to change the angle and direction of incident light, as shown in Figure 2.15.

Let us now consider the *dispersion* of light waves caused by a prism. When a beam of sunlight strikes a triangle prism, the light rays emerges on the other side as a rainbow of colors, as shown in Figure 2.16. In previous sections, we have dealt only with monochromatic light (i.e., light with a single wavelength) when

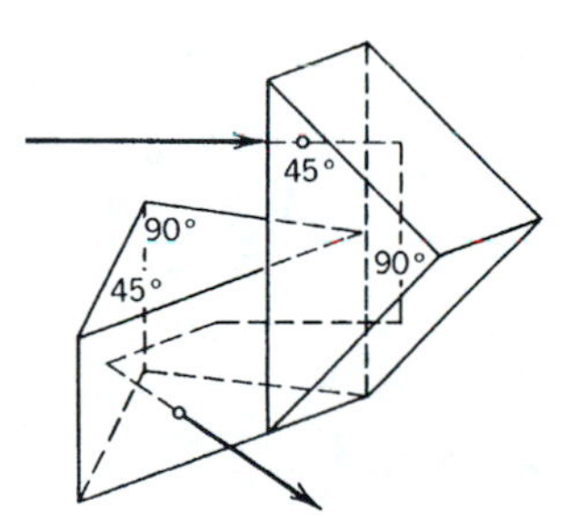

FIGURE 2.15 Reflection by two Porro prisms.

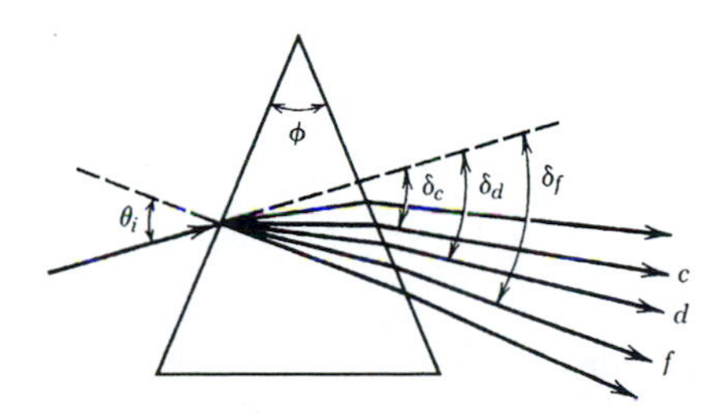

FIGURE 2.16 Effect of light dispersion. The δ_d represents the mean angle of deviation.

discussing the phenomenon of refraction. Now we shall discuss the refraction caused by different wavelengths of light. Although the velocities of light in a vacuum are the same, the wave propagation in a material is dependent on its wavelength. Thus, the refractive index of the material must be wavelength-dependent. The relation between wave propagation and wavelength is known as the *dispersion effect*.

To demonstrate the dispersion of light, let us consider a light ray that strikes the surface of a triangular prism at an angle θ_i, as shown in Figure 2.16. The prism's apex angle is denoted by ϕ, and its refractive index is denoted by η. The medium on either side of the prism is air, and the *angle of deviation* δ can be determined by using Snell's law of refraction. If the angle of incidence is decreasing, the angle of deviation δ can be shown first to decrease, and then to increase. There is a minimum angle of deviation δ_m if the light ray passes the triangular prism symmetrically. We can also show that the refractive index η of the prism satisfies the following equation:

$$\eta = \frac{\sin[(\phi + \delta_m)/2]}{\sin(\phi/2)}. \tag{2.29}$$

Furthermore, for small angles, $\sin\theta \approx \theta$, the preceding equation can be approximated by

$$\eta \simeq \frac{\phi + \delta_m}{\phi}, \tag{2.30}$$

in which we have

$$\delta_m \simeq (\eta - 1)\phi. \tag{2.31}$$

This is a very useful approximation.

Assume that a ray of white light strikes the triangular prism shown in Figure 2.16. Since a larger angle of deviation corresponds to a larger refractive index (see Figure 2.17), violet light would emerge with the largest angle of deviation, and red light would emerge with the smallest angle. Hence, the light emerging from the prism would disperse into a spectrum of rainbow colors wherein each color corresponds to a different angle of deviation. The difference between the angles of deviation of *any* two rays is called the *angle of dispersion*.

The *dispersive power* of a given material is defined as

$$W \triangleq \frac{\eta_f - \eta_c}{\eta_d - 1}, \tag{2.32}$$

where η_c, η_d, and η_f are the refractive indices of the material with respect to the red, yellow, and blue wavelengths. Notice that these wavelengths are arbitrarily selected for reference.

We further note that this dispersive power is the *measure* of the refractive property of a given material to bend spectral lines of light in different angles. A prism (composed of a given material) has this ability to separate polychromatic light into spectral lines of colors.

Consider a narrow beam of polychromatic visible light that is incident to a triangular prism with a narrow apex angle. The minimum angles of deviation for red, yellow, and blue light are denoted by δ_c, δ_d, and δ_f, respectively. From Eq. 2.31 we have

$$\delta_c = (\eta_c - 1)\phi, \tag{2.33}$$

$$\delta_d = (\eta_d - 1)\phi, \tag{2.34}$$

$$\delta_f = (\eta_f - 1)\phi, \tag{2.35}$$

where δ_d is the *mean angular deviation* of the spectrum. From these equations we see that

$$\delta_c - \delta_f = (\eta_c - \eta_f)\phi.$$

Dividing by Eq. 2.34, we have

$$\frac{\delta_f - \delta_c}{\delta_d} = \frac{\eta_f - \eta_c}{\eta_d - 1} = W, \tag{2.36}$$

which is the dispersive power defined by Eq. 2.32. From this we see that the dispersive power is also equal to the ratio of the dispersion between the blue and red angles of deviation (i.e., $\delta_f - \delta_c$), to the mean deviation δ_d of the entire spectrum.

Example 2.8

Figure 2.17 shows the refractive indices of silicate flint glass and borate flint glass as functions of wavelength. We use Eq. 2.32 to determine their corresponding dispersive powers, and we assume that the spectral lines of 475 nm, 550 nm, and 650 nm can be used to determine the reference indices of the materials.

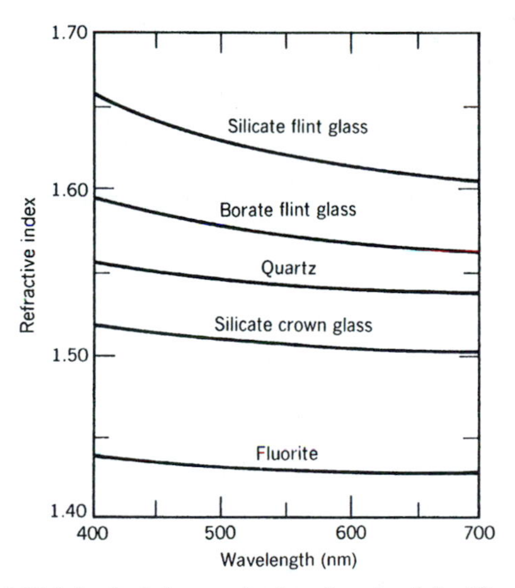

FIGURE 2.17 Refractive indices as a function of wavelength for different media.

From Figure 2.17 we obtain the following refractive indices for silicate flint glass: $\eta_f = 1.635$, $\eta_d = 1.625$, and $\eta_c = 1.618$. Substituting these into Eq. 2.36, we obtain the dispersive power for silicate flint glass:

$$W = \frac{1.635 - 1.618}{1.625 - 1} = 0.0272.$$

Using this same method, we can obtain the refractive indices for borate flint glass: $\eta_f = 1.58$, $\eta_d = 1.57$, and $\eta_c = 1.565$. Thus, the dispersive power for silicate crown glass is

$$W = \frac{1.58 - 1.565}{1.57 - 1} = 0.0175.$$ ■

■ Example 2.9

Given an apex angle of 9°, calculate the mean angular deviation and the power of dispersion produced by a prism composed of silicate flint glass and one composed of borate flint glass. Since the apex angle is small, Eq. 2.31 can be applied.

For silicate flint glass:

$$\text{Mean deviation} = \delta_d = (\eta_d - 1)\phi = (1.62 - 1)9° = 5.58°.$$

$$\text{Dispersion} = W\delta_d = 0.0272 \times 5.58° = 0.15°.$$

For borate flint glass:

$$\text{Mean deviation} = \delta_d = (\eta_d - 1)\phi = (1.57 - 1)9° = 5.13°.$$

$$\text{Dispersion} = W\delta_d = 0.0175 \times 5.13° = 0.089°.$$ ■

REFERENCES

1. F. W. SEARS, *Optics*, Addison-Wesley, Reading, Mass., 1949.
2. B. ROSSI, *Optics*, Addison-Wesley, Reading, Mass., 1957.
3. F. A. JENKINS and H. E. WHITE, *Fundamentals of Optics*, fourth edition, McGraw-Hill, New York, 1976.
4. E. HECHT and A. ZAJAC, *Optics*, Addison-Wesley, Reading, Mass., 1974.

PROBLEMS

2.1 A light ray is incident on a plane surface separating two transparent media of refractive indices 1.65 and 1.35. If the angle of incidence is 35° and the light ray originates in the medium of higher refractive index,

(a) Compute the angle of refraction.

(b) Repeat part a, assuming that the light ray originates in the medium of lower refractive index.

2.2 Consider two identical beakers, one filled with water ($\eta = 1.33$) and the other filled with carbon disulphide ($\eta = 1.63$). If the observer views them from above,

(a) Which beaker appears to have a greater depth of liquid?

(b) Calculate the ratio of the apparent depths.

2.3 A beaker contains a layer of ether alcohol ($\eta = 1.36$) 3 cm deep, which floats on a layer of water ($\eta = 1.33$) 4 cm deep. If an observer views the beaker from above, calculate the apparent distance from the surface of the ether to the bottom of the water layer.

2.4 A glass plate about 4 cm thick is held about 5 cm above a printed page. If its view is from above, calculate the location of the printed page. We assume that the refractive index of the glass is $\eta = 1.5$ and that for air it is $\eta = 1$.

2.5 Consider that a light ray is obliquely incident at an angle of 45° on the surface of a 4 cm thick glass plate. We assume that air ($\eta = 1$) is on both sides of the plate and that the refractive index of the glass is $\eta = 1.5$. Calculate the transverse displacement as the incident ray emerges through the plate.

2.6 Assume that a layer of oil ($\eta = 1.63$) about 0.5 cm thick is floating on a body of water ($\eta = 1.33$). A light ray originating from the body of water impinges on the boundary surface between the water and the oil at an incident angle of 25°.

(a) Calculate the angle of refraction in the oil.

(b) If the medium above the oil layer is air ($\eta = 1$), will the light ray be totally reflected?

(c) What is the critical angle of incidence between the oil and the air?

2.7 A cone of spherical wavefronts, which originates from a point source S, is incident on a boundary surface between two transparent media, as shown in Figure 2.18. Use Huygens' principle to construct the reflected and the refracted wavefronts.

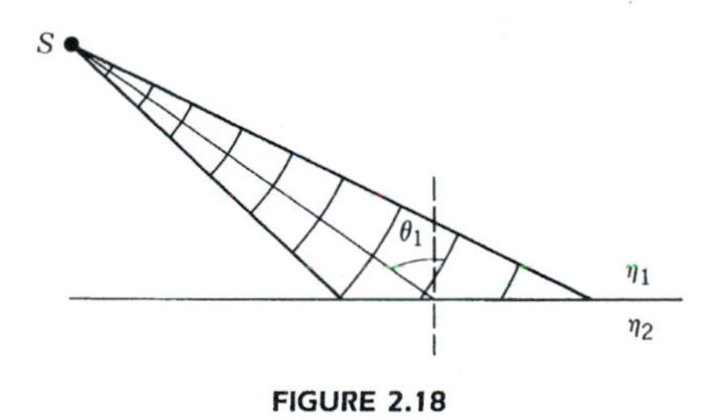

FIGURE 2.18

2.8 An observer is looking perpendicularly through a pond of water. If the bottom of the pond appears to be at a depth of 5 ft, calculate its actual depth. We assume that the refractive index of water is $\eta = 1.33$.

2.9 Suppose a light ray is incident on a rectangular glass plate, which is submerged in clear water, as shown in Figure 2.19. Calculate the light ray's

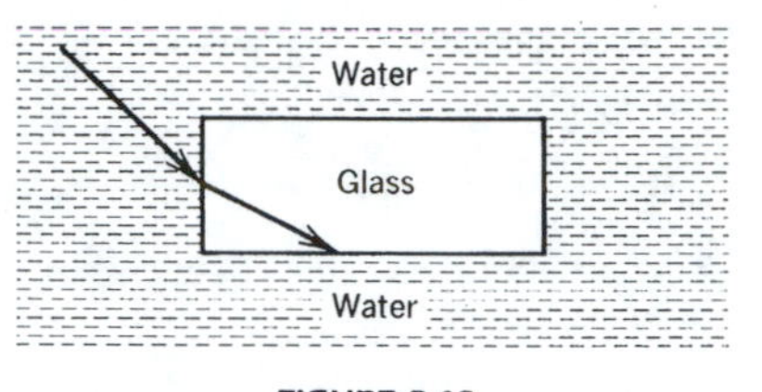

FIGURE 2.19

maximum angle of incidence on the left vertical surface of the glass that will make total internal reflection occur at the top surface of the glass. We assume that the refractive indices of the glass and the water are $\eta = 1.5$ and $\eta = 1.33$, respectively.

2.10 Repeat Problem 2.9 with the same glass plate when it is surrounded by air ($\eta = 1$).

2.11 If a 45°–45°–90° prism is submerged in a body of water ($\eta = 1.33$), calculate the refractive index for the prism that would be required for the normally incident light ray (as shown in Figure 2.13) to be totally reflected.

2.12 Assume that a light ray is normally incident on the shorted face of a 30°–60°–90° prism ($\eta = 1.6$), as depicted in Figure 2.20. If a drop of liquid is placed on the top of the prism, calculate the refractive index of the liquid necessary to produce a total internal reflection.

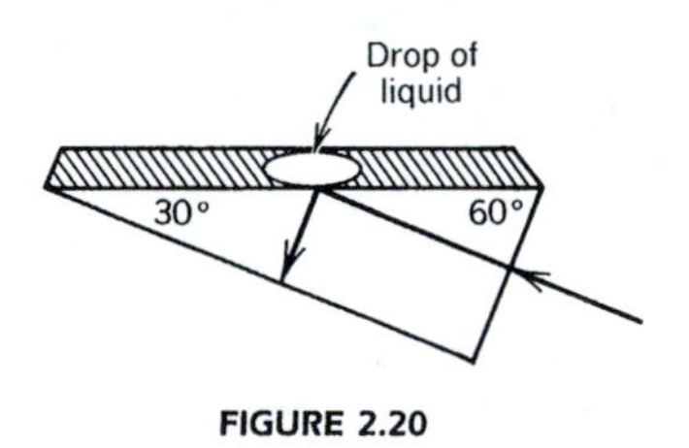

FIGURE 2.20

2.13 Given an equiangular prism made of silicate crown glass, whose index of refraction is that shown in Figure 2.17.

(a) Calculate the angles of minimum deviation for wavelengths of 700 nm and 400 nm, respectively.

(b) Determine the dispersive power of the prism. We assume that spectral lines of 475 nm, 550 nm, and 650 nm (i.e., blue, yellow, and red colors) are used for determining η_f, η_d, and η_c, respectively.

2.14 Given a prism made of silicate crown glass that has an apex angle of 10°, calculate the mean angular deviation and the dispersive power of the prism.

3 LENSES AND ABERRATIONS

A *lens* is defined as an optical element consisting of two or more refractive surfaces that share a common optical axis. A lens consisting of two refractive surfaces is called a *simple lens*, and one that has more than two refractive surfaces is a *compound lens*. Hence, a compound lens is made by combining two or more simple lenses. Generally, the space between the refractive surfaces of the lenses is filled either with air or with the appropriate refractive fluid. High-quality lenses are generally compound lenses, although there are some exceptions. Most compound lenses are aberration-free and are called *diffraction-limited.*

When a light ray passes through a simple lens, it is refracted by both surfaces of the lens. If a lens, simple or compound, causes a negligible deviation of the refracted ray, it is regarded as a *thin lens.* On the other hand, if the deviation cannot be ignored, the lens is called a *thick lens.*

3.1 IMAGE FORMATION

Again with some exceptions, the surfaces of all practical lenses are spherical in shape. We now discuss the reflection and refraction of the spherical surface shown in Figure 3.1.

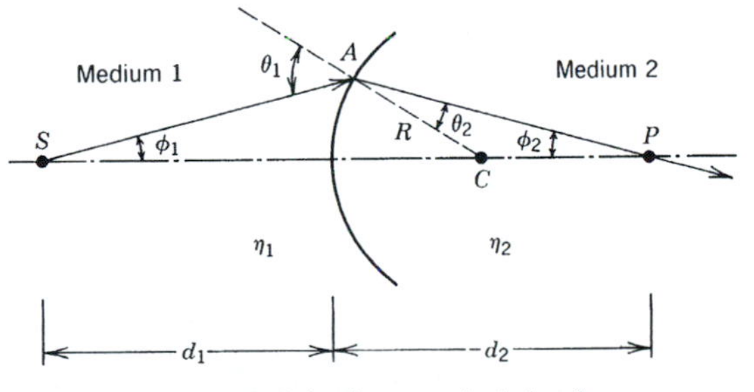

FIGURE 3.1 Refraction at a spherical surface.

In the figure a spherical surface of radius R separates two transparent media. Medium 1 is represented by its index of refraction η_1, and medium 2 is represented by η_2. We assume that all light rays scattered from the object point S will ultimately converge to point P, called the *image* point. The distance d_2 from P to the vertex of the spherical surface is called the *image distance*, and distance d_1 from S to the vertex is called the *object distance*. Thus, by the law of sines, we have

$$\frac{\sin(\pi - \theta_1)}{\sin \phi_1} = \frac{R + d_1}{R}.$$

Since $\sin(\pi - \theta_1) = \sin \theta_1$, the preceding equation reduces to

$$\sin \theta_1 = \frac{R + d_1}{R} \sin \phi_1. \tag{3.1}$$

From triangle SAP we see that

$$\phi_1 + \phi_2 + \theta_2 + (\pi - \theta_1) = \pi,$$

which reduces to

$$\phi_2 = \theta_1 - \theta_2 - \phi_1. \tag{3.2}$$

Using triangle ACP and utilizing the law of sines, we have

$$\frac{d_2 - R}{\sin \theta_2} = \frac{R}{\sin \phi_2}.$$

Thus, the image distance can be written as

$$d_2 = R + R\frac{\sin \theta_2}{\sin \phi_2} = R\left[1 + \frac{\sin \theta_2}{\sin(\theta_1 - \theta_2 - \phi_1)}\right]. \tag{3.3}$$

From this equation we see that the image distance d_2 increases as the emission angle ϕ_1 decreases. This means that the light rays originating from the object point S, do not in actuality all intersect at a common point after the refraction. In other words, the refracted wavefront is no longer spherical in shape, so it creates an image aberration. This type of image aberration is commonly called *spherical aberration*, as will be discussed in Section 3.4.

In addition, we note that if the emission angle ϕ_1 is sufficiently small, the incident light rays can be considered parallel to the optical axis and hence are called *paraxial rays*. Thus, if ϕ_1 is small, the incident and refraction angles of θ_1 and θ_2 will also be small. As a result, the sines of these angles can be approximated by $\sin \phi_1 \simeq \phi_1$, $\sin \phi_2 \simeq \phi_2$, $\sin \theta_1 \simeq \theta_1$, and $\sin \theta_2 \simeq \theta_2$. Thus, from Eqs. 3.1, 3.2, and 3.3, we obtain the following useful relation,

$$\frac{\eta_1}{d_1} + \frac{\eta_2}{d_2} = \frac{\eta_2 - \eta_1}{R}, \tag{3.4}$$

which is known as the *lens maker's equation or the Gaussian formula* for a single spherical surface. The lens maker's equation is derived under the paraxial-ray condition, in which it is independent of the emission angle ϕ_1. Thus, the light rays originating from the object point, under the *paraxial approximation*, will ultimately converge to a common image point after refraction.

It is apparent that if the object is closer to the boundary surface, a divergent refracted wavefront will be obtained, as shown in Figure 3.2. Thus, we see that the image point P located behind the spherical surface, as shown in Figure 3.1, is the *real image*, and the image point P' appearing in front of the spherical surface, as shown in Figure 3.2, is the *virtual image*.

Since a *focal point* is defined as the point through which all the incoming paraxial rays eventually intersect, the *focal length* must be the distance from the vertex of the spherical surface to the focal point. If we assume that the surface is illuminated by a plane wavefront (i.e., parallel rays), the object distance is $d_1 = \infty$ and the image distance is $d_2 = f_2$. By applying the lens maker's equation, Eq. 3.4, we have

$$f_2 = \frac{\eta_2}{\eta_2 - \eta_1} R, \tag{3.5}$$

which is called the *second focal length* of the spherical surface. On the other hand, if we assume that the object point is located at $d_1 = f_1$, the image distance is $d_2 = \infty$. Thus, by Eq. 3.4 we obtain

$$f_1 = \frac{\eta_1}{\eta_2 - \eta_1} R, \tag{3.6}$$

which is known as the *first focal length* of the spherical surface.

The *lateral magnification* of a refractive image can be obtained by using the paraxial ray-tracing technique, as depicted in Figure 3.3. Lateral magnification is defined as

$$M = \frac{h_2}{h_1}, \tag{3.7}$$

where h_1 and h_2 are the separation of the object and image points, respectively.

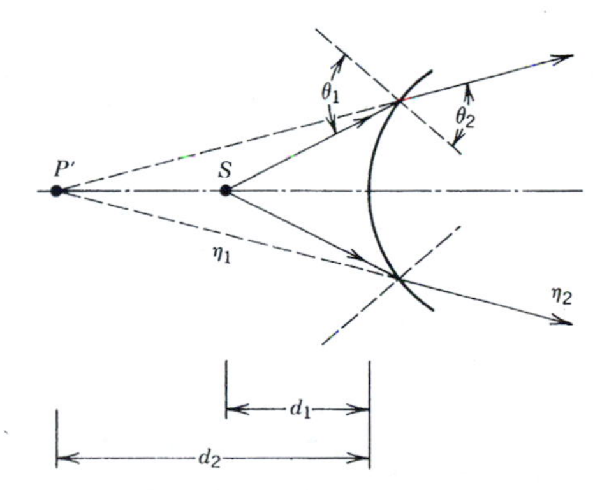

FIGURE 3.2 Divergent refraction at a spherical surface.

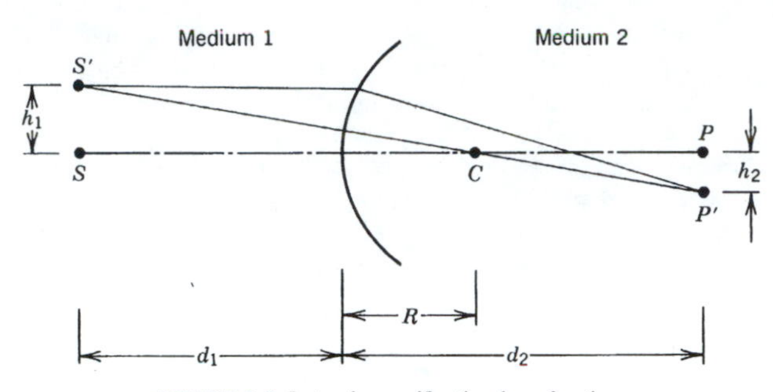

FIGURE 3.3 Lateral magnification by refraction.

We now derive an expression for lateral magnification in terms of the object and image distances and the refractive indices of the media. From the right triangles of SCS' and PCP' in Figure 3.3, we see that

$$M = \frac{h_2}{h_1} = \frac{CP}{CS} = \frac{d_2 - R}{d_1 + R}. \tag{3.8}$$

Substituting Eqs. 3.1 and 3.3 into Eq. 3.8, we have

$$M = \frac{h_2}{h_1} = \frac{\sin\theta_2 \sin\phi_1}{\sin\theta_1 \sin\phi_2}. \tag{3.9}$$

Using Snell's law of refraction, we can reduce this expression to

$$M = \frac{h_2}{h_1} = \frac{\eta_1 \sin\phi_1}{\eta_2 \sin\phi_2}, \tag{3.10}$$

which can be written as

$$h_1\eta_1 \sin\phi_1 = h_2\eta_2 \sin\phi_2. \tag{3.11}$$

This is *Abbe's sine condition*, a well-known relation that must hold to prevent a lens from producing a coma, or pear-shaped spot (see Section 3.4).

Since the emission angle ϕ_1 and the receiving angle ϕ_2 are relatively small, that is, $\sin\phi_1 \simeq h_1/d_1$ and $\sin\phi_2 \simeq h_2/d_2$, Eq. 3.10 can be further reduced to the following form:

$$M = -\frac{\eta_1 d_2}{\eta_2 d_1}. \tag{3.12}$$

■ Example 3.1

Given the spherical convex surface shown in Figure 3.1, the refractive index of medium 1 is $\eta_1 = 1$, of medium 2 it is $\eta_2 = 1.5$, and the radius of the spherical surface is $R = 20$ mm. If an object 2 mm tall is located at a distance of 100 mm from the vertex, calculate the location and the lateral magnification of the image.

By applying these values to Eq. 3.4, we can compute the image distance as

$$\frac{1}{100} + \frac{1.5}{d_2} = \frac{1.5 - 1}{20}$$

$$d_2 = 100 \text{ mm}.$$

Since the image distance d_2 is a positive quantity, a *real* image is formed behind the spherical surface, in medium 2, at a distance of 100 mm from the vertex.

Using Eq. 3.12, we can calculate the corresponding lateral magnification,

$$M = \frac{h_2}{h_1} = -\frac{1 \times 100}{1.5 \times 100} = -0.667.$$

Thus, we see that the image is inverted and appears to be smaller (i.e., demagnified) than the object. ■

■ Example 3.2

Let us continue the preceding problem. If the index of refraction of medium 1 is filled with water, which has a refractive index of $\eta = 1.33$, calculate the location of the refractive image and the corresponding lateral magnification.

Again we use Eq. 3.4 to find the image distance:

$$\frac{1.33}{100} + \frac{1.5}{d_2} = \frac{1.5 - 1.33}{20}$$

$$d_2 = -312.5 \text{ mm}.$$

Since the image distance is a negative quantity, a virtual image is formed in front of the spherical surface in medium 1, at a distance of 312.5 mm from the vertex.

Again from Eq. 3.12, the lateral magnification can be evaluated:

$$M = \frac{h_2}{h_1} = -\frac{(1.33)(-312.5)}{1.5 \times 100} = 2.77.$$

Thus, we see that the image is erected vertically and is about 2.77 times larger (i.e., magnified) than the actual object. ■

3.2 SIMPLE LENSES

We now consider the refraction produced by a simple lens. We assume that the light rays scattered from an object point S are incident on a simple lens L, as shown in Figure 3.4. The refractive index of the lens is η, and R_1 and R_2 are the radii of curvature of the first and second surfaces of the lens, respectively. Using the principles of reflection and refraction, we see that the first surface of the lens produces a virtual image at point S' behind object point S, and the lens produces a real image at point P behind the lens. By applying the lens maker's equation,

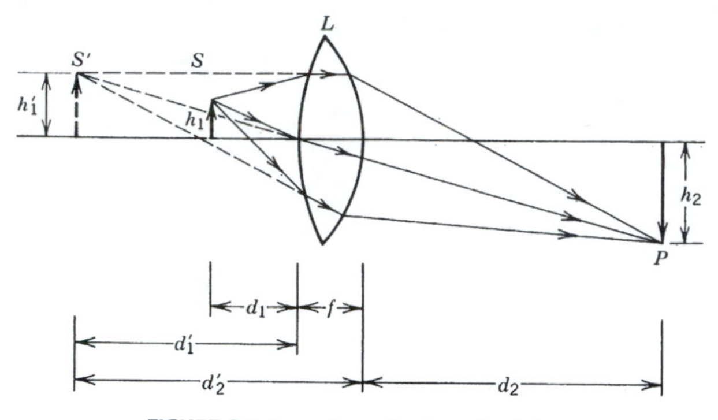

FIGURE 3.4 Image formation by a simple lens.

Eq. 3.4, we can determine the distance of the virtual image,

$$\frac{1}{d_1} + \frac{\eta}{d'_1} = \frac{\eta - 1}{R_1}, \tag{3.13}$$

where R_1 is the radius of curvature of the first surface, and d_1 and d'_1 are the object and the virtual-image distances, as they are measured from the vertex of the first surface.

Similarly, the distance of the real image can be determined for the second surface,

$$\frac{\eta}{d'_2} + \frac{1}{d_2} = \frac{1 - \eta}{R_2}, \tag{3.14}$$

where R_2 is the radius of curvature of the second surface, and d'_2 and d_2 are the virtual-image and the real-image distances, as they are measured from the vertex of the second surface. Notice that Eqs. 3.13 and 3.14 can be used to calculate the image positions and the lateral magnifications for all types of lenses, including compound lenses.

Since a thin lens can be regarded as a plane, a light ray incident on a point on the lens will emerge from the same point behind the lens. To find the focal point of a thin lens, we assume that the object point is located at a distance from the lens, behind which the image point is formed at infinity (i.e., $d_2 = \infty$). Using Eq. 3.14, we obtain

$$\frac{\eta}{d'_2} = \frac{1 - \eta}{R_2}, \tag{3.15}$$

where d'_2 is the virtual-image distance from the vertex of the second surface of the lens.

Since the thickness of a thin lens is negligibly thin, that is, $d'_1 \simeq d'_2$, Eq. 3.15 can also be written as

$$\frac{\eta}{d'_1} = \frac{1 - \eta}{R_2}. \tag{3.16}$$

By substituting Eq. 3.16 in Eq. 3.13, we have

$$\frac{1}{d_1} = (\eta - 1)\left(\frac{1}{R_1} + \frac{1}{R_2}\right), \tag{3.17}$$

where R_1 and R_2 are the radii of curvature of the first and second surfaces of the lens, η is the refractive index of the lens, and d_1 is the distance from the object to the center plane of the lens. By the definition of the first focal length, that is, $f_1 = d_1$, we see that

$$\frac{1}{f_1} = (\eta - 1)\left(\frac{1}{R_1} + \frac{1}{R_2}\right). \tag{3.18}$$

Similarly, the second focal length can be found from

$$\frac{1}{f_2} = (\eta - 1)\left(\frac{1}{R_1} + \frac{1}{R_2}\right), \tag{3.19}$$

which is identical to Eq. 3.18. Thus, when the refractive indices on both sides of the lens are equal, the front and the back focal lengths are also equal. For a *thin lens in the air*, we therefore simply write

$$\frac{1}{f} = (\eta - 1)\left(\frac{1}{R_1} + \frac{1}{R_2}\right), \tag{3.20}$$

which is also known as the *spherical lens maker's equation*, derived from Eq. 3.4.

Let us now derive several useful expressions for thin lenses. We assume that an extended object is located in front of a thin lens, as shown in Figure 3.5. The height of the object is denoted by h_1, and h_2 is the height of the image. We see that the light rays from object point S converge at image point P. We also know that the intersection of any two rays from S is adequate for determining the position of image point P. From triangles $SS'F_1$, F_1CA, BCF_2, and $F_2P'P$, we see that

$$\frac{h_1}{z_1} = \frac{h_2}{f}, \qquad \frac{h_1}{f} = \frac{h_2}{z_2}. \tag{3.21}$$

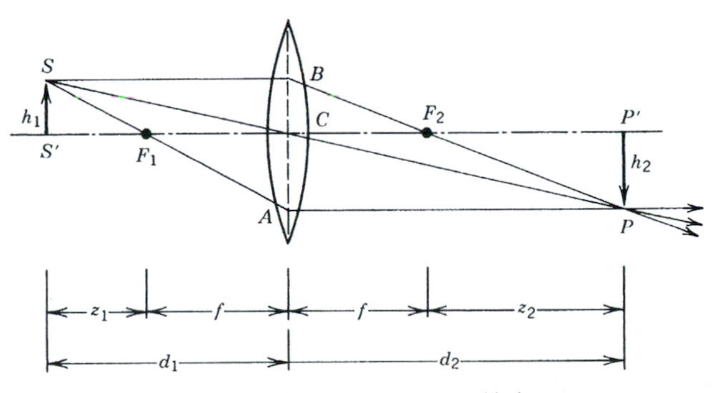

FIGURE 3.5 Image formation by a thin lens.

By dividing these equations, we have

$$f^2 = z_1 z_2, \tag{3.22}$$

which is known as the *Newtonian form* lens equation. Similarly, from triangles SBA, F_1CA, BAP, and BCF_2, we have

$$\frac{h_1 + h_2}{d_1} = \frac{h_2}{f}, \qquad \frac{h_1 + h_2}{d_2} = \frac{h_1}{f}. \tag{3.23}$$

Thus, the lateral magnification is

$$M = \frac{h_2}{h_1} = -\frac{d_2}{d_1}, \tag{3.24}$$

where the negative sign represents an inverted image. Then, by adding Eqs. 3.23, we obtain

$$\frac{1}{d_1} + \frac{1}{d_2} = \frac{1}{f}, \tag{3.25}$$

which is the famous *Gaussian lens equation.* We further note that in the Gaussian lens equation the object distances are measured from the lens, whereas in the Newtonian form lens equation they are measured from the focal points.

Example 3.3

Consider a planoconvex thin lens having a focal length of 20 cm (see Figure 3.6). If the refractive index of the lens is 1.52, calculate the radius of curvature of the convex surface.

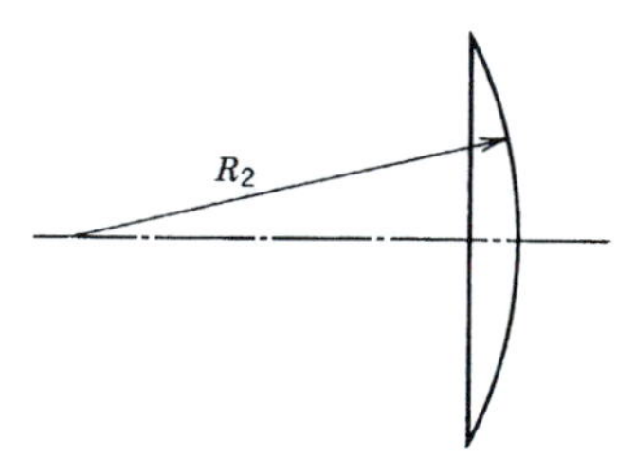

FIGURE 3.6

Since the first surface of the lens is a plane, its radius of curvature is $R_1 = \infty$. By applying the spherical lens maker's equation, Eq. 3.20, we obtain

$$\frac{1}{20} = (1.52 - 1)\left(\frac{1}{\infty} + \frac{1}{R_2}\right)$$

$$R_2 = 10.4 \text{ cm}.$$

Example 3.4

Consider again the planoconvex thin lens shown in Figure 3.6 and described in Example 3.3. If the radius of curvature of the second surface on the planoconvex thin lens is 30 cm, calculate the focal length of the lens.

From the previous example we have $\eta = 1.52$. By applying Eq. 3.20, we obtain

$$\frac{1}{f} = (1.52 - 1)\left(\frac{1}{\infty} + \frac{1}{30}\right)$$

or

$$f = \frac{30}{0.52} = 57.69 \text{ cm.}$$

Notice that if we reverse the planoconvex lens, we still obtain the same focal length. This result shows that when the indices of refraction between two sites on the lens are the same, the first and second focal lengths are equal. ■

Example 3.5

An object is located 35 cm to the left of a thin convex lens. If the focal length is 20 cm, calculate the position and the lateral magnification of the image.

In this problem we have $d = 35$ cm and $f = 20$ cm. By applying Eq. 3.25, the Gaussian lens equation, we obtain

$$\frac{1}{35} + \frac{1}{d_2} = \frac{1}{20}$$
$$d_2 = 46.66 \text{ cm.}$$

We see that the image is real and that it is located 46.66 cm behind the lens.

Moreover, from the Newtonian lens equation, Eq. 3.22, we have

$$(20)^2 = (35 - 20)z_2$$
$$z_2 = 26.66 \text{ cm,}$$

which is in agreement with the result obtained with the Gaussian lens equation.

In computing the lateral magnification, we use Eq. 3.24,

$$M = -\frac{46.66}{35} = -1.33.$$

Thus, we see that the image is inverted and is about 1.33 times larger than the object. ■

3.3 PHASE RETARDATION BY THIN LENSES

An important property of thin lenses is *phase retardation*, or *delay*. As we have illustrated in the preceding chapter, light rays that propagate in a dielectric medium can be described by *wave theory*. Thus, a wavefront that propagates through a transparent dielectric medium (e.g., a lens) exhibits severe wavefront deformation.

Let us consider a light ray entering at a point on one side of a thin lens and emerging from the same point on the other side of the lens. Since the lens is assumed to be thin, the transversal displacement of the light ray inside the lens is negligible. Therefore, only the phase is retarded when a wavefront passes through a thin lens. We stress that the amount of phase retardation is proportional to the variation in thickness of the lens.

In order to calculate the phase retardation of a lens, let us first consider the variation in thickness of a convergent (i.e., positive) lens, as shown in Figure 3.7. The phase retardation $\phi(x, y)$ on the wavefront as it passes through (i.e., is refracted by) the lens can be written as

$$\phi(x, y) = k[\Delta t + (\eta - 1)t(x, y)], \tag{3.26}$$

where $t(x, y)$ is the variation in thickness of the lens, Δt is its maximum thickness, η is its refractive index, $k = 2\pi/\lambda$ is the wave number, λ is the wavelength, and x and y are the spatial coordinates of the lens. We note that the quantity

$$\phi_1(x, y) = k\eta t(x, y)$$

represents the phase retardation caused by the lens alone, and

$$k[\Delta t - t(x, y)]$$

represents the phase retardation caused by free space (or air) alone. Notice that the *phase transform* function of the lens can be represented by the following:

$$T(x, y) = \exp[i\phi(x, y)] = \exp\{ik[\Delta t + (\eta - 1)t(x, y)]\}. \tag{3.27}$$

If the thin lens is illuminated by a monochromatic wavefront $u_i(x, y)$ at plane P_1 (see Figure 3.7), the wavefront that emerges through the lens can be expressed as

$$u_o(x, y) = u_i(x, y)T(x, y), \tag{3.28}$$

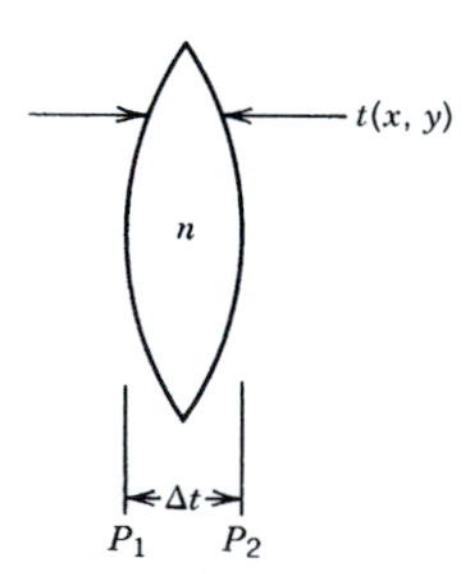

FIGURE 3.7 Variation in the thickness of a convex lens.

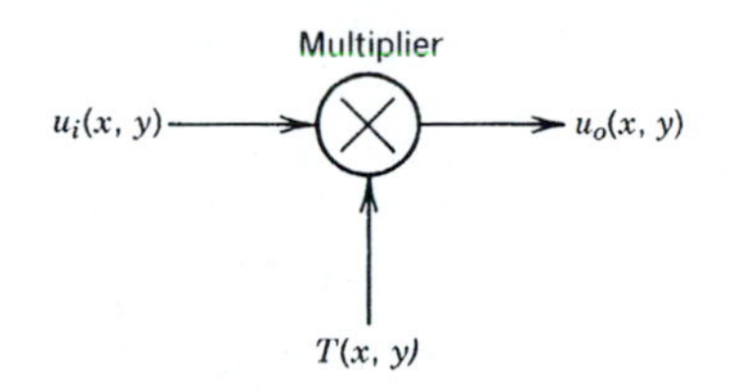

FIGURE 3.8 System analog diagram of a lens transformation.

where $T(x, y)$ is the phase transform of the thin lens. Thus, we see that the output wavefront is the product of the input wavefront multiplied by the phase transform function of the lens. For convenience, an analog system representation of Figure 3.7 is given in Figure 3.8.

To derive the phase transform of a lens, we first consider the variation in thickness of a positive lens. If we divide the lens into left and right halves, as shown in Figure 3.9, the variation in thickness of the left half can be written as

$$\begin{aligned} t_1(x, y) &= \Delta t_1 - [R_1 - (R_1^2 - \rho^2)^{1/2}] \\ &= \Delta t_1 - R_1\left\{1 - \left[1 - \left(\frac{\rho}{R_1}\right)^2\right]^{1/2}\right\}, \end{aligned} \tag{3.29}$$

where Δt_1 is the maximum thickness variation of the left half of the lens, R_1 is the radius of curvature, and

$$\rho^2 = x^2 + y^2.$$

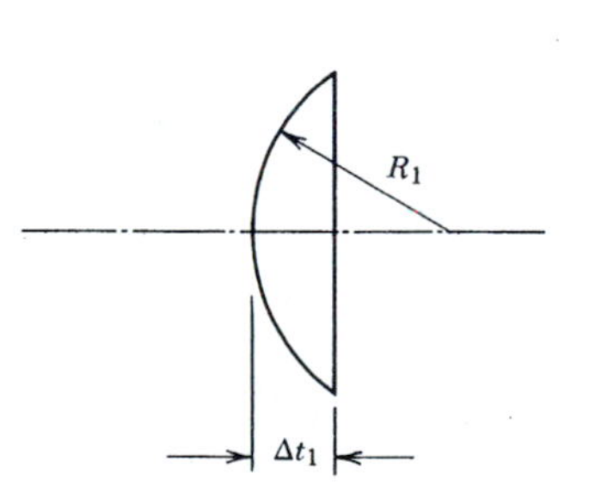

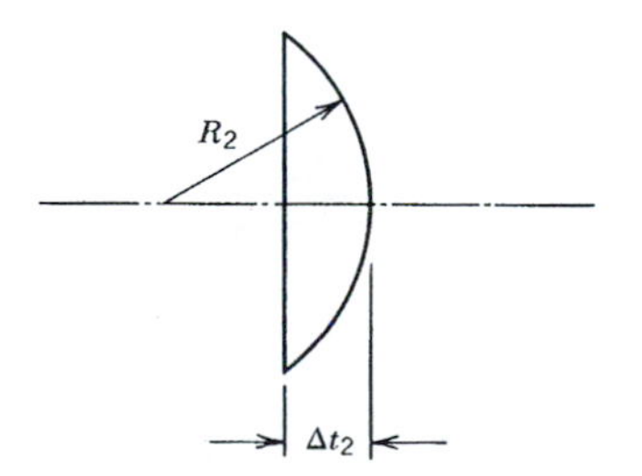

FIGURE 3.9 Left and right halves of a convex lens.

Similarly, the variation in thickness in the right half is

$$t_2(x, y) = \Delta t_2 - [R_2 - (R_2^2 - \rho^2)^{1/2}]$$
$$= \Delta t_2 - R_2\left\{1 - \left[1 - \left(\frac{\rho}{R_2}\right)^2\right]^{1/2}\right\}. \tag{3.30}$$

Thus, the overall thickness variation of the lens is

$$t(x, y) = t_1(x, y) + t_2(x, y)$$
$$= \Delta t - R_1\left\{1 - \left[1 - \left(\frac{\rho}{R_1}\right)^2\right]^{1/2}\right\} - R_2\left\{1 - \left[1 - \left(\frac{\rho}{R_2}\right)^2\right]^{1/2}\right\}, \tag{3.31}$$

where $\Delta t = \Delta t_1 + \Delta t_2$.

To simplify Eq. 3.31 we shall consider only the relatively small region of the lens that lies near the optical axis. In this case, we have

$$R_1 \gg \rho, \qquad R_2 \gg \rho.$$

Using the binomial series expansion, we can make the following approximations:

$$\left[1 - \left(\frac{\rho}{R_1}\right)^2\right]^{1/2} \simeq 1 - \frac{1}{2}\left(\frac{\rho}{R_1}\right)^2 \tag{3.32}$$

and

$$\left[1 - \left(\frac{\rho}{R_2}\right)^2\right]^{1/2} \simeq 1 - \frac{1}{2}\left(\frac{\rho}{R_2}\right)^2. \tag{3.33}$$

These approximations are known as *paraxial approximations*. Thus, the overall thickness of the positive lens can be approximated as

$$t(x, y) \simeq \Delta t - \frac{\rho^2}{2}\left(\frac{1}{R_1} + \frac{1}{R_2}\right). \tag{3.34}$$

It should also be noted that the paraxial approximations of Eqs. 3.32 and 3.33 provide the same result by replacing the spherical surfaces of the lens with *parabolic* surfaces.

Thus, the phase transform function of a positive lens can be written as

$$T(x, y) = e^{ik\eta \Delta t} \exp\left[-ik(\eta - 1)\frac{\rho^2}{2}\left(\frac{1}{R_1} + \frac{1}{R_2}\right)\right], \tag{3.35}$$

and its focal length can be identified as

$$f = \frac{R_1 R_2}{(\eta - 1)(R_1 + R_2)}. \tag{3.36}$$

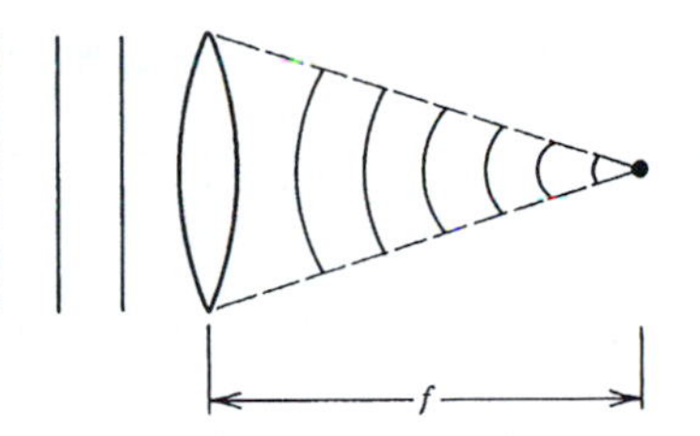

FIGURE 3.10 Convergent effect of a positive lens.

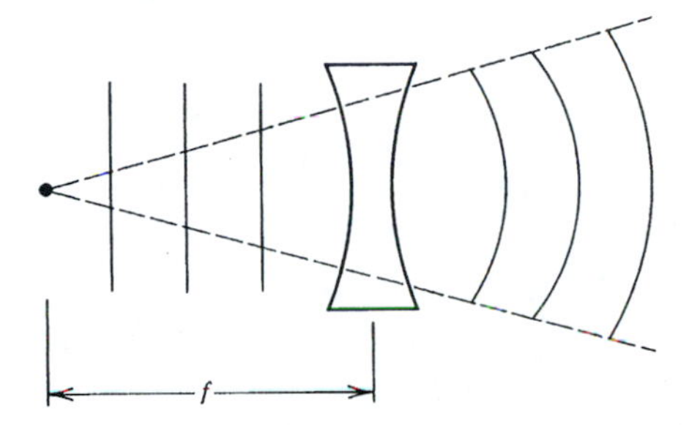

FIGURE 3.11 Divergent effect of a negative lens.

By substituting Eq. 3.36 into Eq. 3.35, we have

$$T(x, y) = C_1 \exp\left(-i \frac{k}{2f} \rho^2\right), \tag{3.37}$$

where $C_1 = e^{ik\eta\Delta t}$ is a complex constant, f is the focal length of the lens, and $\rho = \sqrt{x^2 + y^2}$.

We stress that Eq. 3.37 is very useful in the processing of optical signals. The negative exponent expresses the convergent effect of a wavefront after it is refracted by the lens. In other words, if parallel light rays (in the form of a plane wavefront) are normally incident on the lens, the wavefront emerges as a spherical wavefront after being refracted by the lens, as illustrated in Figure 3.10.

Furthermore, the phase transform function of a divergent (or negative) lens can also be derived by the same approximation. Thus, we obtain

$$T(x, y) = C_2 \exp\left(i \frac{k}{2f} \rho^2\right), \tag{3.38}$$

where C_2 is a complex constant. Notice that Eq. 3.38 is essentially identical to Eq. 3.37, except that it contains a *positive* quadratic phase factor rather than a negative one. Thus, a plane wavefront will emerge as a divergent wavefront after passing through a negative lens, as shown in Figure 3.11.

Example 3.6

Consider a positive thin lens for which the refractive index is $\eta = 1.5$ and the radii of curvature of the front and back surfaces are $R_1 = 10$ cm and $R_2 = 30$ cm, respectively. Calculate the focal length of the lens.

Using Eq. 3.36, we find that the focal length is

$$f = \frac{(10)(30)}{(1.5 - 1)(10 + 30)} = 15 \text{ cm}.$$ ■

■ Example 3.7

Consider two positive lenses of focal length f_1 and f_2 that are cascaded, as shown in Figure 3.12*a*. If this set of lenses is illuminated by monochromatic wavefront $u_i(x, y)$,

(a) Draw an analog system diagram to represent its input–output optical setup.
(b) Calculate the resultant focal length of this set of lenses.

Answers

(a) The analog system diagram of the optical system is shown in Figure 3.12*b*, where the phase transforms of the lenses are

$$T_1(x, y) = C \exp\left[-i \frac{k}{2f_1}(x^2 + y^2)\right]$$

and

$$T_2(x, y) = C \exp\left[-i \frac{k}{2f_2}(x^2 + y^2)\right].$$

(b) The overall phase transform of this setup is

$$T_1(x, y) T_2(x, y) = C \exp\left[-i \frac{k}{2}(x^2 + y^2)\left(\frac{1}{f_1} + \frac{1}{f_2}\right)\right].$$

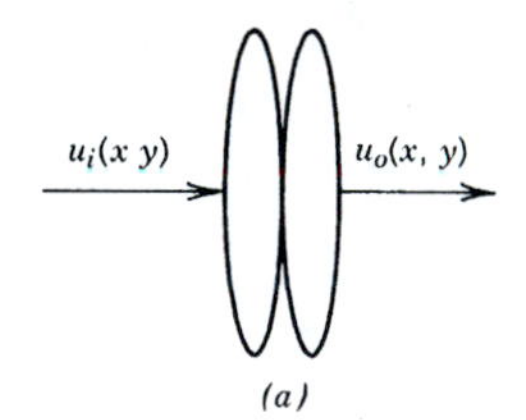

(*a*)

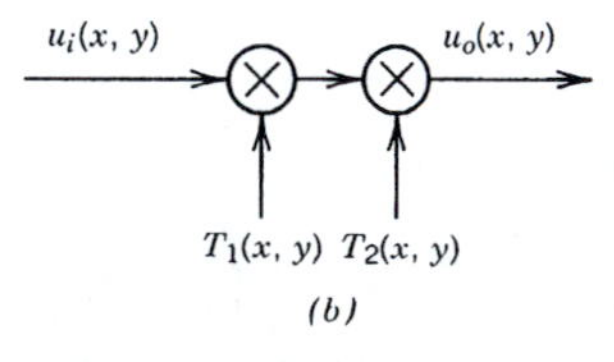

(*b*)

FIGURE 3.12

Thus, the resultant focal length is

$$f = \frac{f_1 f_2}{f_1 + f_2}.$$ ■

3.4 PRIMARY ABERRATIONS

We shall now discuss the primary aberrations of lenses. In the preceding sections we have shown several useful relations—of object and image distances, of focal lengths, of refractive indices, and of radii of curvatures. However, these relations were primarily derived under the assumption of paraxial approximation, which means that they can be applied only to the light rays that have originated from an axial object and are making small incident angles.

However, the expressions derived in the preceding sections may not hold for all light rays that reach a lens either from axial object points or from object points located away from the optical axis. Because of the finite size of the lens, the cone of light rays that forms the image point should not be treated as a very small cone. Thus, the nonparaxial rays from an object point will not in general intersect at the same image point after refraction by the lens.

In addition, from Eq. 3.36, the focal-length equation, we see that the focal length depends on the refractive index of the lens, and that the refractive index varies with the wavelength of the incident light. Strictly speaking, a broad spectrum of color images of different sizes will be in different locations in the image space of the lens. Thus, any deformation or distortion of the image that is caused by the lens and that makes it different from the *first-order approximation* is generally called a *lens aberration*. Aberrations caused primarily by the wavelength of the light source are called *chromatic aberrations*. If the light source is monochromatic, however, these aberrations are called *monochromatic aberrations*. Notice that the lens aberration has nothing to do with any physical imperfection of the lens, but is due to the refraction of light rays at the spherical surfaces.

One of the most common aberration in lenses is spherical aberration. Picture a bundle of light rays originating from an axial object point S and impinging on a positive lens, as shown in Figure 3.13. The light rays that are incident near the rim of the lens are imaged at point P', which is closer to the lens than point P where the paraxial rays are imaged. The light rays incident in the intermediate zone of the lens are imaged between points P and P'. Thus, there exists no clear, sharp image behind the lens. If we insert an observation screen between image

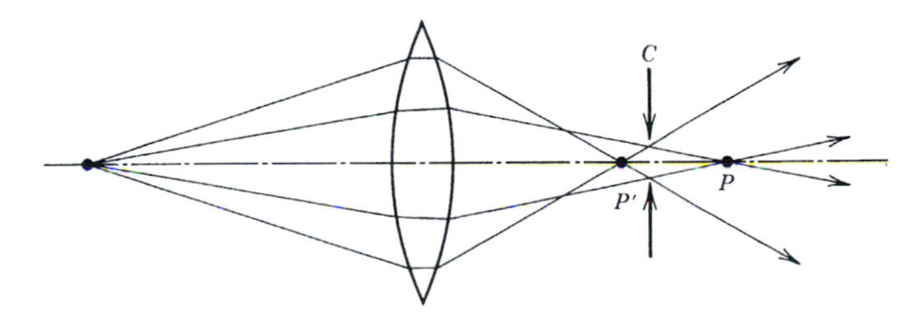

FIGURE 3.13 Spherical abberration of a lens.

points P and P', we find that the image consists of a circular disk image because the cones of light rays from various axial image points overlap. However, there is a location, known as the *circle of least confusion* or *best focus*, where the cross-sectional area of the circular disk image is smallest. The phenomenon of confusion and imprecision in image formation is generally called *spherical aberration* of the lens.

Spherical aberration of a lens can be reduced by inserting a circular diaphragm to block the light rays near the rim and allow only the center portion of the light rays to pass through the lens. However, the decrease in light transmission reduces image resolution, a subject we will discuss in the next section. We stress that spherical aberration of lenses can be minimized by designing the surface curvatures of the lenses properly, for example, by grinding the surfaces to appropriate *aspherical* shapes.

Another primary aberration of the lens is known as *coma*. Coma is similar to spherical aberration in being produced by light rays that are not incident near the optical axis of the lens. However, coma differs from spherical aberration in that the object point images as a pear-shaped spot rather than as a disk, as illustrated in Figure 3.14. The aberration is named for the coma of a comet, the gaseous envelope surrounding its nucleus.

Let us discuss qualitatively how a coma is formed. For simplicity, we assume that coma is the only lens aberration present, because if other aberrations are considered, the coma will be further enhanced. Consider an object point S that is located slightly away from the optical axis of the lens, as shown in Figure 3.14. A very small pencil cone of light rays from the object point that passes through the center zone of the lens converges to image point P. However, a narrow hollow cone of light rays from S, the shaded zone, will be imaged as a circle above image point P. The hollow cones of light rays from within the shaded zone will image as smaller circles below the circle from the shaded zone. Hollow cones that are larger than the shaded zone will image as larger circles above this circle. The total effect of these superpositions of circular images takes the blurred shape of a comet.

Like spherical aberration, coma may be corrected by the proper design of the surface curvature of the lens. Unfortunately, the curvature necessary to correct coma is generally not the curvature that corrects spherical aberration. Therefore, a lens designed for minimum spherical aberration cannot be totally free from coma.

We shall now consider the aberrations caused by *astigmatism* and *curvature of field*. Since these two aberrations come from basically the same effect, we discuss them simultaneously.

Basically, astigmatism is similar to coma except for the spread of the image in relation to the optical axis. Unlike coma, whose spread is along the optical axis,

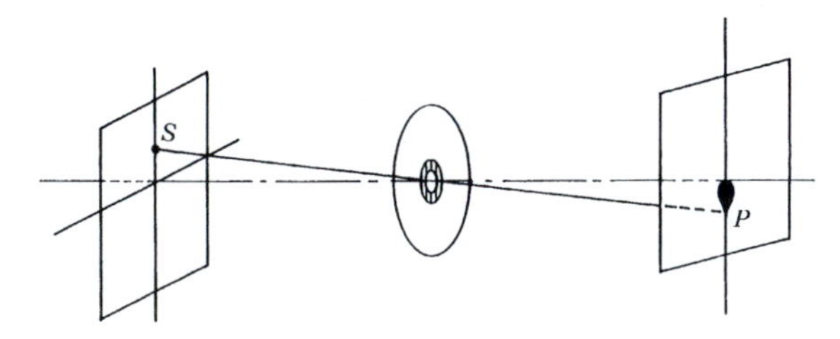

FIGURE 3.14 Coma.

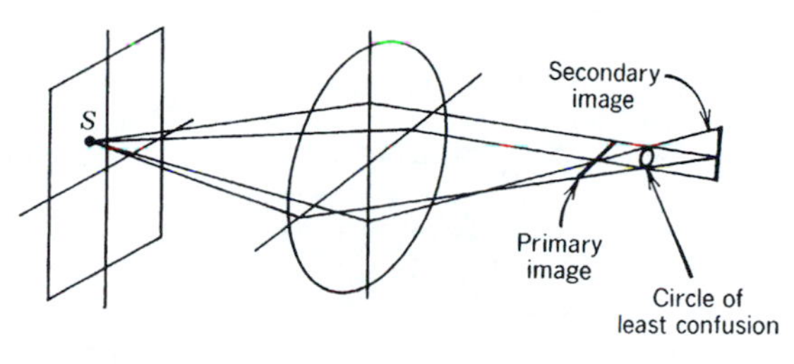

FIGURE 3.15 Astigmatism.

that of astigmatism is over a plane that is perpendicular to the optical axis. As in coma, however, the image is distorted because the object point is not located on the optical axis of the lens.

Referring to Figure 3.15, we see that after being refracted by the lens, the light rays from the object point S converge to a horizontal line image called the *primary image*, after which they converge to a vertical line image called the *secondary image*. The cross section of the refracted light beam is generally elliptical in shape, except at the primary and secondary images. Somewhere between the two line images it becomes a circle. The circular image at this location is called the *least-confused* image or the *best-focused* image.

Furthermore, if the object point is moving on a plane that is perpendicular to the optical axis of the lens, the loci of the primary and secondary images describe the *primary image surface* and the *secondary image surface*, respectively, as illustrated in Figure 3.16. The *surface of best focus* is the locus of least confusion, since all three of these surfaces are tangent at the same point on the optical axis. Thus, when the object point is located on the optical axis, the primary image, the secondary image, and the point of least confusion all converge to a point on the optical axis. Since the surface of best focus is generally a curved surface, the aberration caused by this curve is known as *curvature of field*, and the aberration caused by the noncoincident primary and secondary image surfaces is called *astigmatism* of the lens.

The curvatures of the primary image surface, the secondary image surface, and the surface of best focus are all affected by the curvatures of the lens surfaces. For both curvature of field and astigmatism to be eliminated, the primary and secondary image surfaces must be in the same plane. In practice, however, it is not possible to make them coincident and planar. In other words, it is not possible

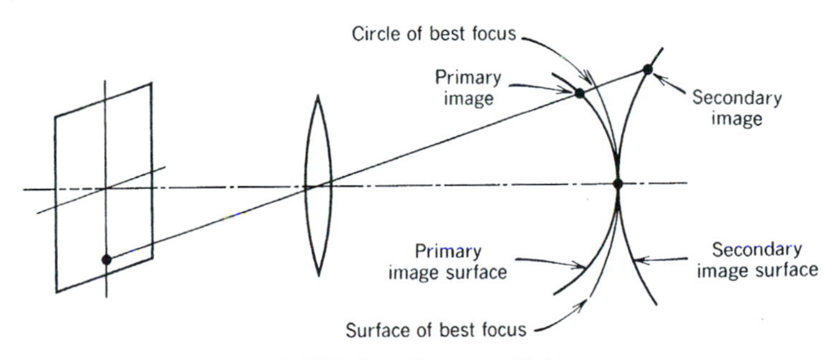

FIGURE 3.16 Curvature of field.

to eliminate simultaneously both curvature of field and astigmatism of a lens. Nevertheless, it is possible to eliminate either one or the other of these two aberrations by properly locating aperture stops near the lens surface.

Apparently, spherical aberration, coma, astigmatism, and curvature of field are aberrations in which an object point fails to image into an image point. We shall now discuss one of the last monochromatic aberrations, namely *distortion.* Distortion is an aberration that is caused not by the best focus of the image, but rather by different magnifications at different distances of the object point from the axis. For example, if the magnification decreases with increasing distance of the object point from the axis, the outer portion of the image becomes smaller. Say a circular screen is imaged by the lens; the image has laterally bulging sides, as shown in Figure 3.17*a*. This distortion is *barrel-shaped.* On the other hand, if the magnification increases as the distance of the object point from the axis increases, the image has concave sides or a *pincushion* distortion, as shown in Figure 3.17*b*. Again, these distortions can be corrected by properly locating stops near the surface of the lens.

Spherical aberration, coma, astigmatism, curvature of field, and distortion are generally known as the *five primary aberrations* of the lens. All five are primarily due to curvature of the refractive surfaces of the lens.

We now briefly discuss the *chromatic aberration* of a lens. Referring to Eq. 3.36, we notice that the focal length varies with the refractive index of the lens. Since the refractive index of the lens varies with the wavelength (see Figure 2.17), a lens will image a chromatic object into a series of images of different colors, at different locations and magnifications. The variation of image distances in relation to the refractive index of the lens is called *axial* or *longitudinal chromatic aberration.* The variation in image size caused by the refractive index of the lens is called *lat-*

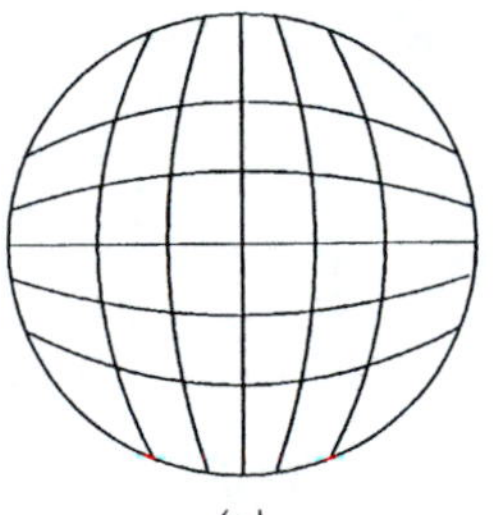

(*a*)

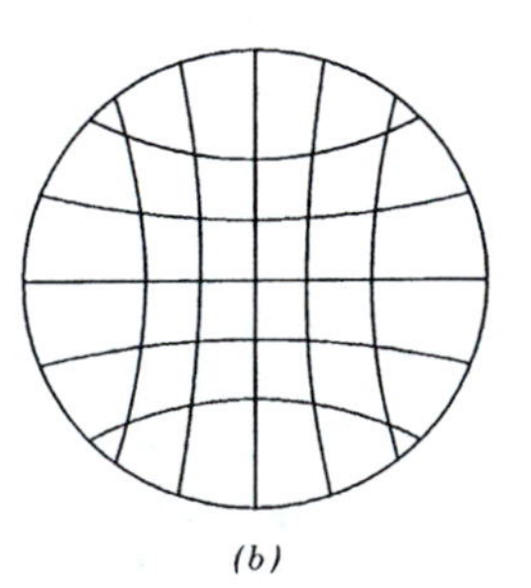

(*b*)

FIGURE 3.17 Distortion. (*a*) Barrel distortion. (*b*) Pincushion distortion.

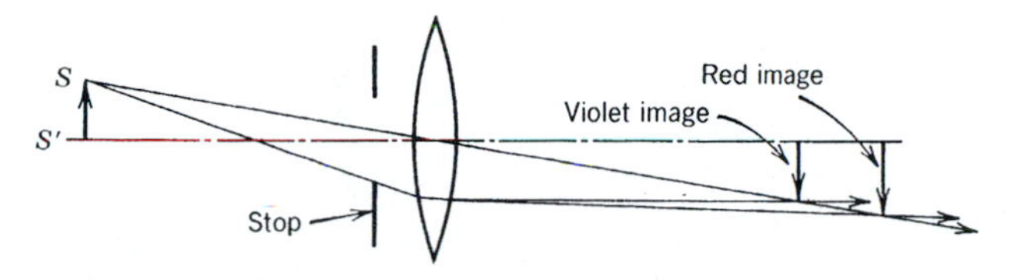

FIGURE 3.18 Chromatic aberrations of a lens.

eral chromatic aberration. Examples of these chromatic aberrations appear in Figure 3.18. The violet image is the closest to the lens and the smallest in size, whereas the red image is the farthest from the lens and the largest. The wavelengths of the other colors are imaged between the violet and red, and their sizes are correspondingly larger and smaller, depending on which of these two colors images they are closer to.

Chromatic aberration may be corrected to some degree by combining several thin lenses with different refractive indices. One such combination, called an *achromatic doublet lens,* consists of two thin lenses that have different refractive indices and are in contact. The doublet lens is designed so that the two lenses have identical focal lengths for two colors of light (e.g., red and violet), thus offsetting the chromatic aberrations of one lens by those of the other.

Example 3.8

Consider a positive lens that is made of silicate flint glass and whose surfaces have radii of curvature of $R_1 = R_2 = 30$ cm. Compute the focal lengths of the lens if it is illuminated by $\lambda_1 = 650$ nm, red light, and $\lambda_2 = 450$ nm, blue light, respectively.

Since the refractive index of the lens is a function of wavelength, that is, $\eta(\lambda)$, the focal length varies as the wavelength changes. Using Eq. 3.36, we have

$$f(\lambda) = \frac{R}{2[\eta(\lambda) - 1]}.$$

From Figure 2.17 we see that $\eta(\lambda_1) = 1.618$ and $\eta(\lambda_2) = 1.64$ for silicate flint glass. Thus, the focal lengths for λ_1 and λ_2 are

$$f(\lambda_1) = \frac{30}{2(1.618 - 1)} = 24.27 \text{ cm}$$

and

$$f(\lambda_2) = \frac{30}{2(1.64 - 1)} = 23.44 \text{ cm}.$$

■

Example 3.9

Consider a doublet lens, as shown in Figure 3.19. We assume that the positive lens is made of silicate flint glass and that the planoconcave lens, the negative lens, is made of silicate crown glass. If the radius of curvature of the surfaces is $R = 30$ cm,

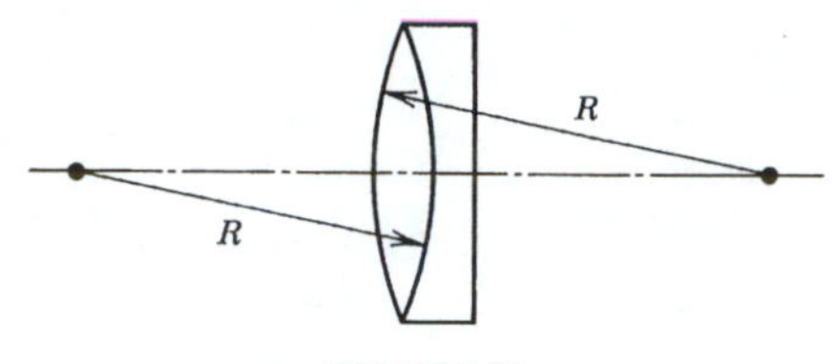

FIGURE 3.19

calculate the focal lengths of this doublet lens under red light ($\lambda_1 = 650$ nm) and blue light ($\lambda_2 = 450$ nm) illuminations.

From Eq. 3.20 we see that the focal lengths of the positive lens and the plano-concave lens can be computed, respectively, as

$$f_1 = \frac{R}{2[\eta(\lambda) - 1]}$$

and

$$f_2 = \frac{R}{\eta(\lambda) - 1}.$$

Again from Figure 2.17, we see that $\eta(\lambda_1) = 1.618$, $\eta(\lambda_2) = 1.64$ for silicate flint glass, and $\eta(\lambda_1) = 1.505$, $\eta(\lambda_2) = 1.515$ for silicate crown glass. Thus, the corresponding focal lengths are

$$f_1(\lambda_1) = \frac{30}{2(1.618 - 1)} = 24.27 \text{ cm},$$

$$f_1(\lambda_2) = \frac{30}{2(1.64 - 1)} = 23.44 \text{ cm},$$

$$f_2(\lambda_1) = \frac{30}{1.505 - 1} = 59.4 \text{ cm},$$

and

$$f_2(\lambda_2) = \frac{30}{1.515 - 1} = 58.25 \text{ cm}.$$

Since the two lenses are in close contact, the resultant focal lengths can be calculated by the following equation,

$$\frac{1}{f} = \frac{1}{f_1} - \frac{1}{f_2} \quad \text{or} \quad f = \frac{f_1 f_2}{f_2 - f_1},$$

where the minus sign reflects the effect of the negative lens. Therefore, the resultant focal lengths under red and blue light illuminations are, respectively,

$$f(\lambda_1) = \frac{f_1(\lambda_1) f_2(\lambda_1)}{f_2(\lambda_1) - f_1(\lambda_1)} = \frac{(24.27)(59.4)}{59.4 - 24.27} = 41.03 \text{ cm}$$

and

$$f(\lambda_2) = \frac{f_1(\lambda_2)f_2(\lambda_2)}{f_2(\lambda_2) - f_1(\lambda_2)} = \frac{(23.44)(58.25)}{58.25 - 23.44} = 39.22 \text{ cm.}$$ ■

3.5 RESOLUTION LIMIT

In this section we discuss the resolution limit of lenses from the standpoint of *diffraction*. The phenomenon of diffraction is discussed in greater detail in Chapter 6. Lens aberrations aside, the image of an object point is not simply the intersection of the light rays that are refracted by the lens; rather, it is a *diffraction pattern*. Thus, there is an ultimate limit in the resolution of a lens. A lens or an optical system is said to be able to *resolve* two object points if the corresponding diffraction patterns are sufficiently separated. In the following, we discuss Baron Rayleigh's widely adopted resolution criterion, which he discovered in 1888.

Consider an optical imaging system or a simple lens, one that is used to image two closely spaced objects, as shown in Figure 3.20. We assume that the lens is *stigmatic*—that is, any light ray from an object point S impinging on the lens will eventually pass through image point P—and that the *optical lengths* (lengths measured in wavelengths of light) of all paths from S to P are equal. Although the geometrical length of the light ray passing through the center of the lens is shorter than the lengths of other rays from S to P, the number of wavelengths is the same, since the lens is thicker at the center. Because of the lens's higher index of refraction, wave propagation in the lens is slower than it is in the air.

In addition to the geometrical ray tracing of the refracted light rays, diffraction also takes place. Thus, the lens can be regarded as an open, circular aperture in which there is diffraction at the *far-field condition*. The result is that the images at points P and P' are not very sharp, each having a far-field circular diffraction pattern around it. Therefore, the images of the object points tend to be confused when they are close together. A question is thereby posed. How close can objects be and still be *resolved*?

The *Rayleigh criterion* states that two equally bright object points can be resolved by the lens of an optical system if the center maximum irradiance of the diffraction pattern of one image falls exactly on the first minimum, that is, the

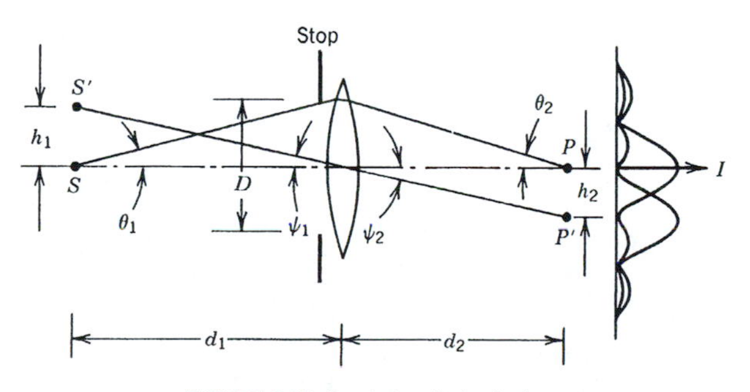

FIGURE 3.20 Resolution limit of a lens.

first dark fringe, of the diffraction pattern of the other. In other words, the distance between the centers of the two circular diffraction patterns should equal the radius of the central bright disk.

Looking again at Figure 3.20, we see that the irradiance curves for image points P and P' are shown at the right. The *minimum resolvable angular* separation can be written as

$$\psi_{\min} = \frac{1.22\lambda}{D}, \tag{3.39}$$

where D is the diameter of the open aperture, and λ is the wavelength of the light source. Therefore, according to Rayleigh's criterion, if the angular separation ψ is smaller than $\psi_{\min}$, the images cannot be resolved.

Moreover, Figure 3.20 shows that the angular separation ψ_1 of the object points equals that of the image points, ψ_2. If the lens is the objective lens of an astronomical telescope, the distance d_1 will be infinitely large, and the minimum resolvable angular separation of Eq. 3.39 will best describe the lens's resolving power. Conversely, for closer observation of the object, as with a microscope, it is better to express Eq. 3.39 in terms of *minimum resolvable separation*; that is,

$$h_{1,\min} = \frac{1.22\lambda d_1}{D}, \tag{3.40}$$

where $\psi = h/d_1$.

When Eq. 3.40 is applied to an optical instrument, it is customary to use Abbe's sine condition of Eq. 3.11:

$$h_1\eta_1 \sin\theta_1 = h_2\eta_2 \sin\theta_2. \tag{3.41}$$

To find the resolving power of the objective of a microscope, we first assume that air is on both sides of the optical system, so that $\eta_1 = \eta_2$. From Figure 3.20 we see that $h_1/h_2 = d_1/d_2$, so that Eq. 3.41 reduces to

$$d_1 \sin\theta_1 = d_2 \sin\theta_2. \tag{3.42}$$

Since the angle θ_2 is very small in microscopic application,

$$\sin\theta_2 \simeq \theta_2 = \frac{D}{2d_2}. \tag{3.43}$$

Substituting Eq. 3.43 into Eq. 3.42, we have

$$\frac{D}{d_1} = 2\sin\theta_1. \tag{3.44}$$

Therefore, from Eq. 3.40, we obtain

$$h_{1,\min} = \frac{1.22\lambda_1}{2\sin\theta_1} = \frac{0.61\lambda_1}{\text{N.A.}}, \tag{3.45}$$

where N.A. = $\sin\theta_1$ is the value called the *numerical aperture* of the objective lens. The N.A. represents the angular radius of the shaft of rays from an object when focused by the lens. Manufacturers usually include the N.A. for purposes of rating the objective of the microscope.

When we consider an oil immersion microscope, $\eta_1 \neq \eta_2$ (i.e., $\eta_2 = 1$), Abbe's sine condition becomes

$$\eta_1 h_1 \sin\theta_1 = h_2 \sin\theta_2. \tag{3.46}$$

Since ψ_1 and ψ_2 are generally not identical, in our present case we have $\eta_1\psi_1 = \psi_2$. Equation 3.39 still holds, provided we express it in terms of ψ_2 and λ_2, that is,

$$\psi_{2,\min} = \frac{1.22\lambda_2}{D} \tag{3.47}$$

or

$$\psi_{1,\min} = \frac{1.22\lambda_2}{\eta_1 D}. \tag{3.48}$$

Thus, the minimum separation of the point object is

$$h_{1,\min} = \frac{1.22\lambda_2 d_1}{\eta_1 D}. \tag{3.49}$$

Since $\eta_1\psi_1 = \psi_2$, we have

$$\frac{\eta_1 h_1}{d_1} = \frac{h_2}{d_2}. \tag{3.50}$$

Substituting Eq. 3.50 into Eq. 3.46, we obtain

$$d_1 \sin\theta_1 = d_2 \sin\theta_2, \tag{3.51}$$

which is the same as Eq. 3.42. If we use Eqs. 3.42 and 3.43, then Eq. 3.49 can be written as

$$h_{1,\min} = \frac{0.61\lambda_2}{\eta_1 \sin\theta_1} = \frac{0.61\lambda_2}{\text{N.A.}}, \tag{3.52}$$

where N.A. = $\eta_1 \sin\theta_1$. Thus, it is possible for an oil immersion microscope to have a numerical aperture greater than unity.

Example 3.10

A lateral object is located at the near point of 250 mm from an adult human eye. The minimum distance of the object for which the eye can accomodate is called the distance of distinct vision, and the point corresponding to the minimum distance is called the *near point*. The near point is the point nearest the eye at which

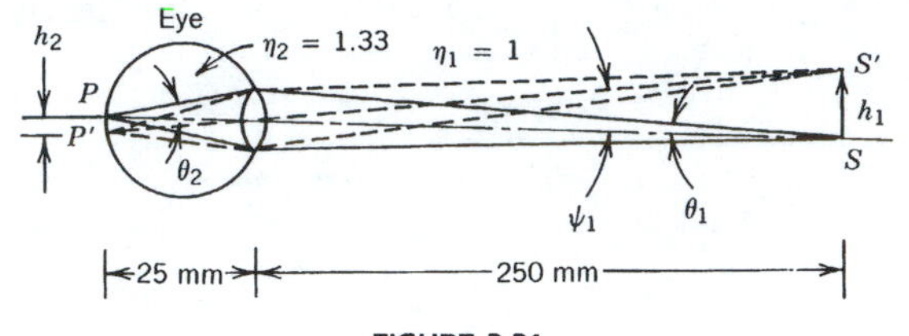

FIGURE 3.21

an object is properly focused on the retina when the maximum degree of accommodation is employed.

Assume that the diameter of the pupil of the eye is 2 mm, the diameter of the eyeball is 25 mm, the refractive index of the vitreous humor is $\eta_2 = 1.33$, and that the object is illuminated by a light source of $\lambda_1 = 550 \times 10^{-6}$ mm. Calculate both the numerical aperture of the eye and the minimum resolution limit of the eye.

Referring to Figure 3.21, we see that the sine of the half-angle θ_1 is

$$\text{N.A.} = \sin\theta_1 = \frac{1}{250} = 0.004.$$

From the Rayleigh resolution criterion of Eq. 3.45, we see that the minimum resolvable separation is

$$h_{1,\text{min}} = \frac{0.61 \times 550 \times 10^{-6}}{0.004} = 0.084 \text{ mm},$$

which is the resolution limit of a normal adult eye. ■

■ Example 3.11

Using the data in the preceding example, calculate both the minimum resolution on the retina in the eye and the minimum angular separation of the lateral object points.

Referring to the image space (i.e., in the eye), we can write the resolution limit on the retina as

$$h_{2,\text{min}} = \frac{0.61\lambda_2}{\sin\theta_2}.$$

Since $\eta_2\lambda_2 = \lambda_1$ (i.e., $\eta_1 = 1$), the minimum resolution on the retina can be calculated as follows:

$$h_{2,\text{min}} = \frac{0.61\lambda_1}{\eta_2 \sin\theta_2}$$

$$= \frac{0.61 \times 550 \times 10^{-6}}{1.33 \times 1/25} = 6.3 \times 10^{-3} \text{ mm}.$$

Referring to Figure 3.21, we see that the minimum angular separation can be determined by Eq. 3.48:

$$\psi_{1,\min} = \frac{1.22\lambda_1}{\eta_1 D}$$

$$= \frac{1.22 \times 550 \times 10^{-6}}{2 \times 10^{-1}} = 3.35 \times 10^{-3} \text{ radian.}$$ ■

REFERENCES

1. F. W. SEARS, *Optics*, Addison-Wesley, Reading, Mass., 1949.
2. B. ROSSI, *Optics*, Addison-Wesley, Reading, Mass., 1957.
3. F. A. JENKINS and H. E. WHITE, *Fundamentals of Optics*, fourth edition, McGraw-Hill, New York, 1976.
4. E. HECHT and A. ZAJAC, *Optics*, Addison-Wesley, Reading, Mass., 1974.
5. F. T. S. YU, *Optical Information Processing*, Wiley-Interscience, New York, 1983, Sections 3.6 and 6.1.

PROBLEMS

3.1 Look again at the spherical convex surface shown in Figure 3.1. The refractive index of medium 1 is $\eta_1 = 1.5$, that of medium 2 is $\eta_2 = 1$, and the radius of curvature of the surface is $R = 20$ mm. If an object 2 mm tall is located at a distance of 100 mm from the vertex,

(a) Calculate the position and the lateral magnification of the image.

(b) Compare the results with the results of Example 3.1.

3.2 Refer to the spherical convex surface shown in Figure 3.1 and described in Example 3.1.

(a) Calculate the *first* and *second* focal lengths of the spherical surface.

(b) Where are the focal points located?

3.3 Find the first and second focal lengths of the spherical surface in Problem 3.1. Compare your results with those of Problem 3.2.

3.4 If the object of Example 3.1 is located at a distance $d_1 = 20$ mm from the vertex, calculate where the image will be located.

3.5 A small tropical fish is at the center of a spherical fishbowl 30 cm in diameter. Assuming that the indices of refraction of the glass and the water are the same, that is, $\eta = 1.33$, calculate the position and lateral magnification of the image of the fish when the observer views the fish from outside the bowl.

3.6 Using the lens maker's equation, derive the focal point for a concave mirror.

3.7 If an object is located at a distance two times the focal length of a concave mirror,

(a) Compute the location of the image from the vertex of the concave mirror.

(b) Calculate the lateral magnification of the image.

3.8 Find the focal length of a thin lens if both sides of the lens are filled with water. We assume that R_1 and R_2 are the radii of curvature of the first and second surfaces of the lens, and that η and η' are the refractive indices of the lens and the water, respectively.

3.9 If one side of a thin lens is filled with ethyl alcohol and the other side is filled with water, find the first and the second focal lengths of the lens. Here R_1 and R_2 are the radii of curvature of the lens's surfaces, and η_1, η, and η_2 are the refractive indices of the ethyl alcohol, the lens, and the water, respectively.

3.10 Repeat Problem 3.9 for the conditions that one side of the lens is filled with air and the other side is filled with water.

3.11 Given a positive thin lens with a refractive index of $\eta = 1.5$, assume that one side of the lens is filled with air ($\eta = 1$) and that the other side is filled with water ($\eta = 1.33$). The radii of curvature of the lens's surfaces are $R_1 = 10$ cm and $R_2 = 15$ cm, respectively. Calculate the first and the second focal lengths of the lens.

3.12 If an object 3 mm tall is located at a distance 1.5 times the first focal length of Problem 3.11, calculate both the image location and the lateral magnification of the image.

3.13 The object is located at a distance 1.5 times the second focal length (i.e., in the water side) of the lens described in Problem 3.11. Compute both the location and the lateral magnification of the image.

3.14 If an object is located at 0.5 times the focal length of a thin lens, calculate the location and the lateral magnification of the image. Assume that both sides of the lens are filled with air, that the index of refraction of the lens is $\eta = 1.5$, and that the radius of curvature of the surfaces is $R = 15$ cm.

3.15 Find the phase transform of a planoconcave thin lens, as shown in Figure 3.22.

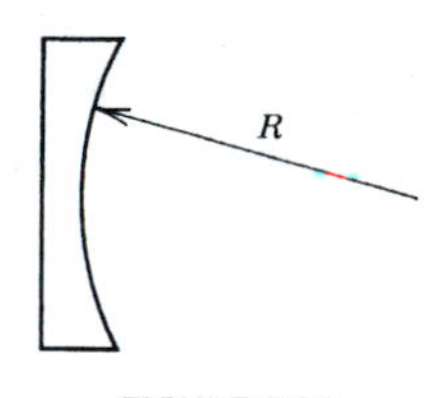

FIGURE 3.22

3.16 Picture an achromatic doublet lens as shown in Figure 3.23; the doublet lens is composed of a positive lens with a refractive index η_1 and a negative lens with a refractive index η_2.

(a) Calculate the phase transform of the doublet lens.

(b) Compute the focal length of the doublet lens.

(c) Draw an analog system diagram of the doublet lens. Assume that the complex light, $u_i(x, y)$, is normally incident at the lens.

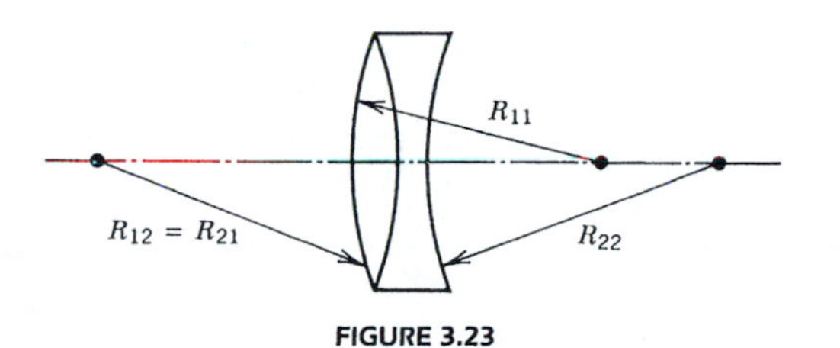

FIGURE 3.23

3.17 Consider two cascaded lenses as given in Example 3.7. If an object is located at a distance $1.2f$, where f is the resultant focal length, calculate the location and the lateral magnification of the image.

3.18 An optical system consists of both a planoconvex and a planoconcave thin lens, as shown in Figure 3.24.

(a) Draw an analog system diagram to represent this optical system. In this diagram $u_i(x, y)$ will represent the complex light illumination.

(b) Compute the overall focal length of the optical system.

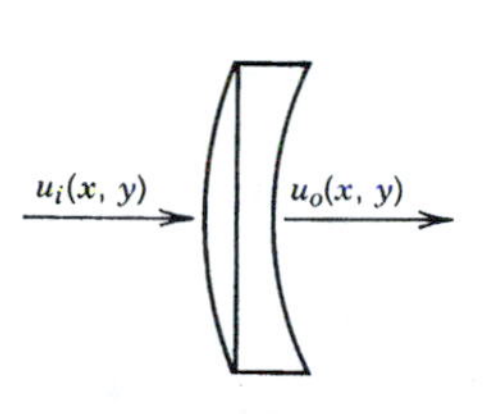

FIGURE 3.24

3.19 A positive lens is made of silicate flint glass and has a radius of curvature of $R = 20$ cm, for both surfaces. If a white-light plane wave is normally incident on the lens,

(a) Calculate the smeared focal length. Assume that the spectral line of the white light varies from 400 to 700 nm. (Note that the refractive indices of the lens can be found in Figure 2.17.)

(b) If an object is located at a distance $d_1 = 30$ cm away from the lens, calculate the positions and the lateral magnifications of the images for $\lambda = 400$ nm and $\lambda = 700$ nm, respectively.

3.20 An optical imaging system is shown in Figure 3.20. Two lateral object points, separated by $h_1 = 0.001$ mm, are located at a distance of $d_1 = 50$ cm from the lens.

(a) If these objects are illuminated by a red light of $\lambda = 650$ nm, calculate the minimum size of lens required to resolve the object.

(b) Repeat part a for violet light illumination of 400 nm. Compare the result with your result for part a.

3.21 An object is located at a distance $d_1 = 100$ cm from an imaging lens. Assume that the diameter of the imaging lens is $D = 1$ cm. If the object is illuminated by a green light of $\lambda = 550$ nm, calculate the minimum resolvable separation of the image (i.e., $h_{2,\min}$). Assume $f = 40$ cm.

3.22 If the object space of Problem 3.21—that is, the left-hand side of the lens in Figure 3.20—is filled with mineral oil of $\eta_1 = 1.4$,

(a) Calculate the minimum angular separation $\psi_{1,\min}$.

(b) Compare the results with those of Problem 3.21.

3.23 If the image space referred to in Problem 3.22 is filled with mineral oil of $\eta_2 = 1.4$, and the object space is filled with air,

(a) Calculate the minimum angular separation $\psi_{1,\min}$.

(b) Compute the numerical aperture of the lens.

(c) Compare your results with those of Problem 3.22.

3.24 If both the object and the image spaces in Figure 3.20 are filled with transparent fluids of η_1 and η_2, respectively,

(a) Calculate the resolution limit of the optical system.

(b) Determine the numerical aperture of the lens.

4 DETECTORS AND DEVICES

In this chapter, we discuss some basic optical instruments, optical detectors, modes of detection, photographic films, and various types of electrooptic devices. Traditional geometrical optical instruments, for example, telescopes and microscope, are based on the principles of reflection and refraction of light and the functions of lenses. These instruments ultimately use the human eye or photographic film to produce the image (or vision of images). There are, however, a large class of modern optical detectors that are electronic in nature. These detectors, for example photomultipliers and photodiodes, are very sensitive and in general require special electronic circuits for photodetection.

Throughout this chapter we will employ electromagnetic units, since they are scientifically well defined. Photometry is a branch of optical detection wherein the units of energy and power are based on the sensitivity of the human eye, which peaks at yellow and drops off sharply at the ultraviolet and infrared regions. Although these photometric units are still used in some photographic processes, in general their use in engineering is not appropriate.

4.1 THE HUMAN EYE

Needless to say, the human eye is perhaps the most important photodetector human beings ever possess. It is a product of millions of years of evolution and has several important characteristics built into it. Figure 4.1 is a schematic of the human eye. The image is formed by the simple process of imaging with a positive (convex) lens. What is interesting, of course, is that the lens in the human eye is a flexible one. With a twitch of the muscle holding it, the lens can either fatten to decrease its focal length (i.e., increase its focal power) or flatten to increase its focal length. It is therefore possible for a person to see both very distant objects and the print in this book, which is only a few centimeters away. The iris in front of the lens can shrink or enlarge to control both the amount of light entering the eye and the field of view. The eyelids act more simply as good shutters.

Images are formed at the retina, which is sensitive to all "visible" light. The wavelengths to which the eye is sensitive range from about 300 to 800 nm. From

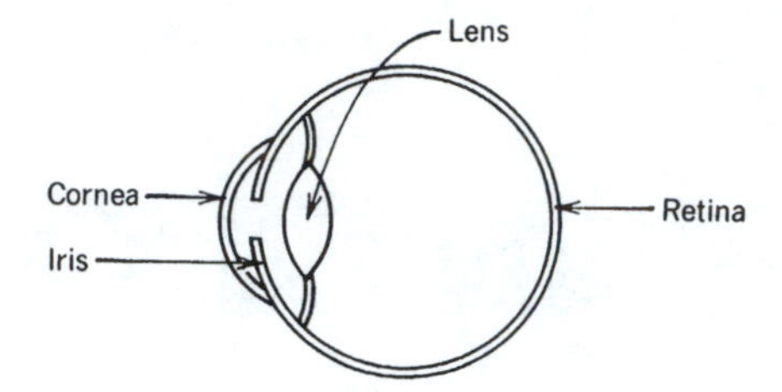

FIGURE 4.1 Schematic of the human eye.

the construction of the eye, it is obvious that the images of objects are formed in an inverted fashion; therefore, the "vision" of right-side-up objects must be accomplished by processing in the brain.

There are, however, a few possible deficiencies in the eye as an imaging system. Two of the most common ones are attributed to faults in an eye's ability to adjust its focus, namely myopia, which is nearsightedness, and hyperopia, which is farsightedness. As we shall see, these deficiencies can be corrected with simple lenses.

1. *Nearsightedness* is the condition in which light rays emanating from distant objects, essentially parallel rays, are brought to focus in front of the retina, as shown in Figure 4.2*a*. The person therefore "sees" a blurred image of the distant objects. As these objects draw closer, they are brought slowly to focus on the retina, as shown in Figure 4.2*b*. The distance at which this happens is called the

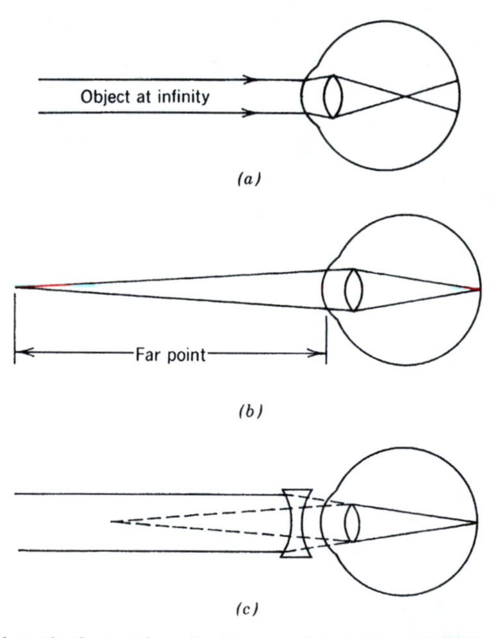

FIGURE 4.2 (*a*) Schematic of a myopic eye focusing on a distant object. (*b*) The far point for a myopic eye. (*c*) Correction of a myopic eye with a negative lens.

far point. The most commonly employed and the most effective way of correcting for myopia is to use a negative lens, as shown in Figure 4.2*c*. The function of the negative lens is simply to refract the light rays from distant objects so that they appear to emanate from the far point. Rays from objects that are nearer will appear to come from points at distances not so faraway as the far point, and can therefore be accommodated by the eye.

Example 4.1

A myopic person cannot see objects clearly at distances greater than 2 m. What type of corrective lens is needed?

As just explained, this person needs a lens that makes rays from distant objects appear to come from distances of less than 2 m. So the image distance is −2 m and the object distance is infinity. The focal length f of the lens is therefore given by

$$\frac{1}{f} = \frac{1}{\infty} + \frac{1}{-2}.$$

Hence, f is −2, which requires a diverging or negative lens. ■

2. *Farsightedness* is almost the opposite of nearsightedness. In this deficiency, however, distant objects are brought to a focus behind the retina, as shown in Figure 4.3*a*. Because of this focusing deficiency, objects closer than a normal

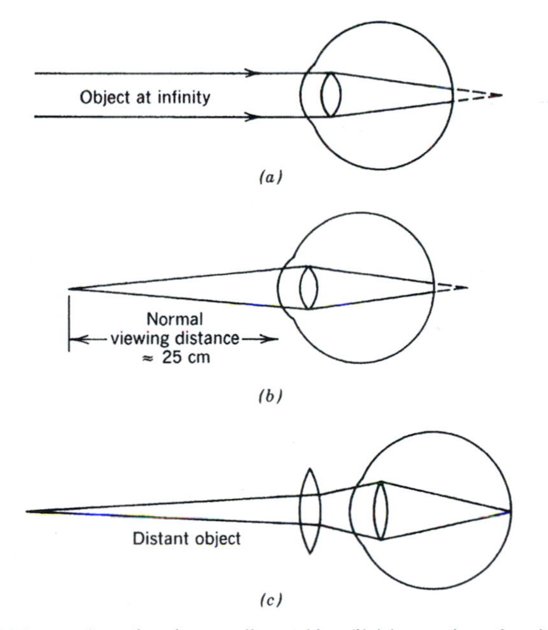

FIGURE 4.3 (*a*) A hyperopic eye focusing on a distant object. (*b*) A hyperopic eye focusing on an object at normal viewing distance. (*c*) Correction of a hyperopic eye with a positive lens.

viewing distance, typically about 25 cm, are focused behind the retina (Figure 4.3*b*). The eye can and must accommodate to see distant objects distinctly. A hyperopic eye cannot, however, adjust to "see" objects nearer than a certain point, the so-called *near point*. Farsightedness can be corrected with a positive lens, which will make objects placed at the normal reading distance from the eye seem to be located at the near point (Figure 4.3*c*).

Example 4.2

Assume that a hyperopic eye can see clearly objects no closer than 100 cm. What should the focal length of the corrective lens be for the hyperopic eye to read this page clearly at 25 cm?

We want the light rays from the object located 25 cm from the eye to appear to come from a distance of 100 cm. Thus, we have

$$\text{Object distance} = 25 \text{ cm}$$

$$\text{Image distance} = -100 \text{ cm}.$$

The minus sign on the image distance is by convention, since the object and the image are both on the same side of the lens. The focal length f of the lens is therefore given by

$$\frac{1}{f} = \frac{1}{25} + \frac{1}{-100} = \frac{3}{100}.$$

Hence, f is 33.3 cm, which requires a positive or convex lens. ■

As a simple lens system, the eye obviously can suffer from any of the lens aberrations discussed in the preceding chapter, for example, coma and astigmatism, in addition to myopia and hyperopia. The corrective lenses required for these aberrations are slightly more complicated than those needed for the common deficiencies.

4.2 MICROSCOPE AND TELESCOPE

There are so many types of telescopes and microscopes that it would require a treatise to discuss them all. In this section we discuss only the most fundamental forms, for the other, more elaborate ones follow essentially the same principles.

Figure 4.4 depicts the basic design of a so-called compound microscope made with two positive lenses. The objective forms a real image of the object near the eyepiece, which in turn forms a virtual image of this real image at a distance of most distinct vision for the eye. There are two magnifications associated with each of the images. The one for the objective is a linear magnification, m_1, and the one for the eyepiece in connection with the eye is an angular magnification, m_2. The overall magnification is the product $m_1 \cdot m_2$.

The linear magnification m is related to the focal length of the objective $m_1 = -x/f$, where x is the distance between the image and the focal point (see Figure 4.5). The angular magnification is given by θ'/θ, where θ' is the angle sub-

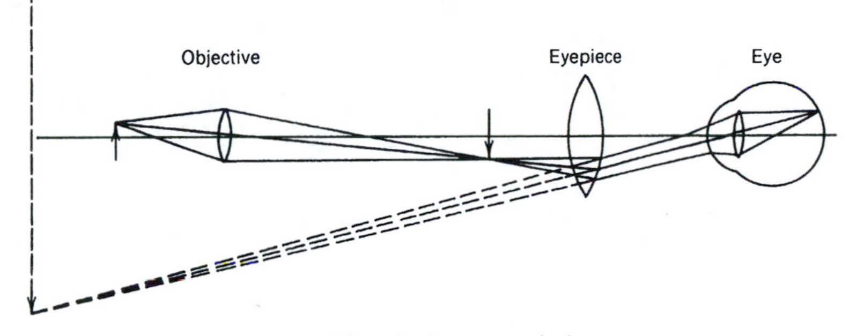

FIGURE 4.4 Schematic of a compound microscope.

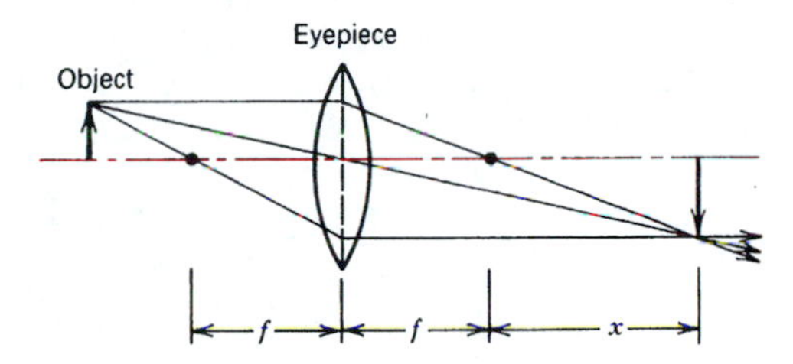

FIGURE 4.5 Linear magnification of the eyepiece.

tended by the image on the eye, and θ is the angle subtended by the object on the eye (see Figure 4.6). Normally, we set the distance of distinct vision, that is, the distance at which the virtual image is formed, at 25 cm. In this instance we can easily show that θ'/θ is given by $25/f_e$, where f_e is the focal length of the eyepiece.

The telescope is essentially a "microscope," but one designed to form the magnified image of a distant object (see Figure 4.7). Angle θ is sometimes called the

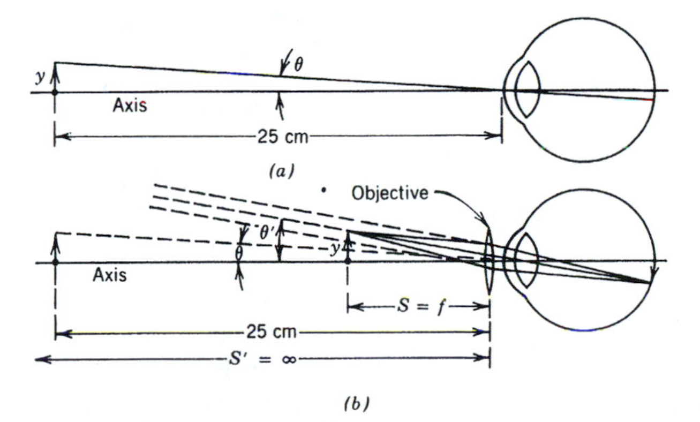

FIGURE 4.6 Angular magnification of the eyepiece.

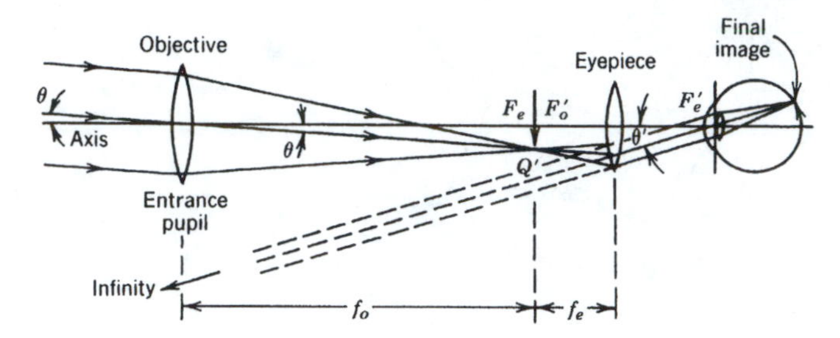

FIGURE 4.7 Schematic of the astronomical telescope.

object field angle; it is also the angle that will be subtended on the naked eye. Angle θ', sometimes called the image field angle, is the angle subtended by the image (located at infinity) on the eye. Magnification factor m is therefore given by $m = \theta'/\theta$. The two lenses are separated by a distance equal to the sum of their respective focal lengths. The first or real image is thus formed at a distance f_o from the objective. This real image, which is at the focal point of the eyepiece, will give rise to a virtual image at infinity from the eyepiece. From Figure 4.6 we have $\tan\theta = h/s$ and $\tan\theta' = -h/s'$, where s is the distance between the telescope objective and the eyepiece, and s' is the distance between the eyepiece and the eye. For small θ and θ', $\tan\theta \sim \theta$ and $\tan\theta' \sim \theta'$. We therefore have $\theta'/\theta - s'/s$. Now since

$$\frac{1}{s} + \frac{1}{s'} = \frac{1}{f_e}$$

and $s = f_e + f_o$, we therefore have

$$\frac{1}{s'} = \frac{f_o}{f_e(f_o + f_e)}.$$

Finally, we get $\theta'/\theta = -f_o/f_e$.

Example 4.3

A telescope consists of a 40-cm focal-length objective and a 2-cm focal-length eyepiece.

(a) What is the angular magnification if both the object and the image are at infinity?

(b) What is the corresponding magnification if the object is 2 m away?

Answers

(a) Since both object and image are at infinity, the angular magnification is given by $\theta'/\theta = -f_o/f_e$, as just discussed. We thus have

$$\frac{\theta'}{\theta} = \frac{-40}{2} = 20\times.$$

(b) If the object is 2 m away, the first image formed by the objective is formed at a distance of slightly more than f_o from the objective. This distance s_o is given by

$$\frac{1}{s_o} + \frac{1}{200} = \frac{1}{40}.$$

Solving for s_o we get $s_o = 50$ cm. The angular magnification is therefore $50/-2 = -25\times$. ■

■ **Example 4.4**

A certain microscope has an objective with a focal length of 4 mm and an eyepiece with a focal length of 25 mm. The distance between the objective and the eyepiece is 150 mm. What is the total magnification?

The total magnification is the product of the linear magnification of the objective and the angular magnification of the eyepiece.

The linear magnification is $m_1 = -\frac{150-25-4}{4} = -30.25\times$. The angular magnification is $m_2 = \frac{250}{25} = 10\times$. Hence, the total magnification is $-30.25 \times 10 = -302.5\times$. ■

The eye, the telescope, and the microscope are but a few examples of the numerous geometrical optical instruments that use lenses. For more information on the more elaborate lens systems (e.g., the camera and prism binocular), consult the technical literature from the manufacturers. Their working principles, however, follow the principles of reflection and refraction discussed in the preceding chapters.

4.3 PHOTODETECTORS

In this section we discuss photodetection in detectors for which the process is essentially electronic and can be discussed in quantitative terms. In comparison with the eye, which responds in fractions of a second, these electronic detectors have very short response times of from 10^{-6} to 10^{-12} sec. Besides being able to detect light in a wide spectral range, from ultraviolet to infrared, these electronic detectors, in combination with the appropriate circuit or electronics, can be extremely sensitive.

4.3.1 Photomultipliers

Figure 4.8 is a schematic of the so-called head-on type of photomultiplier tube, which consists essentially of the cathode, the electron multiplier section, and the anode. The cathode consists of materials with a low *surface work function*, which is the energy needed to eject the electron from the material. Typically, these materials are made of compounds of silver–oxygen–cesium and antimony–cesium. Figure 4.9 is a schematic of the energy required in the photoemission process. The electron in the conduction band (where the electron is free) can be ejected from the surface into the vacuum if it can overcome the work function W. An incident

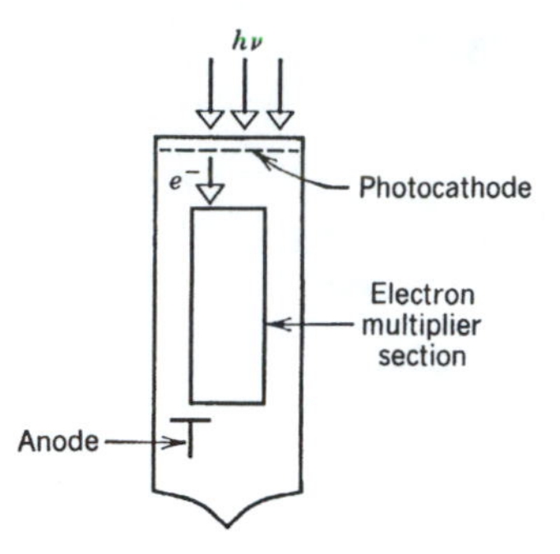

FIGURE 4.8 Schematic of a head-on-type photomultiplier tube.

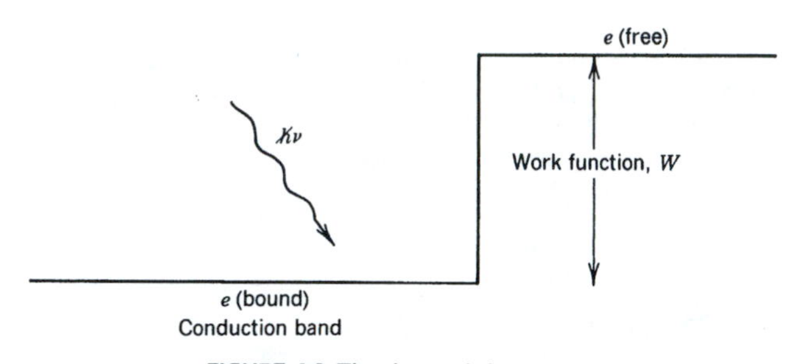

FIGURE 4.9 The photoemission process.

light at frequency ν, with quantized energy unit $h\nu$ (where h is Planck's constant), will be able to eject the electron if $h\nu > W$. Typically, the work functions W for these cathode materials are on the order of 1.5 eV, corresponding to the optical wavelength λ in the ultraviolet (0.1 = μm) to the infrared (1.1 = μm) region. The spectral responses of some commercial phototubes are shown in Figure 4.10.

The electron that is ejected from the photocathode is accelerated to the anode via the electron multiplier, which consists essentially of a series of dynodes at a dc voltage bias of about 100 eV. This kinetic energy enables the electron to cause secondary emission from the dynode. As the electron and the secondary electrons are accelerated down the chain, the process of electron multiplication is repeated, leading finally to substantial current being collected at the anode. Typically, one electron creates on the order of about five secondary electrons. If there are ten dynodes, one electron ejected from the cathode will give rise to 5^{10} ($\approx 10^7$) electrons at the anode.

Photomultiplier tubes are square-law detectors, which means that the response from the photomultiplier tube is proportional to the square of the modulus of the electric field from the incident light. This square-law response is also characteristic of photoconductors and photodiodes (see the following two subsections). The basic mechanism is that the electron involved in this process is induced by the optical field to make a transition from a lower energy state to a higher energy state. In general, the induced transition is proportional to the intensity of the optical field. It is obvious that if two or more fields of different frequencies are incident on the photocathode, the output current will show frequency beating effects. This leads

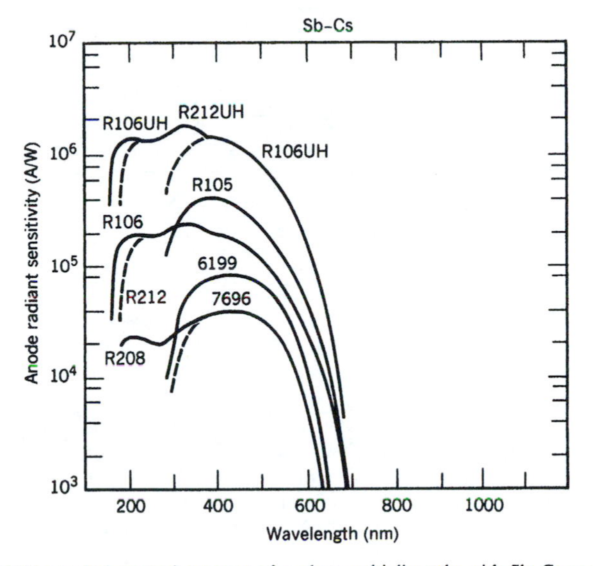

FIGURE 4.10 The typical spectral response of a photomultiplier tube with Sb–Cs as the cathode material.

to, for example, the heterodyne detection scheme, which is discussed in detail in Section 4.4.

Because the electrons involved in this process are only loosely bound to the material, some of them can be excited thermally and break loose from the material. Thus, even in the absence of any applied optical signal, this random electron emission will cause current to flow at the anode, the so-called *dark current*. Typically, the dark current is on the order of 10^{-11} to 10^{-12} A. For example, the dark current can be reduced by more than two orders of magnitude by cooling the tube with dry ice.

The response times of the photomultiplier are typically on the order of nanoseconds to subnanoseconds. The times depend primarily on the dynode material used for each gain stage and the geometrical design of the multiplier structure, including the number of gain stages.

Example 4.5

The energy associated with light at a frequency ν is given by $h\nu$, where h is Planck's constant ($h = 6.626 \times 10^{-34}$ J. sec). Calculate the maximum wavelength of the light that is energetic enough to eject electrons from a photocathode which has a work function of 1.5 eV.

The frequency ν of light is given by c/λ, where λ is the wavelength. In order to eject electrons bound to the cathode with a work function of 1.5 eV, we need $hc/\lambda > 1.5$ eV. This means that

$$\lambda > \frac{hc}{1.5 \text{ eV}}.$$

Note that 1.5 eV = $1.5 \times (1.6 \times 10^{-19})$ J. Then

$$\lambda > \frac{(6.6.26 \times 10^{-34})(3 \times 10^{10}\ \text{cm})}{1.5 \times 1.6 \times 10^{-19}},$$

$$> 8.28 \times 10^{-5}\ \text{cm (828 nm)}.$$ ■

4.3.2 Photoconductive Detectors

Photoconductive detectors are basically semiconductors (*n*-type or *p*-type) that are appropriately doped. Figure 4.11 is a schematic of the detector along with an external circuit. An incident light with sufficient ionizing energy excites electrons into the conduction band (*n*-type) or, in *p*-type semiconductors, into holes in the valence band (see Figure 4.12). This excitation lowers the resistance across the semiconductor, and therefore increases the voltage across the output resistor *R*. The energy needed for these processes varies from material to material, the lowest requiring about 0.04 eV, which corresponds to an optical wavelength of about 32 μm (i.e., infrared). Clearly, the principal advantage of photoconductors over photomultipliers is their ability to detect long wavelength radiation. However,

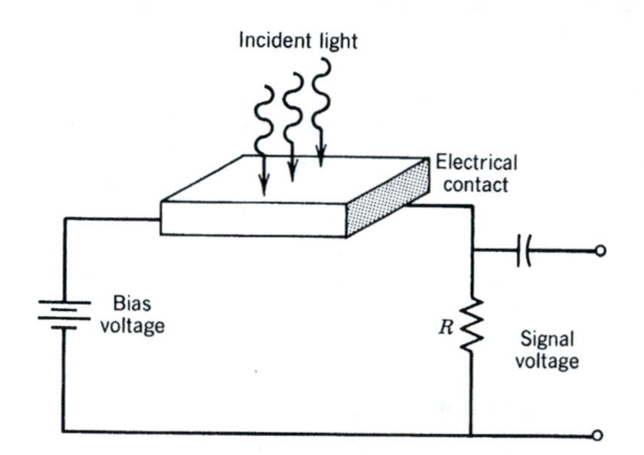

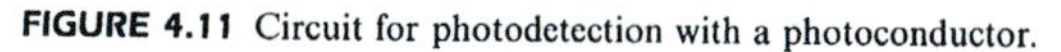

FIGURE 4.11 Circuit for photodetection with a photoconductor.

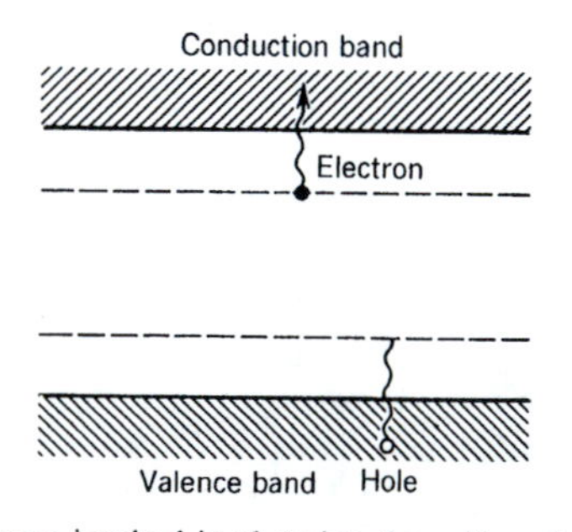

FIGURE 4.12 The two processes involved in photodetection with a photoconductor. Top: electron excitation (*n*-type); bottom: hole excitation (*p*-type).

there are serious problems associated with the rather low energy at which photoelectrons are created. It is imperative, for instance, that the semiconductor be kept at a very low temperature. Moreover, there is no current amplification process to augment the signal level.

The principal noise mechanism in a cooled photoconductor is the randomness of the generation of electrons (or holes) by the incident field. There are two time scales: the drift time for the charge carrier across the length of the semiconductor, td; and the lifetime τ of the photoexcited charge carrier. The statistical randomness of these times causes shot noise (see Section 4.4.1). The response times of photoconductors, which depend on a large number of parameters, including dopant densities, sample dimension, and the recombination rate, are typically in the microsecond range.

4.3.3 Photodiodes

Photodiodes are basically semiconductor *p-n* junctions, and they are often termed *junction photodiodes.* We will discuss *p-n* junctions in more detail in Chapter 7 when we study semiconductor lasers. For the present, it suffices to show the energy band diagram of a *p-n* junction (see Figure 4.13). There are basically three types of charge carrier flows taking place within the diffusion length of the depletion layer, or in the depletion area. If the photon is absorbed on the *p* side, creating an electron–hole pair, the electron will drift toward the depletion layer, cross the junction, and contribute a charge, -*e*, to the external circuit. Similarly, the hole created by photoabsorption on the *n* side will drift in the opposite direction and contribute to the current. Within the depletion layer the electron–hole pair will drift away from the junction in opposite directions, and both will contribute to the current flow.

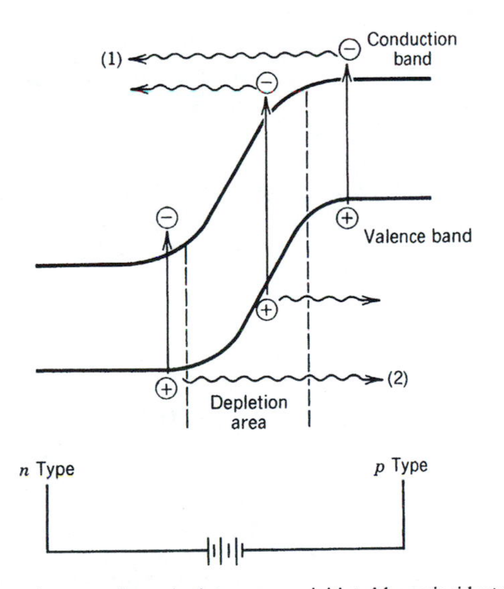

FIGURE 4.13 Various charge carrier excitation processes initiated by an incident light in the vicinity of the *p-n* junction.

In pin photodiodes an intrinsic high-resistivity layer (i) is sandwiched between the p and n semiconductors. The voltage drop across the p and n regions occurs mostly across this layer. If this layer is made wide enough so that most of the incident light is absorbed within it, the long diffusion times associated with processes 1 and 2 as depicted in Figure 4.13 are avoided. Therefore, pin photodiodes respond much faster than do p-n junction diodes. Typically, the response time of a pin diode is on the order of nanoseconds, whereas the response time for junction photodiodes is in tens of nanoseconds or longer.

The energy gap, E_g, of most semiconductors is on the order of 1 eV. This corresponds to an optical wavelength of about 1 μm. In general, therefore, photodiodes can respond to light ranging from the ultraviolet to the infrared.

It is clear from the preceding discussion that there is no electron multiplication process or gain in photodetection by photodiodes. However, if the bias voltage across a p-n junction is increased, the field in the depletion layer may be large enough to accelerate the electron, providing it with sufficient kinetic energy to cause other electrons in the depletion layer to make valence-conduction band transitions. This is similar to the ejection of a secondary electron from the dynode by a high-velocity electron accelerated from the photocathode in the photomultiplier. The process of electron multiplication is often called *avalanche multiplication*. The newly generated electron–hole pairs can in turn create new charge carriers as they are accelerated across the depletion layer. This causes a signal voltage increase of $M^2 \sim 100$, with a typical multiplication factor M of about 10.

Example 4.6

What is a square-law detector?

All the electronic detectors discussed so far—photomultipliers, photoconductors, and photodiodes—are of the class called *square-law detectors*, a term derived from the fact that they respond to the square of the electric field associated with light. Because we will discuss the wave nature of light in more detail in Chapter 5, it is enough hère to state that light consists of electromagnetic waves traveling at a speed of $v = 1/\sqrt{\mu\varepsilon}$. The electric field associated with the light waves will interact with the electrons in the material—the photocathode or the p-n junction—and cause these electrons to be ejected or make transitions to higher energy states. The rate of occurrence, R, of these processes for a square-law detector is proportional to the square of the electric field of the light, that is, $R \propto E^2_{\text{light}}$. ■

4.4 MODES OF DETECTION

The electronic detectors discussed in the previous section are very versatile in terms of applications and, under suitable conditions, can be extremely sensitive. Being essentially electronic in nature, their ultimate sensitivity is dependent on the noise source. Although some noise sources are unique to a particular device, many are common to all detectors. There are also various modes of detection, depending on whether we are interested in detecting an ultrafast signal (e.g., a picosecond laser plus) or a very weak signal (e.g., light from a distant star). In this section we discuss some aspects of these modes of detection, the corresponding sensitivity, and noise sources. First, we take a detailed look at those aspects associated with detection with a photomultiplier. This general description can be extended to other detectors. The features of other detectors that are not similar to those of the photomultiplier will be pointed out at the end of the discussion.

There are essentially two modes of detection, video and heterodyne. In video detection the optical wave to be detected is modulated at some low frequency ω_m before it impinges on the photocathode. As a result of the square-law-detector characteristics of the photomultiplier tube, the photocathode current is given by

$$i_{\text{pmt}}(t) \propto E_c^2(1 + m\cos\omega_m t)^2 \cos^2\omega_c t, \tag{4.1}$$

where m is the amplitude of modulation of the carrier wave (at frequency ω_c) associated with the frequency-modulated optical wave (at ω_m).

Usually, ω_c is very high, and the detector averages the $\cos^2\omega_c t$ dependence over several cycles. Equation 4.1 gives a detector photocathode current of

$$i_{\text{pmt}}(t) = CE_c^2\left[\left(1 + \frac{m^2}{2}\right) + 2m\cos\omega_m t + \frac{m^2}{2}\cos 2\omega_m t\right]. \tag{4.2}$$

The proportional constant C can be evaluated by noting that in the absence of the modulation wave (i.e., $m = 0$), the constant current associated with the carrier wave is

$$i_{\text{pmt}}(t) = CE_c^2. \tag{4.3}$$

If we denote the power associated with the carrier wave as P, the average value of the photocathode current i_{pmt} is given by

$$i_{\text{pmt}} = \frac{Pe\eta}{h\nu_c}, \tag{4.4}$$

where η is the efficiency of converting a photon (light) to an electron, assuming that each photon at frequency ν_c contributes to the ejection of an electron. Hence,

$$CE_c^2 = \frac{Pe\eta}{h\nu_c},$$

and we have

$$i_{\text{pmt}}(t) = \frac{Pe\eta}{h\nu_c}\left[\left(1 + \frac{m^2}{2}\right) + 2m\cos\omega_m t + \frac{m^2}{2}\cos 2\omega_m t\right]. \tag{4.5}$$

As a result of the amplification of the electron in the phototube by a gain factor, g, the output signal from the photomultiplier is

$$i_s(t) = \frac{gPe\eta}{h\nu_c}(2m\cos\omega_m t), \tag{4.6}$$

if we are interested in the signal at the frequency ω_m.

4.4.1 Noise Sources

At this juncture, it is perhaps appropriate to ask what the noise sources are that also contribute to the photomultiplier output current. There are two types of noise, one is associated with the random generation of charge carriers and is called *shot noise.* The other, the so-called *Johnson noise* or Nyquist noise, is associated with fluctuations in the voltage across the dissipative circuit element; here it is output load R of the photomultiplier tube.

Shot noise has a mean square current amplitude given by (see reference 4, Chapter 10)

$$\overline{i_{\text{sn}}^2}(\nu) = 2e\bar{I}\,\Delta\nu, \tag{4.7}$$

where $\bar{I}$ is the average current associated with the electron and any charge carriers generating the shot noise, and $\Delta\nu$ is the bandwidth limit of the detector. For the photomultiplier shot noise is caused by the dark current generated at the cathode. These photoelectrons experience a gain g as they are accelerated through the amplifying stages. Therefore, for the photomultiplier tube,

$$\overline{i_{\text{sn}}^2}(\nu) = 2g^2e(\bar{i}_{\text{pmt}} + \bar{i}_d)\,\Delta i, \tag{4.8}$$

where $\bar{i}_{\text{pmt}}$ is the average signal current, and $\bar{i}_d$ is the average dark current.

The derivation of Johnson noise is quite complicated, yet the result is very simple. Johnson noise has a mean square current amplitude given by

$$\overline{i_{\text{jn}}^2}(\nu) = \frac{4K_BT\,\Delta\nu}{R}, \tag{4.9}$$

where K_B is Boltzmann's constant ($K_B = 1.38 \times 10^{23}$ J/K). This obviously means that the electric power fluctuation across R, that is, $R\overline{i_{\text{jn}}^2}$, is on the order of the thermal energy K_BT within the bandwidth interval $\Delta\nu$.

The signal-to-noise ratio is therefore given by

$$\frac{\text{S}}{\text{N}} = \frac{\bar{i}_{s2}}{\overline{i_{\text{sn}}^2} + \overline{i_{\text{jn}}^2}} = \frac{2(Pe\eta/h\nu_c)}{2g^2e(\bar{i}_s + \bar{i}_d)\,\Delta\nu + (4KTe/R)\,\Delta\nu}. \tag{4.10}$$

Typically, $g \sim 10^6$ for the photomultiplier. The cathode shot noise is, however, much greater than the Johnson noise. Thus, we have

$$\left.\frac{\text{S}}{\text{N}}\right|_{\text{pmt}} \sim \frac{P^2e\eta^2}{h^2\nu_c^2(i_s + i_d)\,\Delta\nu}. \tag{4.11}$$

4.4.2 Sensitivity

In many applications requiring a high level of sensitivity, the photomultiplier tubes are thermoelectrically cooled, and the dark current is reduced until it almost disappears. Under this condition and with S/N set as one, as the acceptable limit on detection, we get

$$1 = \frac{P^2e\eta^2}{h^2\nu_c^2(Pe\eta/h\nu_c)\,\Delta\nu}, \tag{4.12}$$

where, from Eq. 4.4, we have $i_s = Pe\eta/h\nu_c$ and we let $m = 1$. Solving for P, we get, as the optical power for S/N = 1,

$$P_{\min} = \frac{h\nu_c \Delta\nu}{\eta}. \tag{4.13}$$

In heterodyne detection the optical radiation is frequency-modulated. The oscillating signal, $E_s \cos \omega_s t$, is combined with the local oscillator field, $E_L \cos(\omega_s + \omega_L)t$, where $\omega_L \ll \omega_s$. The total field on the photomultiplier tube is thus

$$E_{\text{total}} = E_s \cos \omega_s t + E_L \cos(\omega_s + \omega_L)t. \tag{4.14}$$

The square-law detection characteristic of the photomultiplier again means that the output current from the cathode is

$$i_{\text{pmt}}(t) = C_2(E_L^2 + E_s^2 + 2E_L E_s \cos \omega_L t), \tag{4.15}$$

where C_2 is the proportionality constant to be evaluated. As in the previous section, this evaluation can be made by setting the modulating signal wave to zero, when

$$i_{\text{pmt}} = \frac{P_L e\eta}{h\nu_L},$$

where P_L is the power of the local oscillating field and ν_L is the frequency.

In most applications requiring a high level of sensitivity, that is, detection of low-powered signals, the amplitude of the local oscillator field E_L (derived from a laser) is a great deal larger than that of the oscillating signal E_s. To first order in E_s/E_L, we get

$$i_{\text{pmt}}(t) = \frac{P_L e\eta}{h\nu_L}\left(1 + 2\frac{E_s}{E_L}\cos \omega_L t\right). \tag{4.16}$$

As in the previous section, the dominant noise is cathode shot noise, which is given by

$$\overline{i_{\text{sn}}^2} = 2eg^2\left(id + \frac{P_L e\eta}{h\nu_L}\right)\Delta\nu. \tag{4.17}$$

On the other hand, when we account for gain factor g, the time-averaged signal current from $i_{\text{pmt}}(t)$ is

$$\begin{aligned} \overline{i_s^2} &= \left(\frac{P_L e\eta}{h\nu_L}\right)^2 (4)\langle\cos^2 \omega_L t\rangle \frac{E_s^2}{E_L} g^2 \\ &= 2g^2\left(\frac{P_L e\eta}{h\nu_L}\right)^2 \frac{P_s}{P_L}, \end{aligned} \tag{4.18}$$

where we have used P_s/P_L for E_s/E_L.

The signal-to-noise ratio is therefore given by the ratio of Eq. 4.18 to Eq. 4.17:

$$\frac{\text{S}}{\text{N}} = \frac{e(\eta/h\nu)^2 P_s P_L}{(i_d + P_L e\eta/h\nu)\Delta\nu}. \tag{4.19}$$

One advantage of heterodyne detection is that instead of employing various means to suppress the dark current or other sources of noise, we can increase the power, P_L, of the local oscillator until the noise associated with it dominates over that from all other sources. Then we obtain

$$\frac{\mathrm{S}}{\mathrm{N}} = \frac{P_s \eta}{h\nu\,\Delta\nu}.$$

If we again set S/N as one, we get the lowest level of optical power detected,

$$P_{s,\min} = \frac{h\nu\,\Delta\nu}{\eta},$$

which is identical to the lowest level given in Eq. 4.13. It is important to emphasize that Eq. 4.21 is obtained by requiring that all other noise sources be absent or negligible, an impossible situation in practice.

These two modes of optical signal detection, video and heterodyne, are also applied in photoconductors and photodiodes. In photoconductors the main noise source is the shot noise associated with the random generation of photoelectrons and their recombination with ionized impurities in the photoconductor. In photodiodes the chief noise source is also the shot noise associated with the random generation of carriers. In general, the lowest levels of optical powers detected by photomultipliers and photoconductors are on the order of 10^{-19} W, and those for photodiodes are on the order of 10^{-7} W. Nevertheless, photodiodes are relatively inexpensive and respond to a very wide spectral range, from ultraviolet to infrared.

Example 4.7

Assuming a bandwidth $\Delta\nu$ of 1 Hz, calculate the minimum level of power detected for yellow light using a photomultiplier. Assume also that η is unity.

The central portion of the yellow light has a wavelength λ of about 5900 Å; hence,

$$\nu = \frac{c}{\lambda} = \frac{3 \times 10^{10}}{5900 \times 10^{-8}\ \mathrm{cm}} = \frac{3}{5.9} \times 10^{15}/\mathrm{sec}.$$

Therefore,

$$P_{\min} = \frac{6.26 \times 10^{-34}\ \mathrm{J \cdot sec} \times (3 \times 10^{15}/\mathrm{sec})}{5.9}$$

$$\sim 3 \times 10^{-19}\ \mathrm{W}.$$

■

Example 4.8

Typically, the dark current i_d in an uncooled phototube is about 10^{-11} A. At what power level of the local oscillator (a laser, for example) is the cathode shot noise coming from the local oscillator equal to that from the dark current?

The right-hand side of Eq. 4.17 indicates the contribution to the shot noise of i_d, the dark current, and of $P_L e\eta/h\nu$ from the local oscillator. For these to be equal, we have

$$P_L \frac{e\eta}{h\nu} = i_d.$$

Therefore

$$P_L = \frac{(10^{-11}\ \text{C/sec})(6.626 \times 10^{-34}\ \text{J}\cdot\text{sec})(3/5.9 \times 10^{15}/\text{sec})}{1.6 \times 10^{-19}\ \text{C}}$$

$$= 2 \times 10^{-11}\ \text{W}.$$

■

■ Example 4.9

As a matter of interest, how many photons per second are associated with one watt of light in the yellow region?

Each photon is associated with an energy $h\nu$. For yellow light,

$$h\nu = 6.626 \times 10^{-34} \times \frac{3}{5.9} \times 10^{15}\ \text{J}$$

$$\simeq 3.3 \times 10^{-19}\ \text{J}.$$

Hence, the number of photons per second and in one watt of light is

$$n\ (\text{per second}) = \frac{1\ \text{J/sec}}{0.33 \times 10^{-19}}$$

$$= 3 \times 10^{18}\ \text{photons}.$$

This number is very useful to remember. ■

4.5 PHOTOGRAPHIC FILMS

Photographic film has served as an important optical element since its discovery. Although it is primarily used as a recording medium, it can also be used in the synthesis of complex spatial filters, holograms, and signal transparencies. There are other optical materials whose optical properties are similar to those of photographic film; however, these materials are still in the research and development stage. It is indeed doubtful whether these materials will ever totally replace photographic film.

Photographic film is generally composed of a base, made of a transparent glass plate or acetate film, and a layer of photographic emulsion, as shown in Figure 4.14. The emulsion consists of a large number of tiny photosensitive silver halide particles, which are suspended more or less uniformly in a supporting gelatin. When the photographic emulsion is exposed to light, some of the silver halide grains absorb optical energy and undergo a complex physical change. Some of

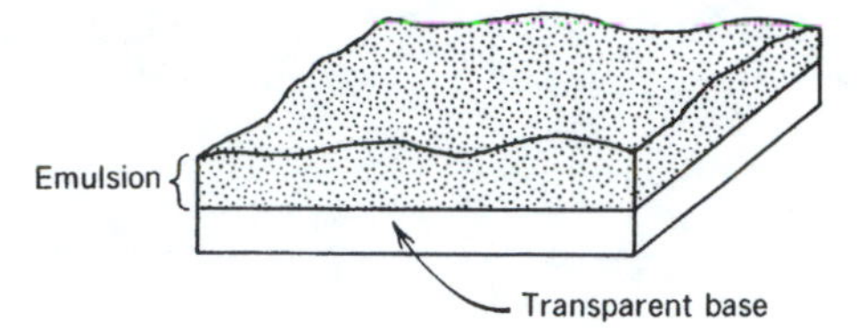

FIGURE 4.14 Section of a photographic film. The emulsion is composed of silver halide particles suspended in gelatin.

the grains that absorb sufficient light energy are immediately reduced, forming tiny metallic silver particles. These are the so-called *development centers*. The reduction to silver is completed by the chemical process of *development*. The grains that were not exposed or that have not absorbed sufficient optical energy will remain unchanged. If the developed film is then subjected to a chemical *fixing* process, the unexposed silver halide grains are removed, leaving only the metallic silver particles in the gelatin. These remaining grains are largely opaque at optical frequencies, so the transmittance of developed film depends on their density. The relation of intensity transmittance to the density of the developed grains was first demonstrated in 1890 by F. Hurter and V. C. Driffield, who showed that the photographic density, the density of the metallic silver particles per unit area, is proportional to $-\log T_i$,

$$D = -\log T_i, \tag{4.22}$$

where T_i is intensity transmittance, which is defined as

$$T_i(x, y) = \left\langle \frac{I_o(x, y)}{I_i(x, y)} \right\rangle. \tag{4.23}$$

The angle brackets represent the localized ensemble average, and $I_i(x, y)$ and $I_o(x, y)$ are the input and output irradiances, respectively, at point (x, y).

One of the most commonly used descriptions of the photosensitivity of a given photographic film is that given by the Hurter–Driffield curve, or the H-and-D curve, as shown in Figure 4.15. This curve is the plot of the density D of the developed grains versus the logarithm of the exposure E. The plot shows that if the exposure is below a certain level, the photographic density is quite independent of the exposure; this minimum density is usually referred as *gross fog*. As the exposure increases beyond the toe of the curve, the density begins to increase in direct proportion to $\log E$. The slope of the straight-line portion of the H-and-D curve is usually referred to as the *film gamma*, γ. If the exposure is increased beyond the straight-line portion of the H-and-D curve, the density saturates, after an intermediate region called the *shoulder*. In the saturated region there is no further increase in the density of the developed grains as the exposure increases.

Conventional photography is usually carried out within the linear region of the H-and-D curve. A film with a high-value gamma is called a *high-contrast film*, whereas one with a low-value gamma is referred as a *low-contrast film*, as illustrated in Figure 4.16. The value of γ is, however, affected not only by the type of photographic emulsion used but also by the chemical of the developer and the time taken by the developing process. In practice it is therefore possible to achieve

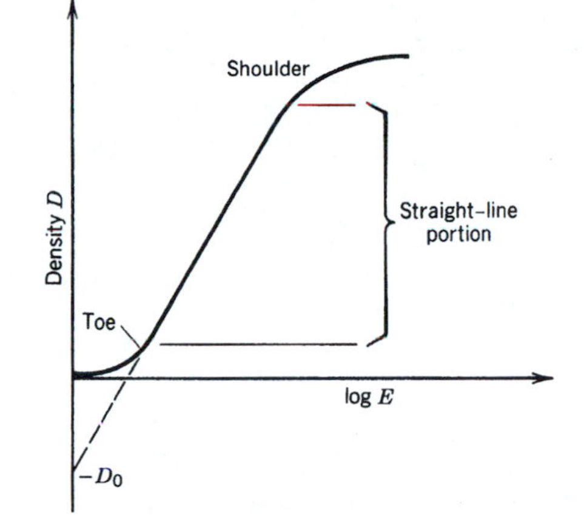

FIGURE 4.15 The Hurter–Driffield (H-and-D) curve.

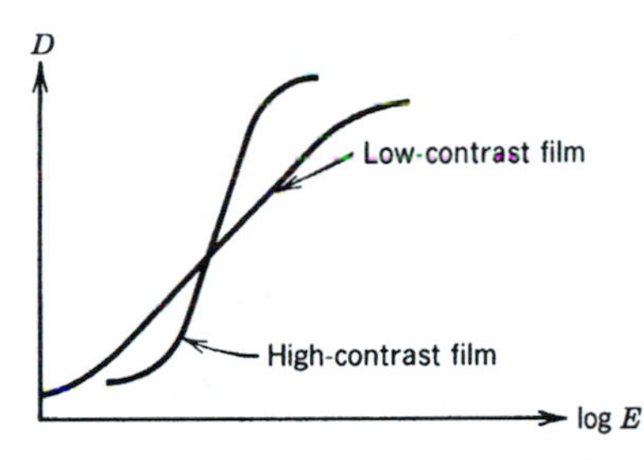

FIGURE 4.16 High- and low-gamma (γ) films.

a prescribed value of γ, with a fair degree of accuracy, by using suitable film, developer, and developing time.

If a given film is recorded in the straight-line region of the H-and-D curve, the photographic density may be written as

$$D = \gamma_n \log E - D_0, \tag{4.24}$$

where the subscript n means that a negative film is being used, and $-D_0$ is the point where the projection of the straight-line portion of the H-and-D curve intercepts with the density ordinate, as shown in Figure 4.15. By substituting Eq. 4.22 in Eq. 4.24, we get

$$\log T_{in} = -\gamma_n \log(It) + D_0, \tag{4.25}$$

where I is the incident irradiance and t is the exposure time. (Note that the exposure of the film is given by $E = It$.) Equation 4.25 can be written as

$$T_{in} = K_n I^{-\gamma_n}, \tag{4.26}$$

where $K_n = 10^{D_0} t^{-\gamma_n}$, a positive constant.

It is apparent that intensity transmittance is highly nonlinear with respect to the incident irradiance. It is, however, possible to obtain a positive exponent relation between intensity transmittance and the incident irradiance. To do so requires a two-step process called *contact printing*.

In the first step a negative film is exposed, and in the second step another negative film is laid under the developed film. An incoherent light is then transmitted through the first film to expose the second film. By developing the second film to obtain the prescribed value of γ, we can obtain a positive transparency with a linear relation between intensity transmittance and the incident irradiance.

To illustrate this two-step process, let the resultant intensity transmittance of the second developed film (i.e., the positive transparency) be

$$T_{ip} = K_{n2} I_2^{-\gamma_{n2}}, \tag{4.27}$$

where K_{n2} is a positive constant, and the subscript $n2$ denotes the second-step negative. The irradiance I_2 that is incident on the second film can be written as

$$I_2 = I_1 T_{in}, \tag{4.28}$$

where

$$T_{in} = K_{n1} I^{-\gamma_{n1}}.$$

Here the subscript $n1$ denotes the first-step negative. The illuminating irradiance of the first film during contact printing is I_1, and I is the irradiance originally incident on the first film. By substituting Eq. 4.28 into Eq. 4.27, we get

$$T_{ip} = K I^{\gamma_{n1}\gamma_{n2}}, \tag{4.29}$$

where

$$K = K_{n2} K_{n1}^{-\gamma_{n2}} I_1^{-\gamma_{n2}}$$

is a positive constant. Thus, we see that a linear relation between the intensity transmittance of the positive transparency and the incident recording irradiance may be obtained if the overall gamma is made to be unity, that is, $\gamma_{n1}\gamma_{n2} = 1$.

If the film is used as an optical element in a coherent system, it is more appropriate to use complex amplitude transmittance rather than intensity transmittance. Complex amplitude transmittance can be defined as

$$T(x, y) = [T_i(x, y)]^{1/2} e^{i\phi(x, y)}, \tag{4.30}$$

where T_i is intensity transmittance, and $\phi(x, y)$ represents random phase retardations. Such phase retardations are primarily due to variations in the thickness of the emulsion. These variations are of two sorts: the coarse "outer-scale" variation, which is a departure from the optical flatness of the emulsion and base; and the fine "inner-scale" variation, which is caused by random fluctuations in the density of the developed silver grains. The silver grains do not swell uniformly in the surrounding gelatin. This fine-scale variation in emulsion thickness obviously depends on the exposure of the film.

In most practical applications, phase retardations caused by variations in the thickness of the emulsion can be removed by means of an *index-matching liquid* gate, as shown in Figure 4.17. Such a gate consists of two parallel, optically flat

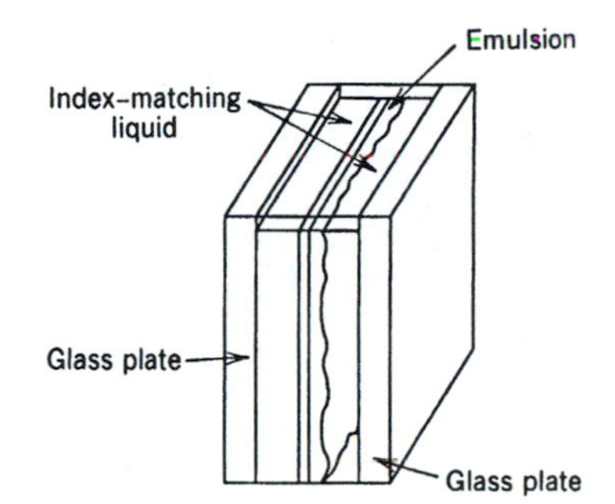

FIGURE 4.17 Refractive-index-matching liquid gate.

glass plates which have the space between them filled with a refractive-index-matching liquid, that is, a liquid whose refractive index is very close to that of the film emulsion. If a developed film is submerged in the liquid gate, the overall complex amplitude transmittance can be written as

$$T(x, y) = [T_i(x, y)]^{1/2}, \tag{4.31}$$

which is a real function, with the random phase retardations removed.

By substituting the negative and positive transparencies of Eqs. 4.26 and 4.29, we have as the amplitude transmittances of these two transparencies

$$T_n = (T_{in})^{1/2} = K_n^{1/2} I^{-\gamma_n/2} = K_n^{1/2}(uu^*)^{-\gamma_n/2} \tag{4.32}$$

and

$$T_p = (T_{ip})^{1/2} = K^{1/2} I^{\gamma_{n1}\gamma_{n2}/2} = K^{1/2}(uu^*)^{\gamma_{n1}\gamma_{n2}/2}, \tag{4.33}$$

where u is the complex amplitude of the incident light field. Thus, the amplitude transmittance of the two-step contact process can be written as

$$T = K_1|u|^{\gamma}, \tag{4.34}$$

where K_1 is a positive constant and $\gamma = \gamma_{n1}\gamma_{n2}$. A linear relation between amplitude transmittance and the amplitude of the recording light field may be achieved by making the overall gamma equal to unity.

In practice, the first gamma is frequently chosen to be less than unity (e.g., $\gamma_{n1} = \frac{1}{2}$), and the second gamma is chosen to be larger than two (e.g., $\gamma_{n2} = 4$). The overall gamma is therefore equal to two. This provides a square-law relation rather than linearity for intensity transmittance versus incident irradiance. As can be seen from Eq. 4.31, however, the relation for amplitude transmittance is linear.

In most coherent systems it is more convenient to use the transfer characteristic directly than to use the H-and-D curve. This direct transfer characteristic is frequently referred to as the T–E curve, of amplitude transmittance versus exposure, as shown in Figure 4.18. As we can see, if the film is properly exposed at an operating point lying well within the linear region of the transfer characteristic of the T–E curve, then within a limited variation in exposure, the recording will offer the best linear amplitude transmittance. If E_Q and T_Q ("quiescent") denote the corresponding bias exposure and amplitude transmittance, then within the linear

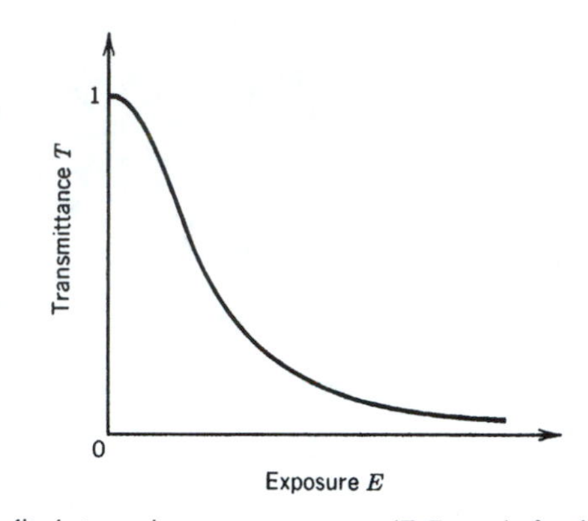

FIGURE 4.18 Amplitude transmittance versus exposure (T–E curve) of a photographic negative.

region of the T–E curve, amplitude transmittance can be written as

$$T \simeq T_Q + \alpha(E - E_Q) = T_Q + \alpha'|\Delta u|^2, \tag{4.35}$$

where α is the slope measured at the quiescent point of the T–E curve, and $|\Delta u|$ is the incremental amplitude variation $\alpha' = \alpha t$, with t the exposure time.

We conclude this section by reiterating that constraint of the signal recorded within the linear region of the T–E curve may not be necessary in some coherent optical systems. This is particularly so in holography, since the holographic construction is mostly a phase-modulated encoding.

Example 4.10

Using the contact printing process, show that it is possible to encode two image transparencies into a photographic film so that the film's amplitude transmittance is the sum of the two image irradiances.

First, we select a low-gamma film (e.g., $\gamma_{n1} < 1$) for the sequential encoding of the two image irradiances, that is, $I_1(x, y)$ and $I_2(x, y)$. We assume that the recording is in the straight-line region of the H-and-D curve, so that the intensity transmittance of the recorded film would be (Eq. 4.26)

$$T_{in}(x, y) = K[I_1(x, y) + I_2(x, y)]^{-\gamma_{n1}}.$$

By contact printing T_{in} onto a high-contrast film (e.g., $\gamma_{n2} > 1$), we obtain for amplitude transmittance of the second film (Eq. 4.33)

$$T_p(x, y) = K_1[I_1(x, y) + I_2(x, y)]^{\gamma_{n1}\gamma_{n2}/2},$$

which is a positive image. Thus, we see that if $\gamma_{n1}\gamma_{n2} = 2$ (e.g., $\gamma_{n1} = \frac{1}{2}$ and $\gamma_{n2} = 4$), then

$$T_p(x, y) = K_1[I_1(x, y) + I_2(x, y)],$$

which is the sum of the two image irradiances. ■

Example 4.11

Use the T–E curve of Figure 4.18.

(a) Draw an input–output block diagram to represent the photographic recording. Assume that recording is limited to the linear region of the T–E curve and that $f(x, y)$ is the complex light illumination.

(b) If the exposure is a phase-modulated function,

$$E(x) = E_0 + E_1 \cos[p_0 x + \phi(x)],$$

where the E's are arbitrary constants, p_0 is the angular carrier frequency, and $\phi(x)$ is the phase variation, show that the recording can be extended beyond the linear region of the T–E curve.

Answers

(a) A block diagram representing the process is shown in Figure 4.19*a*.

(b) Since the instantaneous phase modulation represents the information carrier element, the E_0 (dc bias) crossings preserve $\phi(x)$. Thus, the recording

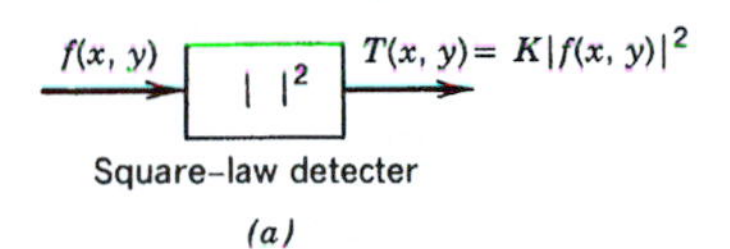

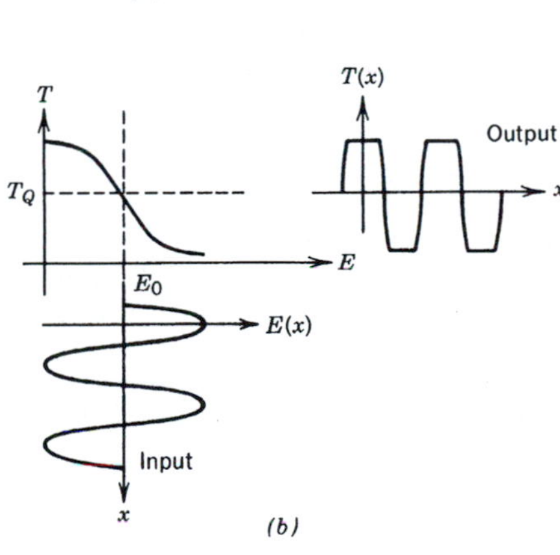

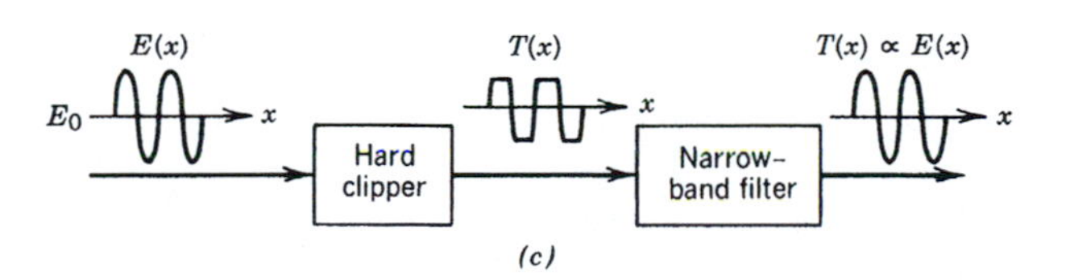

FIGURE 4.19 (*a*) Block diagram of the photographic recording process. (*b*) Recording process extending beyond the linear region of the T–E curve. (*c*) The corresponding block diagram for recording beyond the linear region of the T–E curve.

can be extended beyond the linear region, as sketched in Figure 4.19*b*. Figure 4.19*c* is a block diagram showing the transmittance function proportional to $E(x)$. ■

4.6 ELECTROOPTIC DEVICES

Silver halide photographic films are still the most commonly used and best-developed recording material. However, photographic films have two major drawbacks; the wet development process is cumbersome, and more importantly, the delay encountered in developing films is a major bottleneck in many optical processing systems.

In this section we discuss a few types of electrooptic devices that are commonly used in real-time optical signal processing and computing.

4.6.1 Photoplastic Devices

The basic components of a photoplastic device are shown in Figure 4.20. The device is composed of a clear substrate (usually glass), coated with a transparent conductive layer (tin oxide or indium oxide), on top of which is a layer of photoconductive material, followed by a layer of thermoplastic. For the photoconductor, poly-*n*-vinyl carbazole (PVK) sensitized with trinitrofluorenone (TNF) can be used with an ester resin thermoplastic (Hurculus Floral 105). Before exposure, the device is charged either by a corona discharge or with a charging plate made of another transparent conductive material, as shown in Figure 4.21. The charging plate is separated from the device by a strip of 100-micron Mylar tape. After the charging process, the device can be exposed to the signal light, which causes a variation in charge pattern proportional to the intensity of the input light. The illuminated region displaces the charge from the transparent conductive layer to the photoplastic interface, which in turn reduces the surface potential of the outer surface of the thermoplastic, as shown in the figure. The device is then recharged to the original surface potential and developed by raising the temperature of the thermoplastic to the softening point and then rapidly reducing it to room temperature.

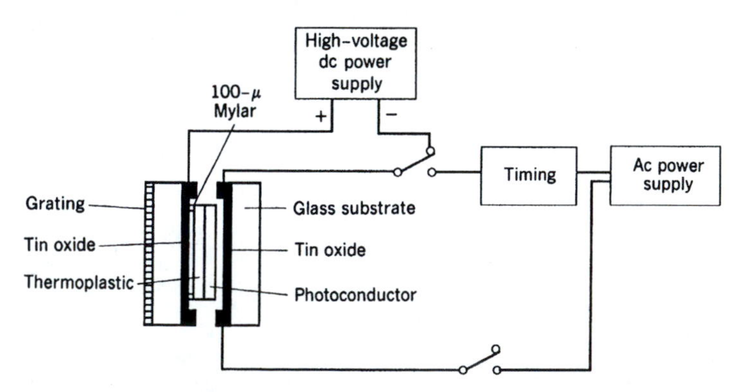

FIGURE 4.20 Composition of a photoplastic device.

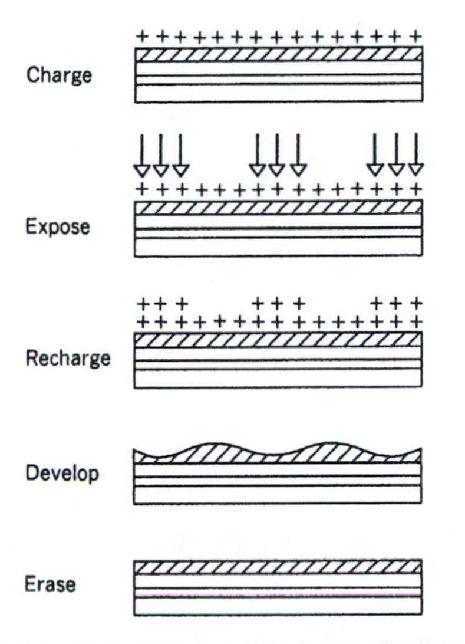

FIGURE 4.21 Operation of the photoplastic device in recording light intensity patterns.

The temperature reduction can be accomplished by passing an ac voltage pulse through the conductive layer. Surface deformation caused by the electrostatic force then produces a phase recording of the intensity of the input light. This recording can be erased by raising the temperature of the thermoplastic above the melting point, thus causing the surface tension of thermoplastic to flatten the surface deformation and erase the recording.

The frequency response of the photoplastic device is poor at low frequencies. To use the device for recording signals with low spatial frequencies, we must modulate the signal with a sinusoidal signal of a spatial frequency corresponding to the peak of the frequency response of the device. We can do this by putting a sinusoidal grating directly in front of the thermoplastic device. A thick thermoplastic layer will make the frequency response of the device peak at 50 lines/mm or below.

The greatest advantage of the photoplastic device is its relatively low cost, which makes it a very practical recording device for use in holography (see Chapter 8). However, its performance is less than ideal in many respects. For example, its cycling time is relatively slow, its lifetime is limited (about 500 cycles, depending on laboratory conditions), and the signal-to-noise ratio is relatively low caused by the random thickness variation of the thermoplastic plate.

4.6.2 Pockels Readout Optical Modulator

Pockels readout optical modulator (PROM) is fabricated from various electrooptic crystals such as ZnS, ZnSe, and $Bi_{12}SiO_{20}$. The basic construction of the PROM device is shown in Figure 4.22. The electrooptic crystal wafer is sandwiched between two transparent electrodes and separated from them by an insulator. The crystal wafer is oriented in such a way that the field applied between the electrodes produces a longitudinal electrooptic effect. The operation of the PROM device is also illustrated in the figure. An applied dc voltage with an erase light

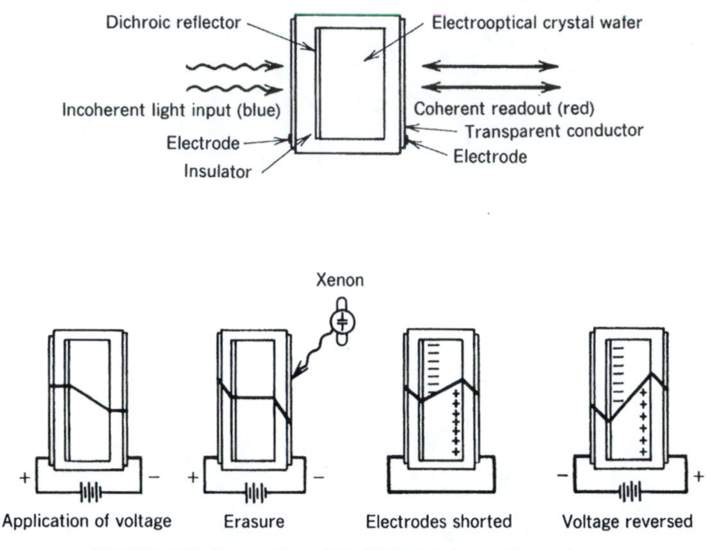

FIGURE 4.22 Composition of the PROM device and its operation.

pulse is used to create mobile carriers that cause the voltage V_0 in the active crystal to decay to zero. The polarity of the applied voltage is brought to zero and then reversed. When a total voltage of $2V_0$ appears across the crystal, the device is exposed to the illumination pattern of blue light. The voltage in the area exposed to the bright part of the input pattern decays, but the voltage in the dark area remains unchanged, thus converting the intensity pattern into a voltage pattern. The relation between input exposure and voltage across the crystal is given by

$$V_c = V_0 e^{-KE}, \tag{4.36}$$

where V_0 is the applied voltage and K is a positive constant. The readout is performed with a red linearly polarized light (e.g., He–Ne laser). For $Bi_{12}SiO_{20}$, the crystal is 200 times more sensitive in the blue region (400 nm) than in the red region (633 nm). Therefore, reading with a He–Ne light source in real time does not produce significant voltage decay over a period of time.

The readout is performed by reflection. In this readout mode, the area of the crystal where the voltage across it has not been affected by the input light intensity acts like a half-wave retardation plate. The angle of polarization of the linearly polarized laser light input reflected by such an area is therefore rotated by 90°. Thus, the light reflected by the area corresponding to the bright region has a polarization perpendicular to that of the dark region. The reflected light is then passed through a polarizer, and the amplitude of the transmitted light is attenuated according to the polarization of the input.

The theoretical curve for amplitude A of output coherent light versus exposure E to input incoherent light is plotted in Figure 4.23. We see that it has a transfer characteristic similar to that of a photographic film, a linear region with a range between $E = 2/K$ and $E = 0$, and a bias point at $E = 1/K$.

The PROM device has some drawbacks, however. For example, the fabrication of the device still needs refinement to improve the uniformity and consistency

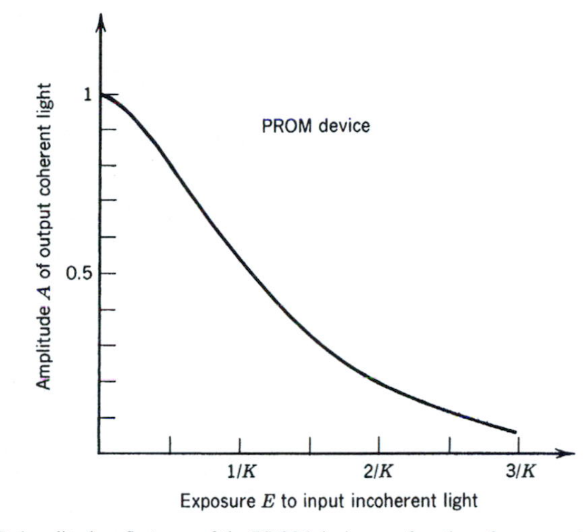

FIGURE 4.23 Amplitude reflectance of the PROM device as a function of exposure to input coherent light.

of the crystal. In addition, the device needs a narrower band-pass in its spectral sensitivity. Any extended exposure to high-intensity readout light causes substantial decay of the crystal voltage, thus reducing the read amplitude of the device. Therefore, the device cannot be read over an extended period of time under strong illumination.

4.6.3 Acoustooptic Modulator

The interaction of light waves with sound waves has been the basis of a large number of devices connected to various laser systems for display, information handling, optical signal processing, and numerous other applications requiring the spatial or temporal modulation of coherent light. The underlying mechanism of acoustooptic interaction is simply the change induced in the refractive index of an optical medium by the presence of an acoustic wave. An acoustic wave is a traveling pressure disturbance which produces regions of compression and rarefaction in a material; these density variations cause corresponding changes in the refractive index of the material.

These interactions between light and sound produce a Bragg diffraction effect, as illustrated in Figure 4.24. The beam of light is incident upon a plane acoustic wave in the acoustooptic (AO) cell. This acoustic wave is emitted by a transducer driven by an electric signal. At a certain critical angle of incidence θ_i, the incident beam produces a coherent diffraction at an angle θ_d. This second angle is known as the Bragg angle of diffraction and is given by

$$\sin\theta_d = \frac{\lambda}{2\Lambda}, \tag{4.37}$$

where λ and Λ are the light and the acoustic wavelengths in the cell, respectively.

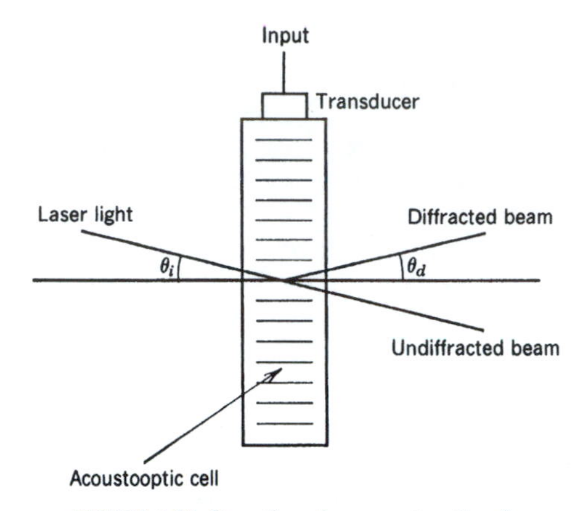

FIGURE 4.24 Operation of an acoustooptic cell.

There are certain physical constraints on the types of signal waveforms that can serve as input and the ways in which AO cells can be used. For example, acoustic waves emitted by the transducer into the acoustic medium usually cannot be seen if they are viewed by conventional optics. Moreover, the electric signals that drive the AO cells must generally be band-pass in nature and have a frequency in the range of 1 MHz to 1 GHz. This restriction on the input signal waveform means that to be processed these signals not originally band-pass in nature must be modulated with a carrier frequency that is suitable for input of the AO cell.

4.6.4 Liquid Crystal Light Valve

Another example of real-time electrooptic spatial light modulators is the liquid crystal light valve (LCLV). The LCLV is capable of converting an incoherent optical image into a coherent image suitable as input for nearly any type of coherent optical processing system (see Chapter 9). Most of us are familiar with liquid crystal materials from their use in the display panels of digital wristwatches and calculators. In fact, liquid crystal displays have found their way into portable laboratory equipment, kitchen appliances, automobile dashboards, and many other applications requiring good visibility, low power consumption, and special display geometries. Most of these applications use the liquid crystal in a twisted nematic cell; a segment of the cell is either opaque or transparent, depending on the magnitude of the applied electric voltage. In applications related to optical signal processing, however, the hybrid field-effect LCLV is more desirable. The LCLV combines the properties of the twisted nematic cell in the "off" state with the property of electrically tunable birefringence to control transmittance over a wide continuous range in the "on" state.

A simplified sketch of a transmission-type twisted nematic cell is shown in Figure 4.25*a*. A thin layer (1 to 20 μm) of nematic liquid crystal is sandwiched between two transparent electrode-coated glass plates. The electrode surfaces are

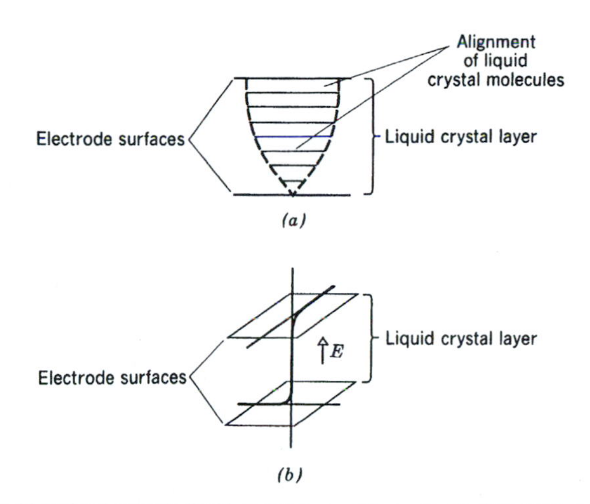

FIGURE 4.25 (*a*) Twisted nematic liquid crystal layer with no electric field applied. (*b*) Molecular alignment in the direction of the applied electric field E.

treated to provide a preferred direction of alignment for the liquid crystal molecules. The plates are so arranged that, with no electric voltage applied across the layer of liquid crystal molecules, the layer is twisted continuously by 90°. A polarizer and an analyzer are placed in front of and behind the sandwiched cell, respectively. The direction in which the light passing through the polarizer is polarized must be the same as the direction of molecular alignment at the front electrode surface. As the light passes through the twisted liquid crystal layer, its direction of polarization is also twisted by 90°. If we make the direction of polarization of the analyzer parallel to the direction of molecular alignment at the back electrode surface, the polarized light will pass through the cell unaffected. If, on the other hand, we make the direction of polarization of the analyzer perpendicular to that of the back surface molecular alignment, the polarized light will not exit the cell, and the device will be in its dark state. With slight modification, this second mode of operating the twisted nematic cell is utilized in the LCLV.

An interesting phenomenon occurs when a low-frequency voltage is applied across the twisted nematic liquid crystal layer. The molecules tend to align themselves in the direction of the applied electric field—that is, perpendicular to the electrode surfaces. The resulting splay and bend of the molecules is shown in Figure 4.25*b*. Thus, if the polarizer and the analyzer have parallel directions of polarization, the polarized light will pass through the liquid crystal layer and the analyzer unaffected. This "on" state behavior is not exhibited in the LCLV, however, because of the reflection-type configuration.

The LCLV takes advantage of the pure birefringence of the liquid crystal material in order to modulate the output laser beam while in the "on" state. Operation of the LCLV is clarified in the side view of Figure 4.26. An incoherent image is focused onto the photoconductor layer to gate the applied alternating voltage to the liquid crystal layer in response to the input intensity at every point in the input space. Laser light illuminating the back of the LCLV is reflected back but modulated by the birefringence of the liquid crystal layer at every point. In order to achieve this effect, the molecular alignment has a 45° twist between the two

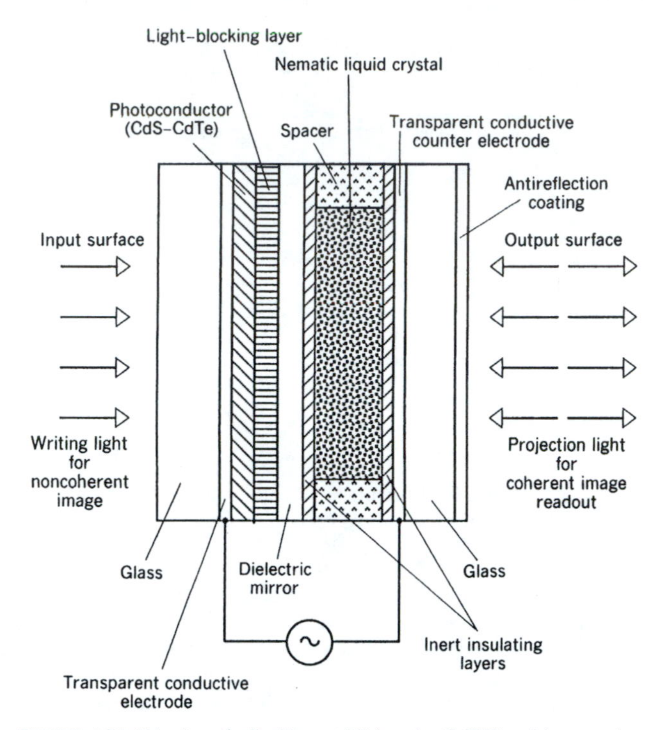

FIGURE 4.26 Side view of a liquid crystal light valve (LCLV) and its operation.

surfaces of the liquid crystal layer. The dielectric mirror plays an important role by providing optical isolation between the input incoherent light and the coherent readout beam. It is worth noting that the resolution of the commercially available LCLV is about 30 lines/mm, and because of the slow response of the liquid crystals, the readout time is about 10 msec.

4.6.5 Magnetooptic Modulator

The magnetooptic modulator consists of a square grid of magnetically bistable mesas (or pixels) that may be used to modulate incident polarized light by the Faraday effect. The device can be electrically switched so that object patterns can be written with a computer. Thus, this device could function as a programmable spatial light modulator.

The basic structure of the magnetooptic spatial light modulator (MOSLM) consists of a bismuth-doped, magnetic iron–garnet film which is epitaxially deposited on a transparent, nonmagnetic garnet crystal substrate. The film is then etched into a square grid of magnetically bistable mesas, and current drive lines are deposited between them. The resulting device is an $n \times n$ matrix of mesas, as depicted in Figure 4.27.

When plane-polarized light is incident on the device, the axis of polarization of the transmitted light will be rotated by the Faraday effect. The plane of polar-

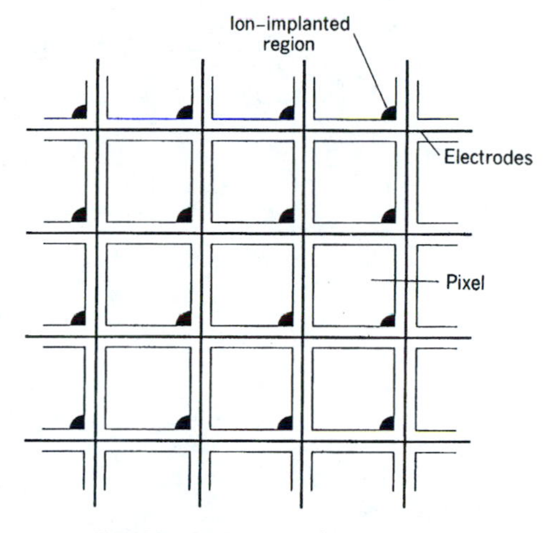

FIGURE 4.27 Structure of MOSLM pixels.

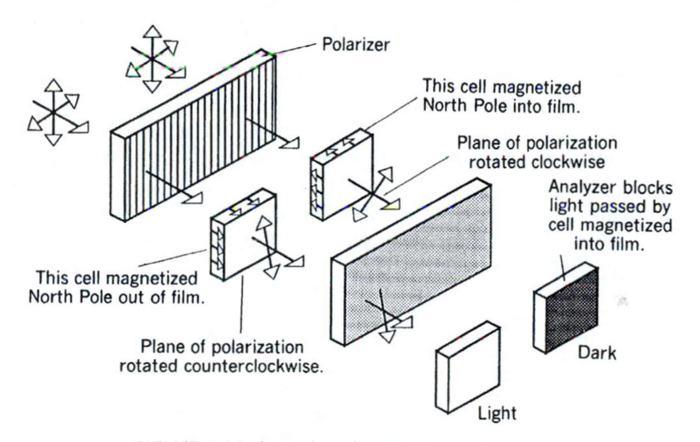

FIGURE 4.28 Operation of MOSLM as a light valve.

ization is rotated in opposite directions for opposite magnetic states, as shown in Figure 4.28. A polarization analyzer then converts this rotation into modulation of image brightness. The magnetization state of the pixel may be changed by sending current to its two adjoining drive lines.

4.6.6 Microchannel Plate Modulator

The microchannel plate spatial light modulator (MSLM) is an electrooptic SLM that has been frequently used in real-time optical signal processing and computing. The MSLM is an optically addressed device (similar to the LCLV) that has a high level of optical sensitivity and high framming speed. The basic structure of an MSLM is illustrated in Figure 4.29. It consists of a photocathode, a microchannel

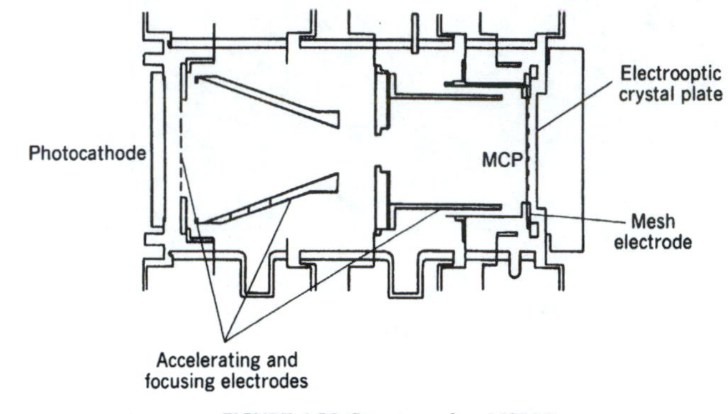

FIGURE 4.29 Structure of an MSLM.

plate (MCP), an accelerating mesh electrode, and an electrooptic crystal plate ($LiNbO_3$) which bears, on its inner side, a high-resistivity dielectric mirror to isolate the readout from the write-in side. Some MSLMs have more than one mesh electrode to improve the electrostatic imaging and focusing. All these components are sealed in a vacuum tube.

The basic operation of an MSLM is shown in Figure 4.30. The write-in light (incoherent image) incident on the photocathode generates a photoelectron image, which is multiplied to about 10^5 times by the MCP, accelerated by the mesh electrode, and deposited on the dielectric mirror of the $LiNbO_3$ crystal plate. This process is somewhat similar to that in an electronic vacuum tube. The resulting charge distribution, in combination with the biasing voltage, creates a spatially varying electric field within the crystal plate in the direction of the optical axis. This field in turn modulates the refractive indices of the crystal plate. Because of

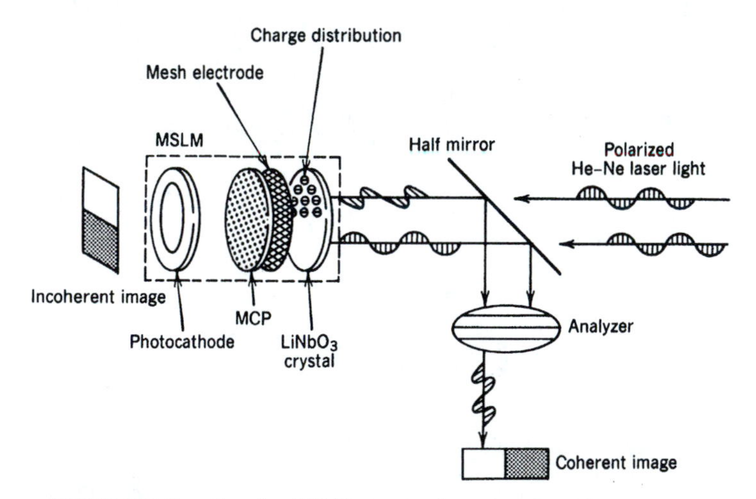

FIGURE 4.30 Operation of an MSLM as an incoherent-to-coherent image converter.

the birefringence property of the $LiNbO_3$ crystal, its refractive indices in the horizontal and vertical planes are modulated differently. Thus, after twice passing the crystal plane, the readout light, which was originally polarized in a plane bisecting the X and Y axes of the crystal plate, will have a relative phase retardation between its x and y components. The higher the charge density, the greater the phase retardation. Thus, the MSLM can be used directly as a phase-only spatial light modulator. In addition, if an analyzer is inserted in the readout light path (as shown in the figure) to detect the phase retardation, a coherent image that is proportional to the input image can be obtained. Generally speaking, not only is the MSLM an incoherent-to-coherent converter, but it can be applied as a wavelength converter, as an input or output transducer, and to image plane or Fourier plane processing.

The versatility of the device comes from its architecture. As just described, photoelectron generation, intensification, transfer, deposit, readout light modulation, and detection are performed by each functional component. It is relatively easy to custom-design the MSLM for different applications. The typical performance specifications of commercially available MSLMs are as follows. The spatial resolution is about 10 lines/mm, the contrast ratio is more than 1000:1, the input sensitivity is around 30 nJ/cm^2, writing time response is about 50 msec, erasing time is about 100 msec, storage time is about one day, and the input window diameter is 15 mm.

4.6.7 Liquid Crystal Television

The basic structure of the commercially available liquid crystal television (LCTV) is illustrated in Figure 4.31*a*. Two polarizing sheets are attached to the substrates with adhesive, one acting as the polarizer and the other as the analyzer. A plastic diffuser provides diffused illumination on one side, and a clear plastic window

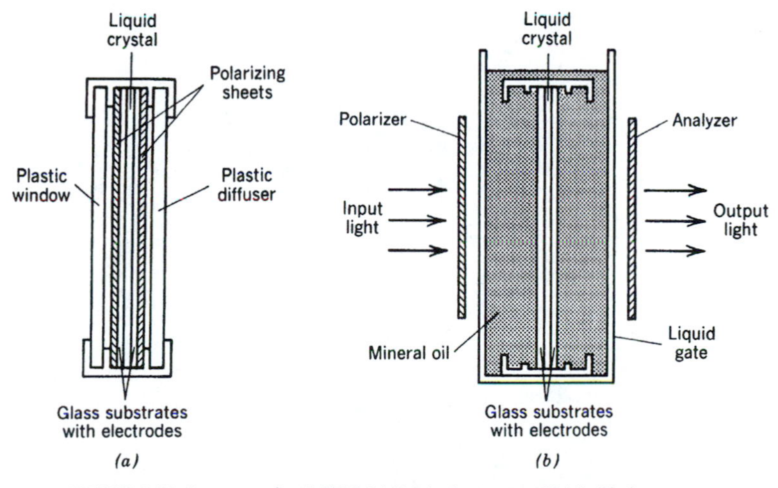

FIGURE 4.31 Structure of an LCTV. (*a*) Original structure. (*b*) Modified structure.

protects the device from dust on the opposite side. The LCTV can be converted to a spatial light modulator as follows.

1. Open the case and break the hinge stops so that the LCTV screen can be opened fully.
2. Disassemble the aluminum frame of the LCTV screen and remove the plastic diffuser and window from the frame.
3. Peel off the two polarizing sheets from the glass substrates.
4. Clean the substrates carefully with acetone.
5. Submerge the modified LCTV screen in an index-matching liquid gate, which will be used as a spatial light modulator, as illustrated in Figure 4.31*b*.

The last step is necessary to remove the phase noise distortion caused by the thin glass substrates.

The operation of an LCTV is illustrated in Figure 4.32. Each liquid crystal cell of the LCTV screen is a nematic liquid crystal twisted 90°. When no electric field is applied, the plane of polarization for linearly polarized light is rotated through 90° by the twisted liquid crystal molecules. Thus, no light can be transmitted through the analyzer, as illustrated in the upper part of the figure. Under an applied electric field, however, the twist and the tilt of the molecules are altered, and the liquid crystal molecules attempt to align parallel with the applied field. This results in partial transmission of light through the analyzer. As the electric field increases further, all the liquid crystal molecules align in the direction of the applied field. The molecules do not affect the plane of polarization, so all the light passes through the analyzer, as shown in the lower part of the figure.

Varying the applied voltage at each liquid crystal cell allows light transmission to be varied. If the applied voltages of the LCTV screen are generated from a computer, a TV camera, or a TV receiver, the LCTV screen can produce gray-scale images.

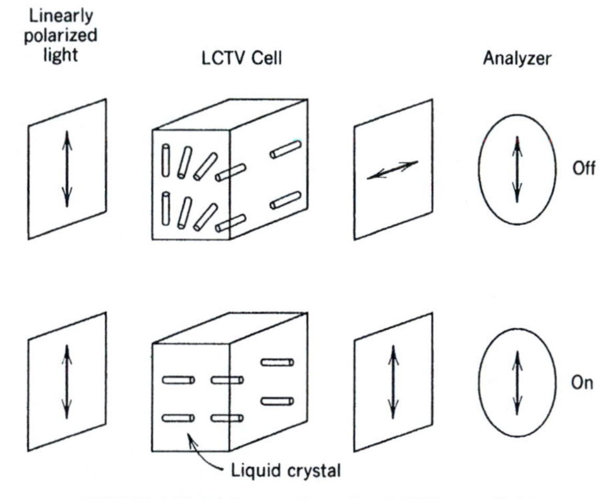

FIGURE 4.32 Basic operation of an LCTV as an SLM.

The major advantages of the LCTV are its low cost and programmability, which make the device a very practical SLM for many optical signal processing and optical computing applications. However, its performance is still far from ideal in several practical aspects. For example, it has low resolution ($\simeq 1.47$ lines/mm), low contrast ($\simeq 20{:}1$), and low speed (the pixels are refreshed at 60 Hz).

4.6.8 Charge-Coupled Devices

An image detector that has been frequently used in optical–digital image processing and optical computing is the charge-coupled device (CCD). The basic structure of a CCD is a shift register formed by an array of closely spaced metal–oxide–silicon (MOS) capacitors.

The MOS capacitor is the basis of many charge-transfer devices. It can be made by depositing a metal electrode on a thermally oxidized *p*-type silicon substrate, as illustrated in Figure 4.33. If a positive voltage is applied to the electrode, the majority carriers (holes) in the silicon are repelled and a potential well is formed at the silicon surface. This well can be considered a bucket, and the minority carriers form the fluid that partially fills the bucket. When a photon is absorbed, an electron–hole pair is produced in the depletion layer of the MOS capacitor. The bucket, the potential well, will contain the charges, the electrons, in direct proportion to the time integral of the incident intensity. These potential wells are organized serially in the CCD, which allows a sequential output charge transfer.

A practical device is produced when individual photosensors (i.e., MOS capacitors) are isolated from the readout register by a transfer gate. When this gate is opened, all the charges integrated in the photosensors transfer in parallel into the readout register, which has one transfer cell opposite each photosensor. After the gate is closed, each photosensor immediately starts integrating the next line, while the previous line is read out along the transfer register. The transfer register is shielded from the incident light, as depicted schematically in Figure 4.34.

Charge-coupled devices are used mostly for image detection and integration applications, since few other devices provide compact, high-resolution, and low-light-level image sensing. High-resolution CCD array detectors are available in

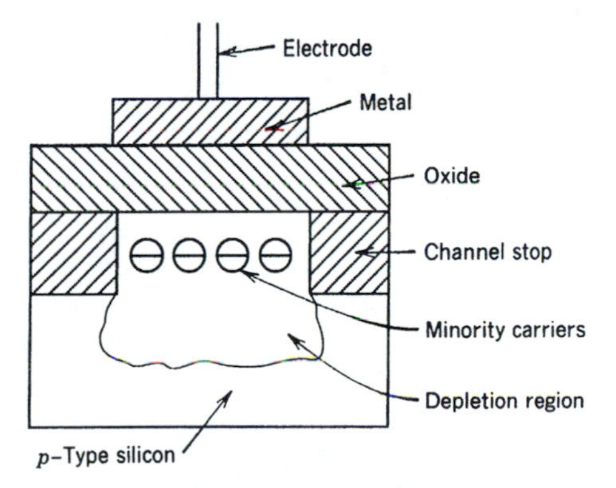

FIGURE 4.33 A MOS capacitor.

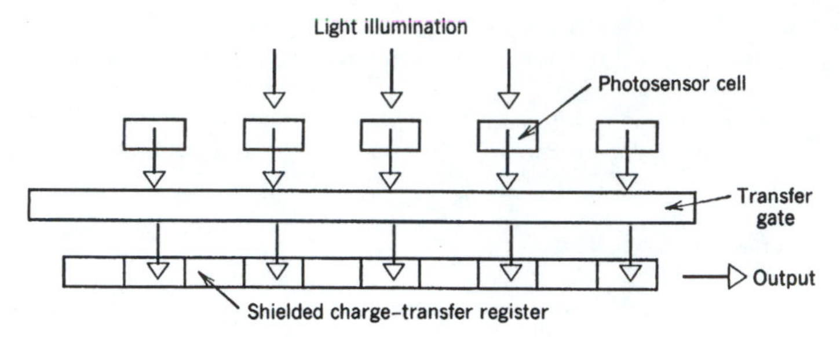

FIGURE 4.34 The operation of a CCD.

256 × 256 to 1024 × 1024 pixels within a 4.8 × 6.4-mm area. Charge-coupled linear devices are single-line arrays that can provide even higher resolutions of up to 6000 pixels in a single line. Charge-coupled cameras can retain a high modulation transfer function (MTF) up to Nyquist limits. At these limits each sensor element receives and detects new data. MTFs of 60 to 75 percent are typical when the illumination spectrum is filtered to simulate human-eye response. Moreover, CCDs have a high level of sensitivity and are capable of integrating and storing data over relatively long periods of time. When the device is cooled, these integration periods can last as long as several minutes before background dark currents affect the quality of the image. New developments also make it possible for CCDs to control exposure shuttering, so that any level of charge integration can be obtained. When adequate light is available, exposure times of less than 1 nsec can be set, thus allowing the capture of high-speed events without blur. Today's CCDs can achieve nonuniformity of less than 5 percent because they have an output signal-to-noise ratio greater than 60 dB (1000:1). In fact, they can be built very small, with a total thickness of only one-quarter inch, and an area of less than one and one-half square inches.

REFERENCES

1. W. G. Driscoll and W. Vaughan, *Handbook of Optics*, McGraw-Hill, New York, 1978.
2. E. Hect and A. Zajac, *Optics*, Addison-Wesley, Reading, Mass., 1979.
3. F. A. Jenkins and H. E. White, *Fundamentals of Optics*, fourth edition, McGraw-Hill, New York, 1976.
4. A. Yariv, *Introduction to Optical Electronics*, Holt, Rinehart and Winston, New York, 1976.
5. M. Born and E. Wolf, *Principles of Optics*, Pergamon, Oxford, 1970.
6. M. Gottlieb, C. L. M. Ireland, and J. M. Ley, *Electro-Optics and Acousto-Optic Scanning and Reflection*, Marcel Dekker, New York, 1983.
7. W. E. Ross, D. Psaltis, and R. H. Anderson, "Two Dimensional Magneto-Optic Spatial Light Modulator for Signal Processing," *Optical Engineering*, Vol. *22*, 485 (1983).
8. C. Warde and J. Thackara, "Operating Modes of the Microchannel Spatial Light Modulator," *Optical Engineering*, Vol. *22*, 695 (1983).

9. L. E. RAVICH, "Charge-Coupled Devices: A Primer," *Laser Focus*, Vol. *23*, No. 6, 166 (1987).

PROBLEMS

4.1 Given the Hurter–Driffield (H-and-D) curve of a certain photographic film, as shown in Figure 4.35,

(a) Evaluate the gamma (γ) of the film.

(b) Write a linear equation to represent the straight-line region of the H-and-D curve.

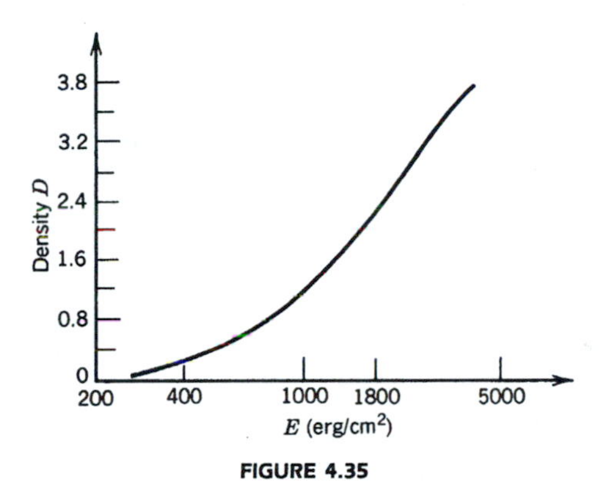

FIGURE 4.35

4.2 Using the film of Problem 4.1, find the linear amplitude transmittance accomplished through the contact printing process. Then calculate the required gamma of the second film.

4.3 Use the H-and-D curve of Problem 4.1.

(a) Plot the amplitude transmittance versus linear exposure (T–E) curve of the film.

(b) Write a linear equation to represent the linear region of the T–E curve.

4.4 We wish to encode two negative image irradiances [i.e., $I_1(x, y)$ and $I_2(x, y)$] in order to obtain a linear (positive) amplitude transmittance.

(a) Describe a procedure to achieve this goal.

(b) Show that output amplitude transmittance would be the sum of the two image irradiances.

4.5 Show that shot noise has the mean square current amplitude given by Eq. 4.7, by considering the charges moving between two electrodes. (*Hint:* A model may be seen in reference 4, Chapter 10.)

4.6 Derive the Bragg scattering angle (Eq. 4.37) in an acoustooptic scattering process.

4.7 In the discussion on magnetooptic devices (Section 4.6.5), we mention the Faraday effect. Using the electromagnetic theory, explain this effect quantitatively in terms of the relevant field parameters.

4.8 Calculate the energy associated with a quanta of light energy for the following typical laser wavelengths: (a) $\lambda = 5145$ Å, (b) $\lambda = 1.06\ \mu m$, (c) $\lambda = 10.6\ \mu m$, and (d) $\lambda = 8000$ Å.

4.9 Conduct a literature search and write a brief report on the practical use of one or two of the real-time detectors discussed in Section 4.6.

4.10 Estimate the amount of current density emitted by a photocathode with a quantum efficiency of $\eta = 0.1$ when a weak He–Ne laser is incident on it with an intensity of 10 mW/cm^2. Using a typical gain of 10^6, estimate the current at the anode.

5
ELECTROMAGNETIC THEORY

In the preceding chapters on lenses, reflection, and refraction, we treated light waves as some rays moving in space. In this chapter we discuss the exact nature of light, namely that it is a traveling electromagnetic field. This electromagnetic nature of light, a very important discovery in science, is credited to James Clerk Maxwell (1831–1879).

Historically, several laws of electrostatics, magnetostatics, and electrodynamics were established during the late nineteenth century. These basic laws, for example, Faraday's laws, Ampère's law, together with the concept of the displacement current, were systematically put together by Maxwell into what came to be called *Maxwell's equations*,

$$\nabla \cdot \mathbf{D} = \rho_f, \tag{5.1}$$

$$\nabla \times \mathbf{E} = -\frac{\partial \mathbf{B}}{\partial t}, \tag{5.2}$$

$$\nabla \cdot \mathbf{B} = 0, \tag{5.3}$$

$$\nabla \times \mathbf{H} = \mathbf{J}_f + \frac{\partial \mathbf{D}}{\partial t}, \tag{5.4}$$

and the *constitutive equations*,

$$\mathbf{D} = \varepsilon_0 \mathbf{E} + \mathbf{P}, \tag{5.5}$$

$$\mathbf{H} = \frac{\mathbf{B}}{\mu_0} - \mathbf{M}. \tag{5.6}$$

These and other electric and magnetic field variables commonly used in the study of optical electronics and their units are summarized in Table 5.1 (see Section 5.8).

The interaction of field variables **E** and **B** with a point charge is given by the Lorentz force equation

$$\mathbf{F} = q(\mathbf{E} + \mathbf{v} \times \mathbf{B}), \tag{5.7}$$

where **v** is the velocity of the charge. Using Eqs. 5.5 and 5.6, we can rewrite Eqs. 5.1 through 5.4 to yield a set of Maxwell equations involving only **E** and **B**:

$$\nabla \cdot \mathbf{E} = \frac{1}{\varepsilon_0}(\rho_f - \nabla \cdot \mathbf{P}), \tag{5.8}$$

$$\nabla \times \mathbf{E} = -\frac{\partial \mathbf{B}}{\partial t}, \tag{5.9}$$

$$\nabla \cdot \mathbf{B} = 0, \tag{5.10}$$

$$\nabla \times \mathbf{B} = \mu_0\left(\mathbf{J}_f + \nabla \times \mathbf{M} + \varepsilon_0 \frac{\partial \mathbf{E}}{\partial t} + \frac{\partial \mathbf{P}}{\partial t}\right). \tag{5.11}$$

In general, the polarization **P** in a material is a function of the electric field **E**, in addition to any permanent polarization associated with the particular physical structure of the material. Roughly speaking, if **P** is a linear function of **E**, we speak of the medium as being linear; otherwise, the medium is said to be nonlinear. In an *isotropic medium*, as opposed to a so-called *anisotropic medium*, the direction of **P** is parallel to the direction of **E**. In some special situations, for example, in optical switching elements or bistable devices, **P** can be a multivalued function of **E**. In an analogous manner magnetization **M** is, in general, a function of magnetic induction **B**.

The electromagnetic energy density associated with fields **E** and **B** is given by

$$U = U_e + U_m = \tfrac{1}{2}\mathbf{D} \cdot \mathbf{E} + \tfrac{1}{2}\mathbf{B} \cdot \mathbf{H}, \tag{5.12}$$

where U_e and U_m denote the electric and the magnetic contributions, respectively.

The flow of electromagnetic energy per unit area is given by *Poynting's vector*:

$$\mathbf{S} = \mathbf{E} \times \mathbf{H}. \tag{5.13}$$

This flow also gives rise to momentum density $\mathbf{g} = \varepsilon_0 \mathbf{E} \times \mathbf{B}$, and angular momentum density $\mathbf{L} = \mathbf{r} \times \mathbf{g}$, where **r** is the displacement from a reference point. The principles behind the definitions and derivations of these functional forms may be found in many standard texts on electromagnetism (see, for example, reference 4). They are summarized here for ease of reference. In their present form they are of limited usefulness; however, when applied to specific problems, the power and the richness of Maxwell's equations will become apparent.

5.1 MAXWELL'S EQUATIONS FOR LINEAR ISOTROPIC HOMOGENEOUS MEDIA

In an isotropic medium the polarization **P** and the magnetization **M** are parallel to **E** and **H**, respectively. That is, we can write

$$\mathbf{P} = \varepsilon_0 \chi_e \mathbf{E} \tag{5.14}$$

and

$$\mathbf{M} = \chi_m \mathbf{H}, \tag{5.15}$$

where χ_e and χ_m are the electric and the magnetic susceptibilities, respectively. In a homogeneous medium these two quantities are spatially constant. The magnitudes of χ_e and χ_m vary widely from one material to another, depending on the frequency (or wavelength) of the fields and on the material's physical structure. The values of the electric and magnetic susceptibilities for particular materials can be obtained from handbooks on physical constants.

Using Eqs. 5.14 and 5.15, we can reexpress **D** and **H** as

$$\mathbf{D} = (1 + \chi_e)\varepsilon_0 \mathbf{E} = \chi_e \varepsilon_0 \mathbf{E} = \varepsilon \mathbf{E} \tag{5.16}$$

and

$$\mathbf{B} = \mu_0(1 + \chi_m)\mathbf{H} = \chi_m \mu_0 \mathbf{H} = \mu \mathbf{H}, \tag{5.17}$$

where

$$\kappa_e = 1 + \chi_e = \text{dielectric constant}, \tag{5.18}$$

$$\varepsilon = \kappa_e \varepsilon_0 = \text{permittivity}, \tag{5.19}$$

and, similarly,

$$\kappa_m = 1 + \chi_m = \text{relative permeability}, \tag{5.20}$$

$$\mu = \kappa_m \mu_0 = \text{permeability}. \tag{5.21}$$

In the same medium the free-current density $\mathbf{J}_f$ is related to **E** by the conductivity σ:

$$\mathbf{J}_f = \sigma \mathbf{E}. \tag{5.22}$$

The set of Maxwell's equations then reduces to

$$\nabla \cdot \mathbf{E} = \frac{\rho_f}{\varepsilon}, \tag{5.23}$$

$$\nabla \times \mathbf{E} = -\frac{\partial \mathbf{B}}{\partial t}, \tag{5.24}$$

$$\nabla \cdot \mathbf{B} = 0, \tag{5.25}$$

$$\nabla \times \mathbf{B} = \mu\sigma\mathbf{E} + \mu\mathbf{J}_f + \mu\varepsilon\frac{\partial \mathbf{E}}{\partial t}, \tag{5.26}$$

where the term $\mu\mathbf{J}_f$ in Eq. 5.26 represents any other free-current density not associated with $\sigma\mathbf{E}$.

Let us now ask two related questions: Do these equations possess any solutions in terms of plane waves? And, if so, what kind of plane waves are they? To answer these questions, we first consider a source-free region.

Assuming that **E** and **B** are in the plane-wave forms

$$\mathbf{E} = \mathbf{E}_0 \exp[i(\mathbf{k} \cdot \mathbf{x} - \omega t)] \tag{5.27}$$

and

$$\mathbf{B} = \mathbf{B}_0 \exp[i(\mathbf{k} \cdot \mathbf{x} - \omega t)], \tag{5.28}$$

then Eq. 5.24 gives

$$i\mathbf{k} \times \mathbf{E} = -i\omega\mathbf{B} \tag{5.29}$$

and Eq. 5.26 gives

$$i\mathbf{k} \times \mathbf{B} = \mu\varepsilon(-i\omega)\mathbf{E}. \tag{5.30}$$

Equations 5.29 and 5.30 yield a solution where **E**, **B**, and **k** are mutually orthogonal to one another, as shown in Figure 5.1. From Eq. 5.29 the ratio of the magnitude of E_0 to B_0 is

$$\frac{E_0}{B_0} = \frac{\omega}{k}. \tag{5.31}$$

On the other hand, Eq. 5.30 gives

$$\frac{E_0}{B_0} = \frac{k}{\omega}\frac{1}{\mu\varepsilon}. \tag{5.32}$$

In conjunction with Eq. 5.31, we finally obtain

$$\frac{1}{\mu\varepsilon} = \frac{\omega^2}{k^2} = v^2, \tag{5.33}$$

where v is the velocity (phase velocity) of the plane waves shown in Eqs. 5.27 and 5.28.

The immediate conclusion we can draw is that Maxwell's equations allow plane-wave solutions where the electric field **E** and the magnetic induction **B** (or **H**, since

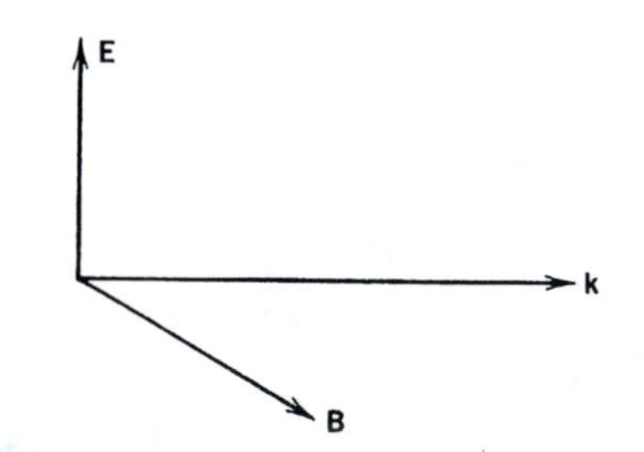

FIGURE 5.1 The electric (**E**) and magnetic (**B**) fields and the propagating vector of a light wave.

$\mathbf{B} = \mu\mathbf{H}$) are mutually orthogonal and are orthogonal to the propagation direction ($\mathbf{k}$) of the wave.

In a vacuum, $\mu \to \mu_0 = 4\pi \times 10^{-7}$ newton/ampere^2, (N/A^2), and $\varepsilon \to \varepsilon_0 = 8.85 \times 10^{-12}$ farad/meter, (F/m). This gives the velocity of the plane electromagnetic wave,

$$v = \frac{1}{\sqrt{\mu_0 \varepsilon_0}} = 3 \times 10^8 \text{ m/sec},$$

which is the speed of light in a vacuum. We shall henceforth denote this quantity by c. That Maxwell's equations correctly describe and establish the electromagnetic nature of light is but one of their many remarkable features.

We can study the electromagnetic-wave nature of light by using the following more formal procedure. In a source-free, nonconducting environment (i.e., $\rho_f = 0$, $J_f = 0$, $\sigma = 0$), Eqs. 5.23 through 5.26 become

$$\nabla \cdot \mathbf{E} = 0, \tag{5.34}$$

$$\nabla \times \mathbf{E} = -\frac{\partial \mathbf{B}}{\partial t}, \tag{5.35}$$

$$\nabla \cdot \mathbf{B} = 0, \tag{5.36}$$

$$\nabla \times \mathbf{B} = \mu\varepsilon \frac{\partial \mathbf{E}}{\partial t}. \tag{5.37}$$

Using the vector identity $\nabla \times (\nabla \times \mathbf{E}) = \nabla(\nabla \cdot \mathbf{E}) - \nabla^2\mathbf{E} = -\nabla^2\mathbf{E}$ (since $\nabla \cdot \mathbf{E} = 0$), and after taking the curl of Eq. 5.35, we get

$$\nabla^2\mathbf{E} - \mu\varepsilon \frac{\partial^2 \mathbf{E}}{\partial t^2} = 0. \tag{5.38}$$

And, similarly, from Eq. 5.37, we get

$$\nabla^2\mathbf{B} - \mu\varepsilon \frac{\partial^2 \mathbf{B}}{\partial t^2} = 0. \tag{5.39}$$

That is, both $\mathbf{E}$ and $\mathbf{B}$ satisfy a wave equation.

The solution of the wave equation depends on what we assume the boundary conditions to be. If we assume that the wave is a plane wave, then $\mathbf{E}$ and $\mathbf{B}$ will be of the form given in Eqs. 5.27 and 5.28, with phase velocity

$$v = \frac{1}{\sqrt{\mu\varepsilon}} = \frac{1}{\sqrt{\mu_0 \varepsilon_0}} \frac{1}{\sqrt{\kappa_m \kappa_e}} = \frac{c}{\eta}, \tag{5.40}$$

where $\eta = (\kappa_m \kappa_e)^{-1/2}$ is the refractive index of the medium. On the other hand, if we are dealing with an optical waveguide, or a beam emerging from some cavity, the form of $\mathbf{E}(\mathbf{r})$ and $\mathbf{B}(\mathbf{r})$ will obviously have some transverse dependence. A special but commonly occurring form of these so-called transverse modes is the Gaussian beam, which is discussed in detail in Chapter 7, when we study the propagation and guidance of a laser beam.

Example 5.1

Given a plane electromagnetic wave propagated in a homogeneous space, and the electric field components

$$E_x = 0,$$

$$E_y = 2\cos\left[\pi \times 10^{15}\left(t - \frac{x}{c}\right) - \frac{\pi}{2}\right] \text{ V/m},$$

$$E_z = 0,$$

where c is the velocity of light.

(a) Calculate the wavelength, the frequency, the direction of propagation, and the amplitude of the electromagnetic wave.

(b) If the wave impinges into the homogeneous medium in which the electric field is given by

$$E_y = 2\cos\left[\pi \times 10^{15}\left(t - \frac{3x}{2c}\right) + \frac{\pi}{2}\right],$$

determine the refractive index of the medium.

Answers

(a) With reference to the electric field component E, we identify the following:

$$k = \frac{2\pi}{\lambda} = \frac{\pi \times 10^{15}}{c}.$$

Hence, the wavelength is

$$\lambda = \frac{2c}{10^{15}} = \frac{2 \times 3 \times 10^{8}}{10^{15}} = 6 \times 10^{-7} = 600 \text{ nm}.$$

The frequency can be determined by

$$\nu = \frac{\omega}{2} = \frac{\pi \times 10^{15}}{2\pi} = 5 \times 10^{14} \text{ Hz}.$$

Propagation is in the direction of the x axis at the amplitude of the electromagnetic wave,

$$A = 2 \text{ V/m}.$$

(b) The refractive index of the medium is

$$\eta = \frac{c}{v} = \frac{c}{2c/3} = 1.5.$$

■

5.2 WAVE EQUATIONS AND POLARIZATION

In the preceding section we assumed that polarization **P** is related to **E** by the simple relation shown in Eq. 5.14. This is true for light propagation in rather uninteresting materials like glass, water, and air, in which the principal effect is a slowing down of the velocity of light. For most materials and processes of interest, the dependence of **P** on **E** remains to be ascertained, either by a vigorous quantum mechanical calculation (e.g., in a simple atom) or by some conceived approximations. It is generally true that **P** will have a linear component that is parallel to **E**, as well as some new and important terms. These new terms may or may not be linear in **E**, but they will contain explicit descriptions of the resonances, nonlinearity, gain, loss, and so forth associated with the processes under consideration.

We designate the first contribution that is linear in **E** by $\mathbf{P}_1$,

$$\mathbf{P}_1 = \varepsilon_0 \chi^{(1)} \mathbf{E}, \tag{5.41}$$

and the second term, as we may call it, by $\mathbf{P}_2(E)$. It therefore follows that the displacement **D** given in Eq. 5.5 now may be written as

$$\begin{aligned} \mathbf{D} &= \varepsilon_0 \mathbf{E} + \mathbf{P}_1 + \mathbf{P}_2 \\ &= \varepsilon \mathbf{E} + \mathbf{P}_2. \end{aligned} \tag{5.42}$$

We can rewrite **B** in Eq. 5.6 or Eq. 5.17 in an analogous way. However, the magnetic effect is usually negligibly small in comparison to its electric field counterpart, except of course for magnetic resonance effects. We shall henceforth simply use relation 5.17 for all the media under study.

In a region without free charges and currents, we have, from Eqs. 5.1 through 5.4 in conjunction with Eq. 5.42, the following equations:

$$\nabla \cdot \mathbf{D} = 0, \tag{5.43}$$

$$\nabla \times \mathbf{E} = -\frac{\partial \mathbf{B}}{\partial t}, \tag{5.44}$$

$$\nabla \cdot \mathbf{B} = 0, \tag{5.45}$$

$$\frac{1}{\mu}(\nabla \times \mathbf{B}) = \varepsilon \frac{\partial}{\partial t}\mathbf{E} + \frac{\partial}{\partial t}\mathbf{P}. \tag{5.46}$$

Taking the curl of Eq. 5.44 and using Eqs. 5.43 and 5.46, we get

$$-\nabla^2 \mathbf{E} = -\varepsilon\mu \frac{\partial^2 \mathbf{E}}{\partial t^2} - \mu \frac{\partial^2 \mathbf{P}_2}{\partial t^2} \tag{5.47}$$

and

$$\nabla^2 \mathbf{E} - \mu\varepsilon \frac{\partial^2 \mathbf{E}}{\partial t^2} = \mu \frac{\partial^2 \mathbf{P}_2}{\partial t^2}. \tag{5.48}$$

Example 5.2

If we have wave propagation in a region with complex susceptibility χ, that is, $\mathbf{P}_2 = \varepsilon_0(\operatorname{Re}\chi - i\operatorname{Im}\chi)\mathbf{E}$, show that it leads either to the amplification or to the attenuation of the wave.

From Eq. 5.42 and $\mathbf{P}_2$ as just given, we get

$$\begin{aligned}\mathbf{D} &= \varepsilon\left(1 + \frac{\varepsilon_0}{\varepsilon}\chi\right)E \\ &= \varepsilon'\mathbf{E},\end{aligned} \tag{5.49}$$

where ε' is the new effective dielectric constant. From Eq. 5.33 for the propagation wave vector, the new effective propagation wave vector k is thus

$$k' = w\sqrt{\mu\varepsilon'} = k\left(1 + \frac{\varepsilon_0}{2\varepsilon}\chi\right), \qquad \text{for} \quad \left|\frac{\varepsilon_0}{2\varepsilon}\chi\right| \ll 1. \tag{5.50}$$

That is, k' now has a real and an imaginary part,

$$k' = \operatorname{Re}k + i\operatorname{Im}k, \tag{5.51}$$

where

$$\operatorname{Re}k = k\left(1 + \frac{\varepsilon_0}{2\varepsilon}\operatorname{Re}\chi\right)$$

and

$$\operatorname{Im}k = -\frac{k\varepsilon_0}{2\varepsilon}\operatorname{Im}\chi.$$

When Eq. 5.51 is substituted back into the plane wave from Eq. 5.27, for light propagation in the x direction, for example, we get

$$\mathbf{E} = \mathbf{E}_0\exp\{i[(\operatorname{Re}k)x + \omega t] - (\operatorname{Im}k)x\}.$$

If $\operatorname{Im}k$ is positive, obviously the electromagnetic wave will attenuate as it propagates along x. On the other hand, it is clear that if $\operatorname{Im}k$ is negative, we have amplification or gain in the electromagnetic wave along x. ■

In Chapter 7, we will show how amplification and attenuation are directly related to the state of the medium in which the light transverses, and how they may lead to laser oscillations.

The preceding discussion of electromagnetic-wave propagation, which was based on the plane-wave approach and assumed an isotropic medium, should of course be properly modified in real life. In anisotropic crystals or birefringent media, in optical waveguides (e.g., fiber or planar waveguides), and in resonators, plane waves are exceptions rather than the rule.

5.3 SCALAR AND VECTOR POTENTIALS

In both classical and quantum mechanical treatments of electromagnetic fields, the so-called *potentials* from which these fields are derived are very useful concepts, for both fundamental and applied problems. In particular, these potentials are found to be very useful in describing the emission of electromagnetic waves, or light, and their interaction with matter. In this section we discuss these potentials from a formal standpoint, but we will wait to discuss their applications in specific problems until we reach the appropriate sections later in this chapter.

From Eq. 5.3 it is obvious that vector **B** is a curl of some vector quantity:

$$\mathbf{B} = \nabla \times \mathbf{A}, \tag{5.54}$$

$$\nabla \times \mathbf{E} = -\frac{\partial}{\partial t}(\nabla \times \mathbf{A}). \tag{5.55}$$

This means that

$$\mathbf{E} = -\nabla\phi - \frac{\partial \mathbf{A}}{\partial t}, \tag{5.56}$$

where **A** is known as the vector potential and ϕ is known as the scalar potential. The reason it is useful to introduce them now is that Maxwell's coupled equations can be reexpressed in terms of uncoupled simple differential equations for **A** and ϕ, with ρ and J as the source terms.

Substituting Eqs. 5.54 and 5.56 into two of Maxwell's equations, we get

$$\nabla^2\phi + \nabla\cdot\frac{\partial \mathbf{A}}{\partial t} = -\frac{\rho_f}{\varepsilon} \tag{5.57}$$

and

$$\nabla^2\mathbf{A} - \mu\sigma\frac{\partial \mathbf{A}}{\partial t} - \mu\varepsilon\frac{\partial^2 \mathbf{A}}{\partial t^2} - \nabla\left(\nabla\cdot\mathbf{A} + \mu\varepsilon\frac{\partial\phi}{\partial t} + \mu\sigma\phi\right) = -\mu\mathbf{J}_f. \tag{5.58}$$

By adding and subtracting the terms $-\mu\sigma(\partial\phi/\partial t)$ and $-\mu\varepsilon(\partial^2\phi/\partial t^2)$, to Eq. (5.57), we get

$$\nabla^2\phi - \mu\sigma\frac{\partial\phi}{\partial t} - \mu\varepsilon\frac{\partial^2\phi}{\partial t^2} - \frac{\partial}{\partial t}\left(\nabla\cdot\mathbf{A} + \mu\varepsilon\frac{\partial\phi}{\partial t} + \mu\sigma\phi\right) = -\frac{\rho_f}{\varepsilon}. \tag{5.59}$$

Equations 5.58 and 5.59 are the two general equations that the vector potential **A** and the scalar potential ϕ obey. Obviously, both **A** and ϕ are coupled together, reflecting correctly the coupling of the fields in Maxwell's equations. The key to untangling these two equations lies in Helmholtz's theorem regarding a vector, which states that a vector field is uniquely specified if both its curl and divergence are specified. For the vector potential **A**, we have specified its curl, $\nabla \times \mathbf{A} = \mathbf{B}$, Eq. 5.54; therefore we are free to choose $\nabla \cdot \mathbf{A}$.

The choice of $\nabla \cdot \mathbf{A}$ is not as arbitrary as it might appear, since it depends very much on our aim. To be more specific, we can choose the rather natural so-called

Lorentz condition

$$\nabla \cdot \mathbf{A} + \mu\varepsilon \frac{\partial \phi}{\partial t} + \mu\sigma\phi = 0. \tag{5.60}$$

This condition immediately reduces Eqs. 5.58 and 5.59 to two uncoupled equations,

$$\nabla^2 \mathbf{A} - \mu\sigma \frac{\partial \mathbf{A}}{\partial t} - \mu\varepsilon \frac{\partial^2 \mathbf{A}}{\partial t^2} = -\mu \mathbf{J}_f \tag{5.61}$$

and

$$\nabla^2 \phi - \mu\sigma \frac{\partial \phi}{\partial t} - \mu\varepsilon \frac{\partial^2 \phi}{\partial t^2} = -\frac{\rho_f}{\varepsilon}. \tag{5.62}$$

We will show in subsequent sections that this condition, and the two resulting equations, are particularly suited for classical radiation theories.

Conversely, we can also adopt the so-called Coulomb condition,

$$\nabla \cdot \mathbf{A} = 0, \tag{5.63}$$

which gives (see, for example, reference 2)

$$\nabla^2 \phi = -4\pi\rho \tag{5.64}$$

and

$$\nabla^2 \mathbf{A} - \frac{1}{c^2} \frac{\partial^2 \mathbf{A}}{\partial t^2} = -\mu \mathbf{J}_f, \tag{5.65}$$

where $\phi = 0$ and $\mathbf{A}$ satisfies the homogenous wave equation. Both sets of conditions have been applied successfully in various electromagnetic problems. The physical laws of electromagnetism (as described by the original set of Maxwell's equations) are of course independent of these choices of conditions. The conditions serve for the most part to remove some of the "difficulties" in the theoretical framework. In the next section we will study the basic process of radiation using the Lorentz condition. Understanding this basic process of dipole radiation will greatly facilitate our understanding of laser oscillations and related coherent optical phenomena.

Example 5.3

The vector potential $\mathbf{A}$ was introduced through the relation $\mathbf{B} = \nabla \times \mathbf{A}$ because of the solenoidal nature of $\mathbf{B}$ (i.e., $\nabla \cdot \mathbf{B} = 0$). In the source-free region, $\nabla \cdot \mathbf{E} = 0$, we define another type of vector potential $\mathbf{A}_e$ for which $\mathbf{E} = \nabla \times \mathbf{A}_e$.

(a) Express $\mathbf{H}$ in terms of $\mathbf{A}$.
(b) Show that $\mathbf{A}$ is a solution of a wave equation.

Answers

(a) Since $\nabla \cdot \mathbf{E} = 0$, we have

$$\mathbf{E} = \nabla \times \mathbf{A}_e.$$

By substituting into the curl of Maxwell's Eq. 5.34, we have

$$\nabla \times (\nabla \times \mathbf{A}) = -\frac{\partial \mathbf{B}}{\partial t},$$

which can be written as

$$\nabla(\nabla \times \mathbf{A}) - \nabla^2 \mathbf{A} = -\frac{\partial \mathbf{B}}{\partial t}. \tag{5.66}$$

From Eq. 5.37, the curl of **H** can be expressed as

$$\nabla \times \mathbf{H} = \varepsilon \frac{\partial \mathbf{E}}{\partial t} = \varepsilon \nabla \frac{\partial \mathbf{A}_e}{\partial t}$$

or equivalently

$$\nabla \times \left(\mathbf{H} - \varepsilon \frac{\partial \mathbf{A}_e}{\partial t} \right) = 0.$$

If we let

$$\mathbf{H} - \varepsilon \frac{\partial \mathbf{A}_e}{\partial t} = -\nabla \phi_m,$$

Eq. 5.66 becomes

$$-\varepsilon\mu \frac{\partial^2 \mathbf{A}_e}{\partial t^2} + \mu \nabla \frac{\partial \phi_m}{\partial t} = \nabla(\nabla \cdot \mathbf{A}_e) - \nabla^2 \mathbf{A}_e. \tag{5.67}$$

By choosing

$$\nabla \cdot \mathbf{A} = \mu \frac{\partial \phi_m}{\partial t},$$

we can express **H** as

$$\mathbf{H} = \varepsilon \frac{\partial \mathbf{A}_e}{\partial t} - \nabla \phi_m.$$

(b) Since

$$\nabla \cdot \mathbf{A}_e = \mu \frac{\partial \phi_m}{\partial t}$$

from Eq. 5.66 we have

$$\nabla^2 \mathbf{A}_e = \varepsilon\mu \frac{\partial^2 \mathbf{A}_e}{\partial t^2}.$$

Thus, we see that $\mathbf{A}_e$ is the solution of a wave equation. ■

5.4 THE CLASSICAL THEORY OF RADIATION

In some practical problems involving lasers or light, it may be necessary to invoke the laws of quantum mechanics. In this chapter we study the details of classical radiation. The basic mechanisms also apply in quantum theory. Except for some numerical constants that modify slightly the expressions describing the various processes, the basic dependence of these processes on electrodynamic parameters, such as the dipole moment, frequency, and the velocity of light, is essentially identical in both quantum and classical descriptions. Although quantum descriptions are undoubtedly correct, the classical picture is more in agreement with our physical intuition. The history of the development of lasers reveals that most of the concepts and reasoning leading to these devices utilizing the natural oscillations of atoms and molecules between energy levels for generating coherent electromagnetic radiation came from analogous concepts in microwave physics. Subsequent developments in coherent phenomena also closely followed the theory and application of classical microwave processes.

We return now to Eqs. 5.61 and 5.62, which were derived in the previous section by using the Lorentz condition. Consider a vacuum region (i.e., $\sigma = 0$, $\mu \to \mu_0$, $\varepsilon \to \varepsilon_0$) outside the current and charge-density sources where we wish to evaluate $\mathbf{A}$ and ϕ (see Figure 5.2). It is obvious that if both $\mathbf{J}$ and ρ are time-independent, we readily recover the static electric and magnetic fields from the solutions for $\mathbf{A}$ and ϕ,

$$\mathbf{A}(\mathbf{r}) = \frac{\mu_0}{4\pi} \int_{v'} \frac{\mathbf{J}(\mathbf{r}')\, d^3\mathbf{r}'}{R} \tag{5.68}$$

and

$$\phi(\mathbf{r}) = \frac{1}{4\pi\varepsilon_0} \int_{v'} \frac{\rho(\mathbf{r}')\, d^3\mathbf{r}'}{R}, \tag{5.69}$$

where v' is the volume enclosing the localized current or charge densities, and $R = |\mathbf{r} - \mathbf{r}'|$. These solutions follow from Poisson's equation, so that both $\mathbf{A}$ and ϕ satisfy

$$\nabla^2 \mathbf{A} = -\mu_0 \mathbf{J}_f \tag{5.70}$$

and

$$\nabla^2 \phi = -\frac{\rho_f}{\varepsilon_0}. \tag{5.71}$$

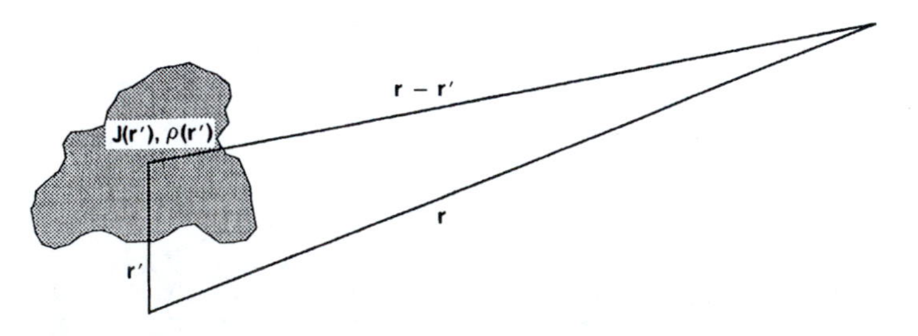

FIGURE 5.2 The potential functions at **r** are related to the current and charge sources at **r**′.

In the time-dependent case, Eqs. 5.61 and 5.62 in the (vacuum) region become

$$\nabla^2 \mathbf{A} - \frac{1}{c^2}\frac{\partial^2 \mathbf{A}}{\partial t^2} = -\mu_0 \mathbf{J}_f \tag{5.72}$$

and

$$\nabla^2 \phi - \frac{1}{c^2}\frac{\partial^2 \phi}{\partial t^2} = -\frac{\rho_f}{\varepsilon_0}. \tag{5.73}$$

These wave equations, whose source terms are on the right-hand side, have the following standard solutions:

$$\mathbf{A}(\mathbf{r}, t) = \mu_0 \int_{v'} \frac{\mathbf{J}(\mathbf{r}', t - R/c)}{R}\, d^3\mathbf{r}' \tag{5.74}$$

and

$$\phi(\mathbf{r}, t) = \frac{1}{\varepsilon_0} \int_{v'} \frac{\rho(\mathbf{r}', t - R/c)}{R}\, d^3\mathbf{r}'. \tag{5.75}$$

The formal mathematics leading to Eqs. 5.73 and 5.75 may be found in other standard texts on electromagnetism. The main difference between Eqs. 5.74 and 5.75 and Eqs. 5.68 and 5.69 is in the integrand; that is, the response at time t at $\mathbf{r}$ is related to the source at $\mathbf{r}'$ at an earlier time, $t - |\mathbf{r} - \mathbf{r}'|/c$. This causal relation is one of the consequences of the wave equations that both $\mathbf{A}(\mathbf{r}, t)$ and $\phi(\mathbf{r}, t)$ satisfy.

The forms of $\mathbf{A}$ and ϕ (and the corresponding $\mathbf{E} = -\nabla\phi - \partial\mathbf{A}/\partial t$ and $\mathbf{B} = \nabla \times \mathbf{A}$) depend on the forms of $\mathbf{J}$ and ρ in the integrands. We now consider containing $\mathbf{J}(\mathbf{r}, t)$ and $\rho(\mathbf{r}, t)$, which are oscillatory terms. For one particular oscillation (at a frequency ω, for example), we have

$$\mathbf{J}(\mathbf{r}', t) = \mathbf{J}_0(\mathbf{r}')e^{-i\omega t} \tag{5.76}$$

and

$$\rho(\mathbf{r}, t) = \rho_0(\mathbf{r}')e^{-i\omega t}. \tag{5.77}$$

Substituting Eqs. 5.76 and 5.77 into Eqs. 5.74 and 5.75, respectively, we get

$$\mathbf{A}(\mathbf{r}, t) = \frac{\mu_0 e^{-i\omega t}}{4\pi} \int_{v'} \frac{e^{ikR}\mathbf{J}_0(\mathbf{r}')\, d^3\mathbf{r}'}{R}, \tag{5.78}$$

and

$$\phi(\mathbf{r}, t) = \frac{e^{-i\omega t}}{4\pi\varepsilon_0} \int_{v'} \frac{e^{ikR}\rho_0(\mathbf{r}')\, d^3\mathbf{r}'}{R}, \tag{5.79}$$

where $k = \omega/c$.

Writing out $R = |\mathbf{r} - \mathbf{r}'|$ explicitly, we get

$$R = |\mathbf{r} - \mathbf{r}'| = [r^2 + (r')^2 - 2rr' \cos\theta']^{1/2}, \tag{5.80}$$

where $\theta' = \mathbf{r} \cdot \mathbf{r}'/rr'$. If we define

$$x = -2\left(\frac{r'}{r}\right)\cos\theta' + \left(\frac{r'}{r}\right)^2,$$

then

$$R = r(1 + x)^{1/2}. \tag{5.82}$$

We assume that the source dimension is small compared to $|\mathbf{r}|$. Of course, this is always true in optical processes for which the sources are of atomic size and the observation distance is macroscopic. (However, if we consider the fields associated with an atomic source as experienced by a nearby atom, this assumption should not be made without some detailed justification.) If the source dimension is small compared to $|\mathbf{r}|$, then

$$\frac{1}{R} \sim \frac{1}{r}(1 - \tfrac{1}{2}x + \tfrac{3}{8}x^2 + \cdots). \tag{5.83}$$

Similarly, the exponent e^{ikR} becomes

$$\begin{aligned} e^{ikR} &= e^{ikr}e^{ik(R-r)} \simeq e^{ikr}e^{ikr(\frac{1}{2}x - \frac{1}{8}x^2)} \\ &\simeq xe^{ikr}\left[1 + ikr(\tfrac{1}{2}x - \tfrac{1}{8}x^2) - \tfrac{1}{2}(kr)^2\left(\frac{x}{2} - \frac{x^2}{8}\right)^2\right] \\ &\simeq e^{ikr}\{1 + \tfrac{1}{2}ikrx - \tfrac{1}{8}[ikr + (kr)^2]x^2\} + \cdots \text{ terms higher than } x^2. \end{aligned} \tag{5.84}$$

Combining Eqs. 5.83 and 5.84 gives

$$\frac{e^{ikR}}{R} \sim \frac{e^{ikr}}{r}\left\{1 + (1 - ikr)\frac{\mathbf{r} \cdot \mathbf{r}'}{r^2} + [1 - ikr - \tfrac{1}{2}(kr)^2]\left(\frac{r'}{r}\right)^2\right\}. \tag{5.85}$$

We can usually separate the terms in Eq. 5.85 into three terms, I_0, I_1, and I_2, according to the increasing order of r'/r:

$$I_0 = \frac{e^{ikr}}{r}, \tag{5.86}$$

$$I_1 = (1 - ikr)\frac{\mathbf{r} \cdot \mathbf{r}'}{r}, \tag{5.87}$$

$$I_2 = \left[1 - ikr - \frac{(kr)^2}{2}\right]\left(\frac{\mathbf{r}'}{r}\right)^2. \tag{5.88}$$

Substituting Eqs. 5.86 through 5.88 into the expression for $\mathbf{A}$, for example, yields the following three terms:

$$A_0 = \frac{\mu_0 e^{i(kr-\omega t)}}{4\pi r}\int_{v'} \mathbf{J}_0(\mathbf{r}')\,d^3\mathbf{r}', \tag{5.89}$$

$$A_1 = \frac{\mu_0(1 - ikr)e^{i(kr-\omega t)}}{4\pi r^2}\int_{v'} (\mathbf{r} \cdot \mathbf{r}')\mathbf{J}_0(\mathbf{r}')\,d^3\mathbf{r}', \tag{5.90}$$

$$A_2 = \frac{\{\mu_0[1 - ikr - \frac{1}{2}(kr)^2]\}e^{i(kr-\omega t)}}{4\pi r^3} \int_{v'} \mathbf{J}_0(\mathbf{r}')r'^2\, d^3\mathbf{r}'. \tag{5.91}$$

In the time-independent case, the integral over $\mathbf{J}_0(\mathbf{r}')$ is vanishing, but when $\mathbf{J}$ is time-dependent, the situation is slightly more complicated. To investigate this further, consider the continuity equation

$$\nabla' \cdot \mathbf{J} + \frac{\partial \rho}{\partial t} = 0, \tag{5.92}$$

where the prime on the ∇' signifies that the derivative corresponds to $\mathbf{r}'$. From the time dependence of $\rho(\mathbf{r}, t)$, Eq. 5.92 gives

$$\nabla' \cdot \mathbf{J} = i\omega\rho_0(\mathbf{r}, t). \tag{5.93}$$

We now borrow a vector identity that relates a surface integral to the integral over the volume that the surface encloses:

$$\int_S \mathbf{A}(\mathbf{B} \cdot d\mathbf{s}) = \int_v [(\mathbf{B} \cdot \nabla)\mathbf{A} + \mathbf{A}(\nabla \cdot \mathbf{B})]\, d^3\mathbf{r}. \tag{5.94}$$

Letting $\mathbf{A} = \mathbf{r}'$ and $\mathbf{B} = \mathbf{J}_0$, and choosing the surface so that $\mathbf{J}_0$ on the surface is vanishing, we get

$$\begin{aligned} \int_{v'} \mathbf{J}_0\, d^3\mathbf{r}' &= -\int_{v'} \mathbf{r}'(\nabla' \cdot \mathbf{J}_0)\, d^3\mathbf{r}' \\ &= -i\omega \int_{v'} \rho_0(\mathbf{r}')\mathbf{r}'\, d^3\mathbf{r}' \\ &= -i\omega \mathbf{d}_0, \end{aligned} \tag{5.95}$$

where $\mathbf{d}_0$ is the electric dipole moment of the charge distribution and is given by

$$\mathbf{d}_0 = \int_{v'} \rho_0(\mathbf{r}')\mathbf{r}'\, d^3\mathbf{r}'. \tag{5.96}$$

Hence, the first term of the vector potential $\mathbf{A}_1$ now becomes

$$A_0 = \frac{-i\mu_0\omega}{4\pi r} \mathbf{d}_0 e^{i(kr-\omega t)}, \tag{5.97}$$

from which the electric and magnetic fields can be deduced once $\mathbf{d}_0$ is known.

Although a little more vector analysis is needed, the evaluation of $\mathbf{A}_1$ and $\mathbf{A}_2$ can be performed in a similar way. Here we simply give the results. It is found that $\mathbf{A}_1$ consists of two terms. One term is

$$A_{1,1} = \frac{\mu_0(1 - ikr)(\mathbf{m}_0 \times \mathbf{r})}{4\pi r^2} e^{i(kr-\omega t)}, \tag{5.98}$$

where $\mathbf{m}_0$ is the magnetic dipole moment,

$$\mathbf{m}_0 = \frac{1}{2}\int_{v'} \mathbf{r} \times \mathbf{J}_0 \, d^3\mathbf{r}', \tag{5.99}$$

and the other term is

$$A_{1,2} = \frac{-\mu_0\omega(1 - ikr)e^{i(kr-\omega t)}}{8\pi r^2}\int_{v'} \mathbf{r}'\left(\frac{\mathbf{r}\cdot\mathbf{r}'}{r}\right)\rho_0(\mathbf{r}')\, d^3\mathbf{r}', \tag{5.100}$$

which is due to the electric quadrupole moment.

Electric dipole radiations are perhaps the most important and dominant atomic (and molecular) processes; others are in comparison relatively small in magnitude. For this reason we shall from now on concentrate only on dipole radiations. Of course, in some particular systems dipole radiations may not be allowed, so we will have to study the higher-order radiation processes by evaluating the magnetic dipole or the electric quadrupole moments.

5.5 ELECTRIC DIPOLE RADIATIONS

Consider the instance in which the dipole moment has a constant amplitude and

$$\mathbf{d}_0 = d_0\hat{z}. \tag{5.101}$$

In the spherical coordinate

$$\mathbf{d}_0 = d_0(\cos\theta\hat{r} - \sin\theta\hat{\theta}). \tag{5.102}$$

From $\mathbf{B} = \nabla \times \mathbf{A}$ and the expression for $\nabla \times \mathbf{A}$ in the spherical coordinate,

$$\begin{aligned}\nabla \times \mathbf{A} &= \frac{r}{r\sin\theta}\left[\frac{\partial}{\partial\theta}(\sin\theta A_\phi) - \frac{\partial A_\theta}{\partial\phi}\right] \\ &+ \frac{\theta}{r}\left[\frac{1}{\sin\theta}\frac{\partial A_r}{\partial\phi} - \frac{\partial}{\partial r}(rA_\phi)\right] \\ &+ \frac{\phi}{r}\left[\frac{\partial}{\partial r}(rA_\theta) - \frac{\partial A_r}{\partial\theta}\right],\end{aligned}$$

we get

$$\mathbf{B} = \frac{-\mu_0 k^2\omega d_0}{4\pi}\left[\frac{1}{kr} + \frac{i}{(kr)^2}\right]\sin\theta e^{i(kr-\omega t)}\hat{\phi}. \tag{5.103}$$

For $\mathbf{E}$ $(= -\nabla\phi - \partial A/\partial t)$ we need to evaluate ϕ, which we can obtain from the Lorentz condition

$$\frac{1}{c^2}\frac{\partial\phi}{\partial t} = -\nabla\cdot\mathbf{A}.$$

From Eq. 5.78 for ϕ, we note that

$$\frac{\partial \phi}{\partial t} = -i\omega \frac{\partial \phi}{\partial t}.$$

Hence we get

$$\phi = \frac{-ic^2}{\omega} \nabla \cdot \mathbf{A}. \tag{5.104}$$

Upon evaluating both $\nabla \phi$ and $\partial \mathbf{A}/\partial t$, we obtain for $\mathbf{E}$

$$\mathbf{E} = \frac{-k^3 d_0}{4\pi\varepsilon_0} \left\{ \left[\frac{2i}{(kr)^2} - \frac{2}{(kr)^3} \right] \cos\theta \hat{r} + \left[\frac{1}{kr} + \frac{i}{(kr)^2} - \frac{1}{(kr)^3} \right] \sin\theta \hat{\theta} \right\} e^{i(kr-\omega t)}. \tag{5.105}$$

Both $\mathbf{E}$ and $\mathbf{B}$ are quite complex as a function of r. Since $\mathbf{E}$ is in the plane defined by θ and r, whereas $\mathbf{B}$ is along $\hat{\phi}$, we note that $\mathbf{E}$ is obviously perpendicular to $\mathbf{B}$, consistent with Maxwell's equation. A particular feature of these fields is that they contain terms in various orders of $1/kr$. If we note that

$$k = \frac{\omega}{c} = \frac{2\pi}{\lambda},$$

where λ is the wavelength of the electromagnetic wave, these various orders in $1/kr$ amount to various orders in λ/r, that is, in the ratio of the wavelength of the electromagnetic wave to the observation distance. For light, ultraviolet through visible to infrared, the wavelength is very small compared to the dimension r under consideration, $r \gg \lambda$. The region of interest is typically labeled the far field or the radiation field zone. The corresponding electric and magnetic fields are given by

$$\mathbf{E} \sim \frac{-k^2 d_0}{4\pi\varepsilon_0 r} \sin\theta e^{i(kr-\omega t)} \hat{\theta} \tag{5.106}$$

and

$$\mathbf{B} \sim \frac{-\mu_0 k\omega d_0}{4\pi r} \sin\theta e^{i(kr-\omega t)} \hat{\phi}, \tag{5.107}$$

i.e., they both contain the spherically outgoing waveform $e^{i(kr-\omega t)}/r$. Both $\mathbf{E}$ and $\mathbf{B}$ vectors are perpendicular to the direction of propagation ($\hat{r}$), and they are orthogonal to one another, that is, they are transverse electromagnetic waves of the type we first encountered in Section 5.2. From Eqs. 5.106 and 5.107 we can see that

$$\mathbf{B} = \frac{1}{c} \hat{\mathbf{r}} \times \mathbf{E}.$$

Poynting's vector **S**, when time-averaged, gives

$$\langle \mathbf{S} \rangle_{\text{time average}} = \frac{1}{2} \operatorname{Re}\left(\mathbf{E} \times \frac{\mathbf{B}}{\mu_0} \right)$$
$$= \frac{\mu_0 \omega^4 |d_0|^2}{32\pi^2 c r^2} \sin^2 \theta \hat{r}. \tag{5.108}$$

That is, the energy flows radially outward.

The total rate of energy flow can be found by integrating **S** over the surface enclosing the dipole. By considering a spherical surface with radius r, we get the total dipole radiation power,

$$P = \int \langle \mathbf{S} \rangle \cdot d\mathbf{s} \int_0^{2\pi} \int_0^{\pi} \langle \mathbf{S} \rangle r^2 \sin\theta \, d\theta \, d\phi$$
$$= \frac{\mu_0 \omega^4 |d_0|^2}{12\pi c}. \tag{5.109}$$

The preceding classical radiation theory points out two simple dependencies that are also borne out in the quantum theory of radiation, namely that the rate of radiation is proportional to the squared amplitude of the dipole moment, and that the rate increases dramatically with the frequency. There are, however, other important points. One subtle one is that we have been assuming that the dipole has a constant amplitude, and therefore that some driving force must supply the energy. The "driving" of this dipole moment and its emission are somehow mixed up in the preceding consideration. In quantum theory, as first quantitatively treated by Albert Einstein, this process is separated into two related processes: spontaneous emission by the excited dipole and stimulated emission by the dipole oscillator when driven by an external field. Furthermore, we must also use quantum mechanics to interpret the dipole moment correctly and to indicate how atomic structure and physical properties influence the dipole moment and the corresponding radiation rate.

5.6 POLARIZATIONS

Knowing that light is a form of electromagnetic wave makes it easier to understand the concept of polarization. Put simply, polarization refers to the *direction* of the electric field vector **E**. Since these field vectors are always perpendicular to the direction of the propagation of light, **k**, the correct specification of the polarization direction is with respect to **k** (see Figure 5.1).

Since there are only two independent axes in a plane orthogonal to **k**, there can be only two definable independent states of polarization. If we designate the direction **k** as the z axis, the set of x and y axes is the obvious choice. As light propagates along the z axis, we can imagine the electric field vector oscillating up and down in the x or y direction, or in a direction in the x, y plane formed by a linear combination of the x and y components.

If polarization is in a particular direction, then as the light propagates along the z axis, we say that the light is *linearly polarized.* If the direction of polarization is not constant, but moves around in the x, y plane, the states of polarization may

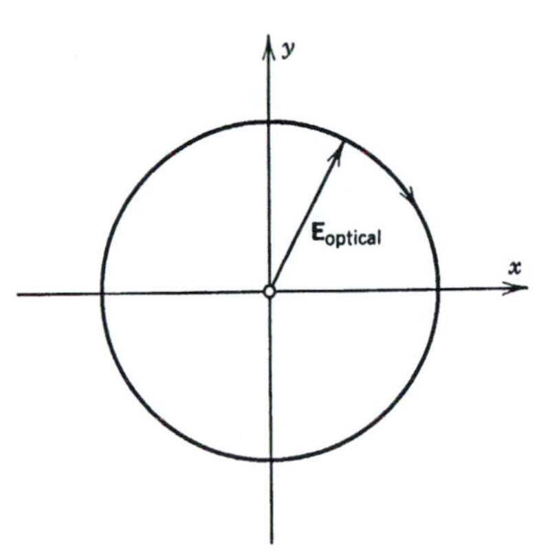

FIGURE 5.3 A right-hand circularly polarized light propagating toward the observer.

be characterized in several ways, depending on the direction of motion or change of polarization vector.

Figures 5.3 and 5.4 illustrate two states of polarization that could happen if the electric field vector has equal amplitudes along the x and y axes, as viewed by an observer looking back along the z axis. If the motion of the electric field vector is such that it sweeps out a circle in a clockwise direction, it is called a *right-hand circularly polarized* light (Figure 5.3), but if it sweeps out a circle in the counterclockwise direction, it is called a *left-hand circularly polarized* light (Figure 5.4). With the unit vector along the x axis designated by $\hat{i}$ and that along the y-axis designated by $\hat{j}$, the right or left-hand circularly polarized light can be represented as a linear combination:

$$\mathbf{E} = \hat{i}E_{0x}\cos(kz - \omega t) + \hat{j}E_{0y}\cos(kz - \omega t + \phi). \tag{5.110}$$

In Eq. 5.110, E_{0x} and E_{0y} are the electric field amplitudes ($E_{0x} = E_{0y}$) and ϕ is the phase of the y component relative to the x component.

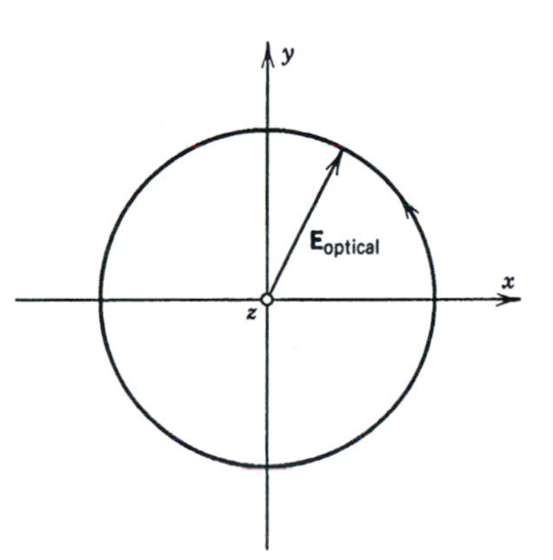

FIGURE 5.4 A left-hand circularly polarized light propagating toward the observer.

Using the representation given in Eq. 5.110, we obtain the right-hand circularly polarized light if $E_{0x} = E_{0y}$ and

$$\phi = -\frac{\pi}{2} \pm 2m\pi, \qquad m = 0, \pm 1, \pm 2, \ldots . \tag{5.111}$$

Equation 5.110 then becomes

$$\mathbf{E}_{\mathrm{RC}} = \hat{i} E_{0x} \cos(kz - \omega t) + \hat{j} E_{0y} \sin(kz - \omega t). \tag{5.112}$$

Conversely, the left-hand circularly polarized light corresponds to the condition $E_{0x} = E_{0y}$ and

$$\phi = \frac{\pi}{2} \pm 2m\pi, \tag{5.113}$$

so Eq. 5.110 becomes

$$\mathbf{E}_{\mathrm{LC}} = \hat{i} E_{0x} \cos(kz - \omega t) - \hat{j} E_{0y} \sin(kz - \omega t). \tag{5.114}$$

If the x and y components of the field vector are not equal (i.e., $E_{0x} \neq E_{0y}$), the addition of these two components to a phase factor ϕ gives rise to what we call an *elliptically polarized* light (see Figure 5.5).

The state of polarization of light is preserved if the medium in which it propagates is linear and isotropic. In an anisotropic medium, in which the responses of the medium are different for different axes, the polarization of the incident light can be very drastically affected. Polarizing crystals, which preferentially transmit

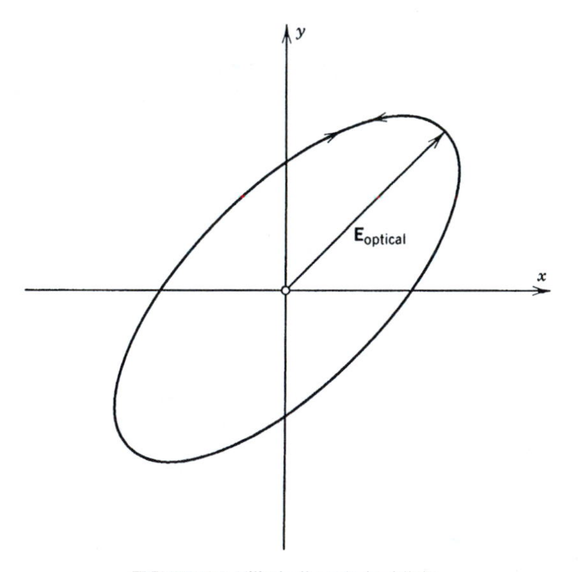

FIGURE 5.5 Elliptically polarized light.

light polarized in a particular direction, are a dramatic but very common example. Incident randomly, or circularly, or elliptically polarized light will be converted by the crystal into linearly polarized light, as the other orthogonal component is extinguished.

Most sources of light produce randomly polarized light. This is so because the atoms or molecules in these sources are free to radiate in all directions. Put in another way, the dipoles associated with these emitting atoms or molecules are randomly oriented. The mechanisms that we use to excite these dipoles, such as electric discharges and combustions, do not orient the dipoles in a specific direction. In lasers (Chapter 7) the emitting system is forced (or induced) to emit light along a particular direction, so polarization occurs in a collective manner, producing highly polarized and directional radiation.

■ Example 5.4

Describe the state of polarization of each of the following electromagnetic waves.

(a) $\mathbf{E} = \hat{i}E_0 \cos(kz - \omega t) - \hat{j}E_0 \cos(kz - \omega t).$

(b) $\mathbf{E} = \hat{i}E_0 \sin(\omega t - kz) + \hat{j}E_0 \sin\left(\omega t - kz - \frac{\pi}{4}\right).$

(c) $\mathbf{E} = \hat{i}E_0 \cos(\omega t - kz) + \hat{j}E_0 \sin\left(\omega t - kz - \frac{\pi}{2}\right).$

Answers

(a) It is a linearly polarized wave because

$$\mathbf{E} = (\hat{i} - \hat{j})E_0 \cos(kz - \omega t).$$

(b) It is an elliptically polarized wave, because

$$\left(\frac{E_x}{E_0}\right)^2 + \left(\frac{E_y}{E_0}\right)^2 - \sqrt{2}\left(\frac{E_x}{E_0}\right)\left(\frac{E_y}{E_0}\right) = \frac{1}{2},$$

where $E_x = E_0 \sin(\omega t - kz)$, and $E_y = E_0 \sin(\omega t - kz - \pi/4)$.

(c) It is a circularly polarized wave, because

$$\left(\frac{E_x}{E_0}\right)^2 + \left(\frac{E_y}{E_0}\right)^2 = 1.$$

■

5.7 REFLECTION AND REFRACTION REVISITED

The fact that light is an electromagnetic field radiating from an atomic or a molecular dipole opens up many questions about all the optical processes that we discussed in earlier chapters. In this section we therefore visit again reflection and refraction phenomena, where the electromagnetic and vectorial aspects of light bring forth new, interesting and practically important effects.

Figures 5.6 and 5.7 show two distinct situations encountered by linearly polarized light as it passes from one medium to another. In Figure 5.6 the electric

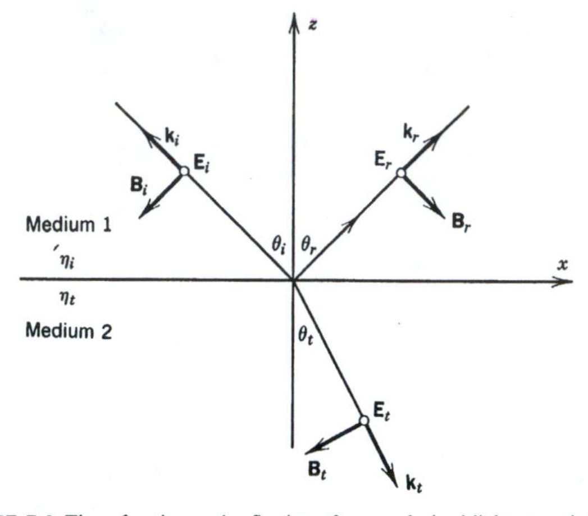

FIGURE 5.6 The refraction and reflection of an *s*-polarized light at an interface.

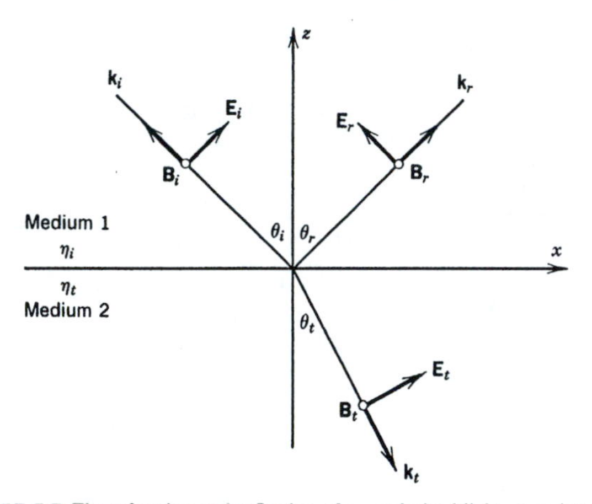

FIGURE 5.7 The refraction and reflection of a *p*-polarized light at an interface.

field vector of the incident light is perpendicular to the plane of incidence. This state of polarization is called, in reflective optics terminology, an *s-polarized* incident light. Conversely, when the electric field vector of the incident light is in the plane of incidence, it is called a *p-polarized* incident light (see Figure 5.7).

The reflection and transmission coefficients of these two forms of incident light are quite different because of this fundamental difference in the direction of polarization. This is a very important revelation, which a scalar theory of light cannot provide.

We denote the incident, reflected, and transmitted light, respectively, by

$$\mathbf{E}_i = \mathbf{E}_{0i} \cos(\mathbf{k}_i \cdot \mathbf{r} - \omega_i t), \tag{5.115}$$

$$\mathbf{E}_r = \mathbf{E}_{0r} \cos(\mathbf{k}_r \cdot \mathbf{r} - \omega_r t), \tag{5.116}$$

and

$$\mathbf{E}_t = E_{0t}\cos(\mathbf{k}_t \cdot \mathbf{r} - \omega_t t). \tag{5.117}$$

We can then apply elementary electromagnetic theories to derive many of the laws of reflection and refraction discussed earlier.

One electromagnetic theory holds that the total amount of tangential components of the electric field **E** (also of **H**) on one side of the interface is the same as the total amount on the other side. Denoting the unit vector normal to the interface by $\hat{i}_n$, we can obtain the tangential components of all the $\mathbf{E}_s$ involved by a vector cross product with $\hat{i}_n$.

$$\hat{i}_n \times \mathbf{E}_i + \hat{i}_n \times \mathbf{E}_r = \hat{i}_n \times \mathbf{E}_t. \tag{5.118}$$

Substituting Eqs. 5.115 through 5.117 in Eq. 5.118, we get

$$\hat{i}_n \times \mathbf{E}_{0i}\cos(\mathbf{k}_i \cdot \mathbf{r} - \omega_i t) + \hat{i}_n \times \mathbf{E}_{0r}\cos(\mathbf{k}_r \cdot \mathbf{r} - \omega_r t) = \hat{i}_n \times \mathbf{E}_{0t}\cos(\mathbf{k}_t \cdot \mathbf{r} - \omega_t t). \tag{5.119}$$

Equation 5.119 holds true at *any instant* and at *any point* on the interface. This means that for any point on the interface (a plane defined by $z = 0$, for example), we have

$$(\mathbf{k}_i \cdot \mathbf{r} - \omega_i t)_{z=0} = (\mathbf{k}_r \cdot \mathbf{r} - \omega_r t)_{z=0} = (\mathbf{k}_t \cdot \mathbf{r} - \omega_r t)_{z=0} \tag{5.120}$$

at any instant. Equation 5.120 holds only if

$$\omega_i = \omega_r = \omega_t \tag{5.121}$$

and

$$\mathbf{k}_i \cdot \mathbf{r}|_{z=0} = (\mathbf{k}_r \cdot \mathbf{r})|_{z=0} = (\mathbf{k}_t \cdot \mathbf{r})|_{z=0}. \tag{5.122}$$

From the first equality in equation 5.122, we have

$$(\mathbf{k}_i - \mathbf{k}_r) \cdot \mathbf{r} = \phi_r, \tag{5.123}$$

which must hold at all points. Clearly, if we choose **r** to be the origin (0, 0, 0), then $\phi_r = 0$. This would also hold for $\phi_t = 0$.

Thus, we have

$$(\mathbf{k}_i - \mathbf{k}_r) \cdot \mathbf{r} = 0 \tag{5.124}$$

and

$$(\mathbf{k}_i - \mathbf{k}_t) \cdot \mathbf{r} = 0. \tag{5.125}$$

From Eq. 5.124 we can say that $\mathbf{k}_i - \mathbf{k}_r$ is perpendicular to **r**, which is a vector in the plane $z = 0$ (the interface). If we choose **r** as the unit vector along x, we have, from Eq. 5.124,

$$k_i \sin\theta_i = k_r \sin\theta_r. \tag{5.126}$$

Since $k_i = k_r$—the reflected and the incident waves are in the same medium—we have, from Eq. 5.126,

$$\theta_i = \theta_r. \tag{5.127}$$

Similarly, from Eq. 5.125, we get

$$k_i \sin \theta_i = k_t \sin \theta_t.$$

Since $k_i = \eta_i(2\pi/\lambda)$ and $k_t = \eta_t(2\pi/\lambda)$, we have

$$\eta_i \sin \theta_i = \eta_t \sin \theta_t, \tag{5.128}$$

which is Snell's law.

The preceding approach can be used to derive more quantitative components of reflection and transmission at an interface. Since there are two unknowns (E_t and E_r), we need two equations. They are obtained by equating the tangential components of $\mathbf{H}(= \mathbf{B}/\mu)$ and $\mathbf{E}$. Using these two sets of equations and the relation between $|\mathbf{B}|$ and $|\mathbf{E}|$ (e.g., Eq. 5.31), we can show that for s polarization the amplitude reflection coefficient is

$$r_s = \frac{E_{0r}}{E_{0i}} = \frac{\dfrac{\eta_i}{\mu_i}\cos\theta_i - \dfrac{\eta_t}{\mu_t}\cos\theta_t}{\dfrac{\eta_i}{\mu_i}\cos\theta_i + \dfrac{\eta_t}{\mu_t}\cos\theta_t}, \tag{5.129}$$

and the amplitude transmission coefficient is

$$t_s = \frac{E_{0t}}{E_{0i}} = \frac{2\dfrac{\eta_i}{\mu_i}\cos\theta_i}{\dfrac{\eta_i}{\mu_i}\cos\theta_i + \dfrac{\eta_t}{\mu_t}\cos\theta_t}. \tag{5.130}$$

For p polarization, we have amplitude reflection coefficient

$$r_p = \frac{E_{0r}}{E_{0i}} = \frac{\dfrac{\eta_t}{\mu_t}\cos\theta_i - \dfrac{\eta_i}{\mu_i}\cos\theta_t}{\dfrac{\eta_i}{\mu_i}\cos\theta_t + \dfrac{\eta_t}{\mu_t}\cos\theta_i} \tag{5.131}$$

and amplitude transmission coefficient

$$t_p = \frac{2\dfrac{\eta_i}{\mu_i}\cos\theta_i}{\dfrac{\eta_i}{\mu_i}\cos\theta_t + \dfrac{\eta_t}{\mu_t}\cos\theta_i}. \tag{5.132}$$

Equations 5.129 through 5.132 can be simplified if we note that for most dielectrics, the values for the permeabilities, μ, are very close to one another, that is, $\mu_i \approx \mu_t \approx \mu_0$, the permeability of free space. Therefore μ disappears from these equations, and the two most important parameters governing reflection and refraction at an interface are the incident angle and the refractive indices, for a given state of polarization. As we remarked earlier, because of the electromagnetic interactions at the interface, the reflection and transmission coefficients are drastically different for the s- and the p-polarized incident waves.

Example 5.5

A linearly polarized light originates in a glass medium and is incident at a 30° angle on the boundary surface between the glass ($\eta_i = 1.5$) and air ($\eta_t = 1.0$). Determine the transmission and reflection coefficients and the percentage of energy that would be transmitted. Assume that the **E** fields are oriented normally to the plane of incidence.

From Snell's law of refraction, we have

$$\sin\theta_t = \frac{\eta_i}{\eta_t}\sin\theta_i = \frac{1.5}{1.0}\sin(30°) = 0.75, \qquad \theta_t = 48.6°.$$

The transmission coefficient can be determined:

$$t_s = \frac{2\eta_i\cos\theta_i}{\eta_t\cos\theta_i + \eta_t\cos\theta_t} = \frac{2\times 1.5\times\cos 30°}{1.5\cos 30° + \cos 48.6°} = 1.325.$$

The reflection coefficient can be calculated:

$$r_s = \frac{\eta_i\cos\theta_i - \eta_t\cos\theta_t}{\eta_i\cos\theta_i + \eta_t\cos\theta_t} = \frac{1.5\cos 30° - 1.0\cos 48.6°}{1.5\cos 30° + 1.0\cos 48.6°} = 0.325.$$

Therefore, the transmitted energy would be

$$T_s = 1 - (r_s)^2 = 1 - (0.325)^2 = 0.894 \quad \text{or} \quad 89.4 \text{ percent.}$$

5.8 SYSTEMS OF UNITS

Several systems of units have been devised to describe nature quantitatively. To date, the two most commonly used in quantum electronics are the centimeter–gram–second or cgs system and the meter–kilogram–second or mks system, which is also known as the Système International or S.I. system of units. We conclude this chapter with a discussion of this issue. In actual applications or devices, many of the interrelations between the two systems of units have to be carefully studied in order to obtain accurate numerical results. Perhaps one convenient place to start the discussion is with Coulomb's and Ampère's laws.

Coulomb's law states that the magnitude of the force between two point charges q and q', separated by a distance R, is given by

$$F = C_e\frac{qq'}{R^2}, \tag{5.133}$$

where C_e is a constant of proportionality.

Ampere's law states that the force per unit length (dF/dl) between two parallel currents I and I', separated by a distance r, is given by

$$\frac{dF}{dl} = C_m \frac{II'}{r}, \tag{5.134}$$

where C_m is a constant of proportionality.

In *the mks system*, which we have used throughout the text, C_e and C_m are given by

$$C_e = \frac{1}{4\pi\varepsilon_0} \tag{5.135}$$

and

$$C_m = \frac{\mu_0}{4\pi},$$

respectively. In this system,

$$\frac{C_e}{C_m} = \frac{1}{\mu_0\varepsilon_0} = c^2,$$

where c is the velocity of light.

On the other hand, *the cgs system* is mixed in the sense that electric quantities are measured in *electrostatic* units, (esu), and the constants C_e and C_m appear as 1 and $1/c^2$, respectively, whereas magnetic quantities are measured in *electromagnetic units* (emu), and C_m and C_e appear as 1 and c^2, respectively. It is immediately obvious that this procedure leads to numerous discrepancies in many of the equations we use in quantum electronics (e.g. Maxwell's equations and the constitutive equations). Using the cgs system, we can proceed with the usual electromagnetic laws and obtain the equivalent equations expressed in the preceding sections. More specifically, Maxwell's equations become

$$\nabla \cdot \mathbf{D} = 4\pi\rho_f, \tag{5.136}$$

$$\nabla \cdot \mathbf{B} = 0, \tag{5.137}$$

$$\nabla \times \mathbf{E} = -\frac{1}{c}\frac{\partial \mathbf{B}}{\partial t}, \tag{5.138}$$

and

$$\nabla \times \mathbf{H} = \frac{4\pi J_f}{c} + \frac{1}{c}\frac{\partial \mathbf{D}}{\partial t}, \tag{5.139}$$

where

$$\mathbf{D} = \mathbf{E} + 4\pi\mathbf{P} \tag{5.140}$$

and

$$\mathbf{H} = \mathbf{B} - 4\pi\mathbf{M}. \tag{5.141}$$

In a linear isotropic medium

$$\mathbf{P} = \chi_e \mathbf{E} \tag{5.142}$$

and

$$\mathbf{M} = \chi_m \mathbf{H}, \tag{5.143}$$

so we have

$$\mathbf{D} = \varepsilon \mathbf{E}, \tag{5.144}$$

$$\mathbf{H} = \frac{\mathbf{B}}{\mu}, \tag{5.145}$$

and

$$\mathbf{J}_f = \sigma \mathbf{E}. \tag{5.146}$$

In terms of the scalar vector potentials ϕ and $\mathbf{A}$, respectively, we have

$$\mathbf{B} = \nabla \times \mathbf{A} \tag{5.147}$$

and

$$\mathbf{E} = -\nabla\phi - \frac{1}{c}\frac{\partial \mathbf{A}}{\partial t}, \tag{5.148}$$

where vectors **E**, **D**, **B**, **H**, **M**, and **P** have the same dimensions.

For converting equations from one system of units to the other, starting with the first principle as outlined earlier, regarding Coulomb's and Ampère's laws, would be rather cumbersome. In Table 5.1 we therefore list some of the more frequently used quantities in both their mks and cgs forms, so that equations written in the mks system can be converted to their cgs counterparts by substituting the symbols. For example, in the mks system, Poynting's vector **S** is defined as

$$\mathbf{S} = \mathbf{E} \times \mathbf{H} \quad \text{(mks)}$$

Using their counterparts $\mathbf{E} \to (4\pi\varepsilon_0)^{-1/2}\mathbf{E}$ and $\mathbf{H} \to (4\pi\mu_0)^{-1/2}\mathbf{H}$, we get

$$\mathbf{S} = \frac{1}{4\pi\sqrt{\mu_0\varepsilon_0}}\,\mathbf{E} \times \mathbf{H}$$

and

$$\mathbf{S} = \frac{c}{4\pi}(\mathbf{E} \times \mathbf{H}) \quad \text{(cgs)}. \tag{5.150}$$

TABLE 5.1
Table for Conversion from Mks to Cgs Units

Quantity	Mks	Cgs
Charge	q (coulomb)	$(4\pi\varepsilon_0)^{1/2}q$ (3×10^9 statcoulomb)
Capacitance	C (farad)	$4\pi\varepsilon_0 C$ (9×10^{11} statfarads)
Conductivity	σ $(\Omega\cdot\text{m})^{-1}$	$4\pi\varepsilon_0\sigma$ (9×10^9 stat$\Omega\cdot$cm)
Dielectric constant	K_e	ε
Permittivity	ε	$\varepsilon\varepsilon_0$
Permeability (relative)	K_m	μ
Dipole moment (electric)	d	$(4\pi\varepsilon_0)^{1/2}d$
Displacement	D (coulomb/m^2)	$(\varepsilon_0/4\pi)^{1/2}D$ ($12\pi \times 10^5$ statvolt/cm)
Electric field	E (volt/m)	$(4\pi\varepsilon_0)^{-1/2}E$ ($\frac{1}{3} \times 10^{-4}$ statvolt/cm)
Magnetic field	H (ampere/m)	$(4\pi\mu_0)^{-1/2}H$ ($4\pi \times 10^{-3}$ oersted)
Magnetic induction	B [weber/(m^2 = tesla)]	$(\mu_0/4\pi)^{1/2}B$ (10^4 gauss)
Magnetization	M (ampere/m)	$(4\pi/\mu_0)^{1/2}M$ (10^{-3} oersted)
Polarization	P (coulomb/m^2)	$(4\pi\varepsilon_0)^{1/2}P$ (3×10^5 statvolt/cm)
Scalar potential	ϕ (volt)	$(4\pi\varepsilon_0)^{-1/2}\phi$ ($\frac{1}{300}$ statvolt)
Speed of light	$(\mu_0\varepsilon_0)^{-1/2}$	c
Vector potential	A	$(\mu_0/4\pi)^{1/2}A$
Susceptibilities	χ_e (κ_m)	$4\pi\chi_e$ (χ_m)

Note: The units in parenthesis after each symbol show the equivalence of one unit of quantity in mks units and the corresponding numerical value in cgs units, for example, 1 coulomb/m^2 (for P) in mks units equals 3×10^5 statvolt/cm (for P) in cgs units; 1 volt units equals $\frac{1}{300}$ statvolt.

REFERENCES

1. M. BORN and E. WOLF, *Principles of Optics*, Macmillan, New York, 1964.
2. D. J. GRIFFITHS, *Introduction to Electrodynamics*, Prentice-Hall, Englewood Cliffs, N.J. 1981.
3. R. K. WANGSNESS, *Electromagnetic Fields*, Wiley, New York, 1979.
4. L. D. LANDAU and E. M. LIFSHITZ, *Electrodynamics of Continuous Media*, Pergamon Press, London, 1965.
5. J. D. JACKSON, *Classical Electrodynamics*, second edition, Wiley, New York, 1975.

PROBLEMS

5.1 Derive the amplitude reflection and transmission coefficients for s- and p-polarized light given in Eqs. 5.129 through 5.132.

5.2 Plot the reflection coefficient $R = r^2$ for p- and s-polarized light as a function of the incidence angles for $\eta_i = 1.5$ and $\eta_t = 1.0$.

5.3 A point charge q is oscillating near the origin with a displacement of $\mathbf{z} = z_0\hat{z}e^{-i\omega t}$ + complex conjugate. By expressing its dipole moment in terms of acceleration $a = d^2z/dt^2$, show that the $\mathbf{E}$ and $\mathbf{B}$ fields of this dipole in the radiation zone are given by

$$\mathbf{E} = \frac{qa \sin\theta}{4\pi\varepsilon_0 c^2 r}\,\hat{\theta}$$

and

$$\mathbf{B} = \frac{qa \sin\theta}{4\pi\varepsilon_0 c^3 r}\,\hat{\phi}.$$

Show that the total power radiated instantaneously is given by

$$P = \frac{q^2a^2}{6\pi\varepsilon_0 c^3}.$$

5.4 Assuming a linear isotropic homogeneous medium, find the differential equations satisfied by the vector potential $\mathbf{A}$ and the scalar potential ϕ if we impose the Coulomb condition $\nabla \cdot \mathbf{A} = 0$.

5.5 A beam of laser is polarized in the $\hat{x}$ direction and propagates along $\hat{z}$. A polarizer P_1 is placed in the beam with its axis at 45° to the $\hat{x}$ axis. Another polarizer P_2 is placed behind P_1, with its axis in the $\hat{y}$ direction. What percentage of the incident light intensity is transmitted?

5.6 A polarizer is placed between two crossed polarizers, and it is rotating at the rate of ω. What is the output intensity from the second polarizer if the intensity emerging from the first polarizer is I_1?

5.7 The polarizer P_1 in Problem 5.4 is replaced by a so-called electrooptic crystal whose refractive indices for light polarized in the $\hat{x}$- and $\hat{y}$ directions are, respectively,

$$\eta_x = \eta_0 - rE_{\text{dc}}(z)$$

and

$$\eta_y = \eta_0 + rE_{\text{dc}}(z),$$

where $E_{\text{dc}}(z)$ is an applied dc field in the z direction. What electric field strength is needed to get the first maximum transmission? What electric field strength is needed to get the first minimum transmission?

5.8 Using Eqs. 5.130 through 5.132, plot the amplitude reflection and transmission coefficients for s- and p-polarized light incident on a glass from the air, assuming that the refractive index of the glass is 1.52. Show that for the p-polarized light, the reflection coefficient becomes zero and the transmission coefficient becomes unity at an angle of $\tan^{-1} 1.55$. This is the so-called *Brewster angle*.

5.9 Assuming that the incident, reflected, and transmitted light are given in the forms shown in Eqs. 5.115 through 5.117, show that at total internal reflection (at an air–glass interface, for example)

(a) The optical electrical field penetrates into the medium to a depth of about one optical wavelength.

(b) There is no transmitted optical power.

5.10 A beam of randomly polarized light is incident on an air–water interface. The refractive index of water is about 1.33. Assuming that the angle of incidence is about 80°, calculate the percentage of reflected light. Is the reflected light randomly polarized, or is it preferentially polarized in some particular direction?

5.11 Prove that Eqs. 5.129 through 5.132 can be expressed in the following alternate forms:

$$r_s = \frac{-\sin(\theta_i - \theta_t)}{\sin(\theta_i + \theta_t)},$$

$$r_p = \frac{\tan(\theta_i - \theta_t)}{\tan(\theta_i + \theta_t)},$$

$$t_s = \frac{2\sin\theta_t\cos\theta_i}{\sin(\theta_i + \theta_t)},$$

$$t_p = \frac{2\sin\theta_t\cos\theta_i}{\sin(\theta_i + \theta_t)\cos(\theta_i - \theta_t)}.$$

5.12 In a practical situation it is more convenient to consider the intensity rather than the amplitude of a light wave. If we define *reflectance* R as the ratio of the reflected power to the incident power, and *transmittance* T as the ratio of the transmitted power to the incident power, show that

$$R = r^2$$

and

$$T = \frac{\eta_t\cos\theta_t}{\eta_i\cos\theta_i}\, t_2,$$

where we assume that $\mu_i \simeq \mu_t = \mu_0$.

6 DIFFRACTION

Suppose that a point source of light is illuminating an opaque object, casting a shadow of the object on an observation screen. If we examine the sharpness of the shadow, we see that the edge of the shadow fades gradually over a short distance rather than changing abruptly. Furthermore, if the point source of light is monochromatic, there will be narrow bands of light, called *fringes*, parallel to the edges of the geometrical shadow. It is obvious that the light that passes the edges of the object deviates from a straight-line propagation. This phenomenon is called the *diffraction* of light.

Historically, it was the observation of diffraction that led to general acceptance of the wave theory of light. Wave theory shows that the magnitude of the diffraction effect, that is, the angle of deviation from straight-line propagation, is directly proportional to the wavelength. Thus, with a sound wave we are usually not conscious of a shadow at all, because it has a long wavelength and the diffraction angle is large. Even with sound, however, shadows can be demonstrated if supersonic frequencies are used. Since the wavelengths of visible light are extremely short, the diffraction angle is also small. Therefore, the straight-line propagation assumed in geometrical optics is only the limit approached as the wavelength approaches zero.

6.1 FRAUNHOFER AND FRESNEL DIFFRACTIONS

In the preceding, we have spoken of the diffraction of light as it passes the edge of an obstacle. Diffraction may be treated more simply, however, if we consider the light as passing through one or more small apertures in a diffraction screen.

It is customary to divide diffraction into two types, each of which has been named after one of the early investigators of diffraction. If the source and viewing point are located at effectively infinite distances relative to the aperture in the diffracting screen, the diffraction is known as *Fraunhofer diffraction.* On the other hand, if either the source or the viewing screen is located at a finite distance relative to the aperture, the diffraction is known as *Fresnel diffraction.* The boundary between these two types is somewhat arbitrary and depends on the accuracy desired. In most instances the Fraunhofer method can be used if the difference in the distances does not exceed one-twentieth of a wavelength. Of course, Fraunhofer diffraction may be achieved without the source being at a great physical distance, if a collimating lens is used to make the rays of light from the source nearly parallel.

Figure 6.1 illustrates the preceding definitions. A point source of monochromatic light is at S, the viewing point is at P, and between them is an opaque screen with a finite number of apertures. A circle C, which is as small as possible while still enclosing all the apertures, has been drawn on the plane of the screen. Circle C is the base of cones that have S and P as vertices. Spherical surfaces Σ_1 and Σ_2, with S and P as their centers, are at the bases of radii r_1 and r_2, which are the shortest distances from S and P to the circle C. If the longest distance from C to Σ_1 and from C to Σ_2 is not more than one-twentieth of the wavelength of the light used, the diffraction is Fraunhofer, and the light falling on the observing screen forms a *Fraunhofer diffraction pattern.*

On the other hand, because of the large size of C, or the shortness of the distance to S or P, the distance between C and Σ_1 or Σ_2 is greater than one-twentieth of the wavelength, the diffraction is Fresnel, and a *Fresnel diffraction pattern* is produced.

The radius of the circle C is denoted by ρ in Figure 6.2, l is the shortest distance from S to the screen, and Δl is the greatest separation between sphere and screen. According to the definition of Fraunhofer diffraction, Δl must be a small fraction of the wavelength. However, ρ may be many wavelengths long (as it can also be in Fresnel diffraction). In the right triangle of Figure 6.2, we have

$$(l + \Delta l)^2 = l^2 + \rho^2, \tag{6.1}$$

and, because of the small size of $(\Delta l)^2$ in comparison with the other quantities, we may make the approximation

$$l \simeq \frac{\rho^2}{2\,\Delta l}. \tag{6.2}$$

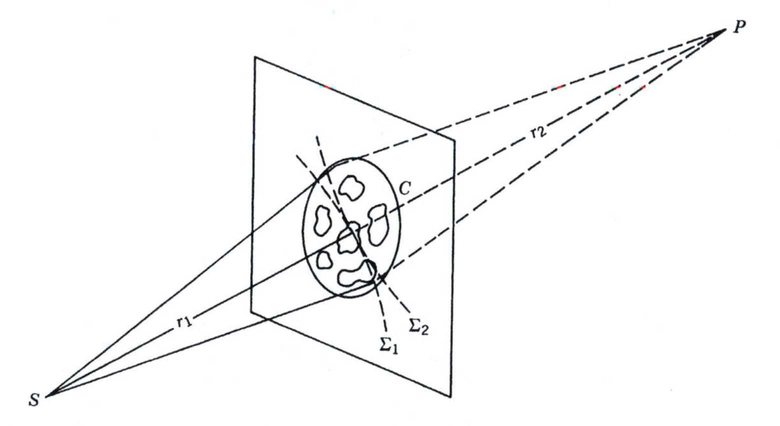

FIGURE 6.1 Geometry for defining Fraunhofer and Fresnel diffractions.

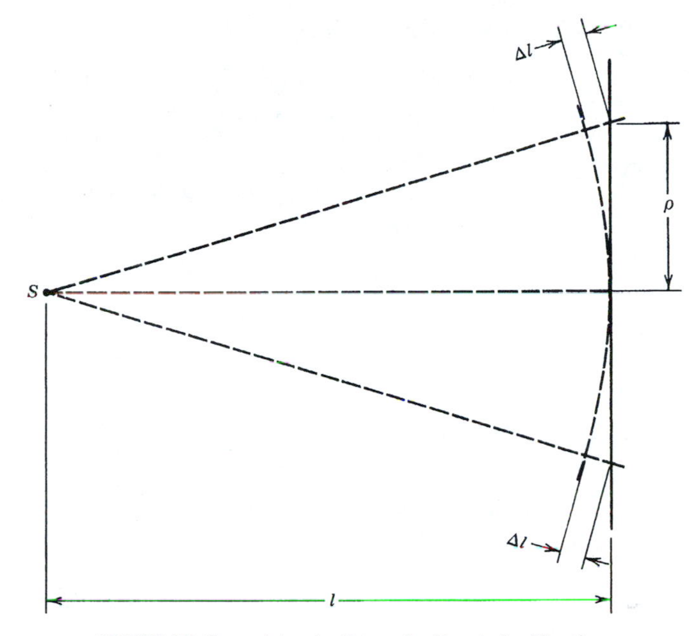

FIGURE 6.2 Determining the distance for Fraunhofer diffraction.

Example 6.1

Let us assume that a diffraction screen, which contains a number of open apertures, is illuminated by a monochromatic point source of visible green light of 5×10^{-5}-cm wavelength. If the open apertures in the diffraction screen are encompassed within a circle of 2-cm diameter, calculate the length of the separation between the light source and diffraction screen (as well as the distance between the diffraction screen and the viewing screen) that is necessary to achieve the Fraunhofer diffraction condition.

To calculate the Fraunhofer diffraction condition, we determine that $\rho = 1$ cm and $\lambda = 5 \times 10^{-5}$ cm. By letting $\Delta l = \lambda/20$, we have

$$\Delta l = \frac{5 \times 10^{-5}}{20} = 2.5 \times 10^{-6} \text{ cm}.$$

Applying Eq. 6.2, we obtain

$$l \simeq \frac{\rho^2}{2\,\Delta l} = \frac{1}{2(2.5 \times 10^{-6})} = 2 \text{ km}.$$

The light source should be about 2 km from the diffraction screen to achieve the Fraunhofer diffraction condition. ■

6.2 THE FRESNEL–KIRCHHOFF INTEGRAL

The Fresnel–Kirchhoff integral can generally be derived from the *scalar wave theory* by the application of Green's theorem. However, this approach is rather tedious and mathematically involved. So in this section we shall approach the Fresnel–Kirchhoff integral by simple *linear system theory.*

According to Huygens' principle in Section 2.2, the complex amplitude observed from a point p' of the coordinate system $\sigma(\alpha, \beta, \gamma)$, caused by a monochromatic light source located in another coordinate system $\rho(x, y, z)$, as shown in Figure 6.3, may be calculated by assuming that each point of the light source is an infinitesimal spherical radiator. Thus, the complex light amplitude $h_l(x, y, z)$ contributed by a point p in the ρ coordinate system must come from an unpolarized monochromatic point source, for which

$$h_l(x, y, z) = -\frac{i}{\lambda r} e^{i(kr - \omega t)}, \tag{6.3}$$

where λ is the wavelength, $k = 2\pi/\lambda$ is the wave number, and ω is the angular time frequency of the point source p, respectively, and r is the distance between the point source p and the point of observation p', which can be written as

$$r = [(l + \gamma - z)^2 + (\alpha - x)^2 + (\beta - y)^2]^{1/2}. \tag{6.4}$$

If the separation l of the two coordinate systems is assumed to be large compared with the regions of interest in the ρ and σ coordinate systems, then the r in the denominator of Eq. 6.3 can be approximated by l, and that in the exponent is replaced by

$$r \simeq (l + \gamma - z) + \frac{(\alpha - x)^2}{2l} + \frac{(\beta - y)^2}{2l}. \tag{6.5}$$

Therefore, Eq. 6.3 can be written as

$$h_l(\alpha - x, \beta - y, \gamma - z) \simeq -\frac{i}{\lambda l} \exp\left\{ik\left[l + \gamma - z + \frac{(\alpha - x)^2}{2l} + \frac{(\beta - y)^2}{2l}\right]\right\}, \tag{6.6}$$

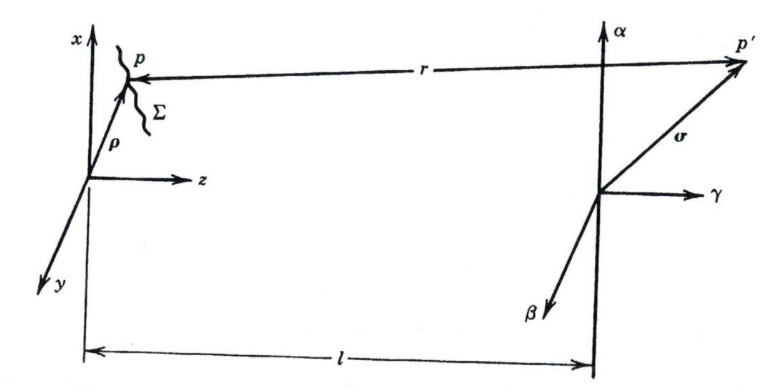

FIGURE 6.3 The Fresnel–Kirchhoff theory.

where the time-dependent part of the exponent, ωt, has been dropped for convenience. Since Eq. 6.6 represents the free-space radiation from a monochromatic point source, it is known as the free-space or *spatial impulse response*. In other words, the complex amplitude produced at the $\boldsymbol{\sigma}$ coordinate system by a monochromatic radiating surface located in the $\boldsymbol{\rho}$ coordinate system can be written in the following abbreviated form,

$$g(\boldsymbol{\sigma}) = \iint_{\Sigma} f(\boldsymbol{\rho}) h_l(\boldsymbol{\sigma} - \boldsymbol{\rho}; k)\, d\Sigma, \tag{6.7}$$

which is essentially a two-dimensional convolution integral, where $f(\boldsymbol{\rho})$ is the complex light field of the monochromatic radiating surface, Σ denotes the surface integral, and $d\Sigma$ is the incremental surface element. We note that the convolution integral of Eq. 6.7 is called the Fresnel–Kirchhoff integral.

For simplicity, we assume that a complex monochromatic radiating field $f(x, y)$ is distributed over the x, y plane. The complex light disturbances at the α, β coordinate plane can be obtained by the following convolution integral,

$$g(\alpha, \beta) = \iint_{-\infty}^{\infty} f(x, y) h_l(\alpha - x, \beta - y)\, dx\, dy, \tag{6.8}$$

where

$$h_l(x, y) = C \exp\left[i \frac{k}{2l} (x^2 - y^2) \right] \tag{6.9}$$

is the spatial impulse response between the spatial coordinate systems (x, y) and (α, β), and

$$C = -\frac{i}{\lambda l} e^{(ikl)}$$

is a complex constant. Equation 6.8 can be represented by the block diagram shown in Figure 6.4.

In addition, if the complex light disturbances at (α, β) are known, the monochromatic radiating field of $f(x, y)$ can be determined in a similar manner,

$$f(x, y) = \iint_{-\infty}^{\infty} g(\alpha, \beta) h_l^*(x - \alpha, y - \beta)\, d\alpha\, d\beta, \tag{6.10}$$

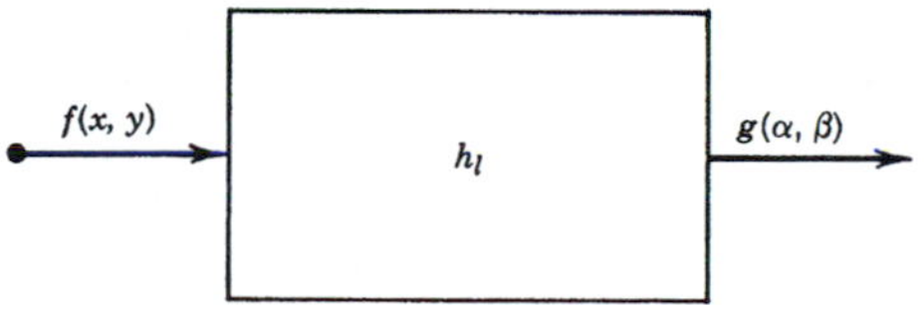

$f(x, y)$ → h_l → $g(\alpha, \beta)$

FIGURE 6.4 A linear representation of Eq. 6.8.

where the superscript asterisk denotes the complex conjugate,

$$h_l^*(\alpha, \beta) = C^* \exp\left[-i\frac{k}{2l}(x^2 + y^2)\right], \tag{6.11}$$

and

$$C^* = \frac{i}{\lambda l} e^{-ikl}.$$

Notice that Eq. 6.11 represents a convergent spherical wavefront instead of a divergent wavefront as described by Eq. 6.9.

Example 6.2

Given a monochromatic point source located at the origin of the (x, y) coordinate system, as shown in Figure 6.5, use the Fresnel–Kirchhoff theory to calculate the complex light field arriving at the α, β plane.

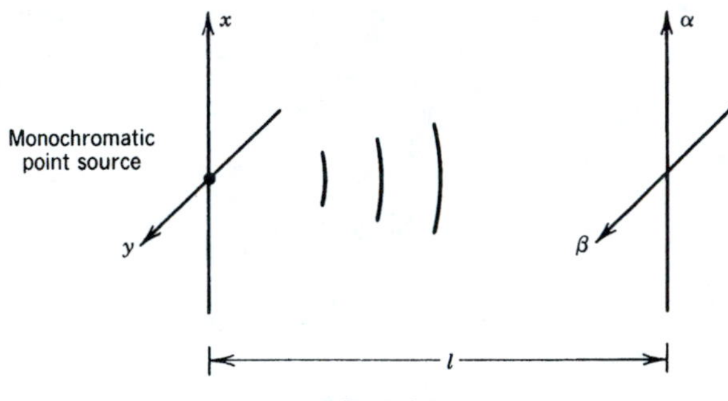

FIGURE 6.5

Since the distance between the x, y and α, β planes is assumed linear and spatially-invariant, the spatial impulse response can be written as

$$h_l(x, y) = C \exp\left[i\frac{k}{2l}(x^2 + y^2)\right],$$

where C is a proportionality complex constant. The complex light distributed over the α, β plane can be computed by

$$\begin{aligned} g(\alpha, \beta) &= \delta(x, y) * h_l(x, y) \\ &= \iint_{-\infty}^{\infty} \delta(x, y) h_l(\alpha - x, \beta - y)\, dx\, dy \\ &= C \exp\left[i\frac{k}{2l}(\alpha^2 + \beta^2)\right], \end{aligned}$$

which is a spherical wavefront. ■

■ Example 6.3

Referring to the final result of Example 6.2, find the complex light field at distance $l' \neq l$ in front of the α, β plane, as shown in Figure 6.6*a*.

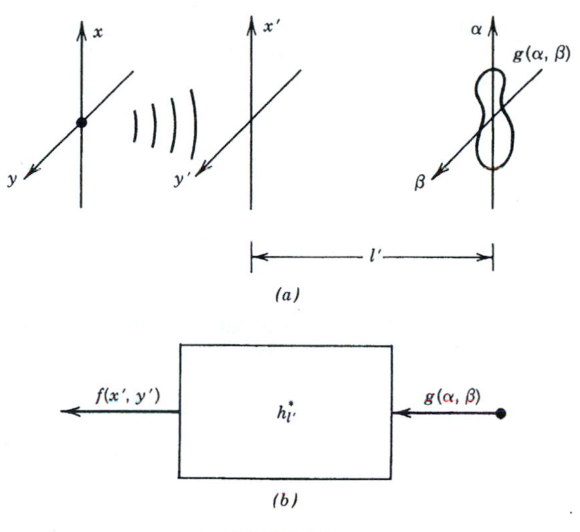

FIGURE 6.6

Let us first draw a block diagram for this problem, as shown in Figure 6.6*b*. We note that the spatial impulse response is written in conjugate form, which represents a convergent spherical wavefront. The complex light distribution over the x', y' plane can be obtained by

$$f(x', y') = g(\alpha, \beta) * h_{l'}^*(\alpha, \beta),$$

where

$$h_{l'}^*(\alpha, \beta) = C^* \exp\left[-i \frac{k}{2l'} (\alpha^2 + \beta^2)\right].$$

Thus, we have

$$f(x', y') = \iint_{-\infty}^{\infty} g(\alpha, \beta) h_{l'}^*(x' - \alpha, y' - \beta)\, d\alpha\, d\beta$$

$$= C \exp\left[i \frac{k}{2l'} (x' + y')\right]. \qquad ■$$

6.3 FOURIER TRANSFORM IN FRAUNHOFER DIFFRACTION

We assume a diffraction screen that is illuminated by a monochromatic point source, as shown in Figure 6.7. To calculate the complex light distribution over the α, β plane, we first draw an analog system diagram to represent this problem, as

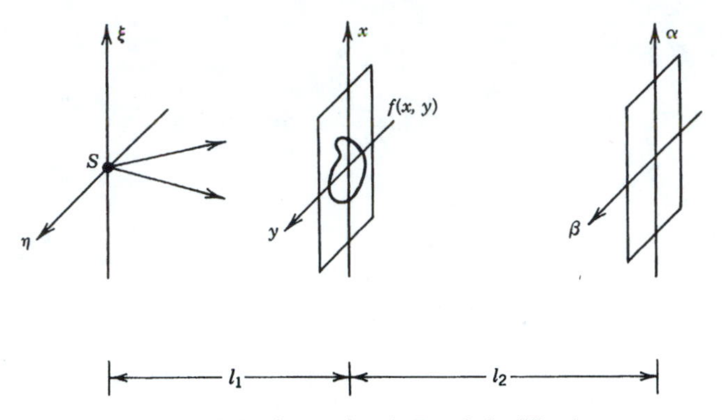

FIGURE 6.7 Fourier transform in Fraunhofer diffraction.

shown in Figure 6.8. Referring to this diagram, we see that the complex light distribution over the output plane can be obtained by

$$g(\alpha, \beta) = [\delta(\xi, \eta) * h_{l1}(\xi, \eta)]f(x, y) * h_{l2}(x, y), \tag{6.12}$$

where the asterisks represent the convolution operations, and

$$h_l(\xi, \eta) = C \exp\left[\frac{ik}{2l}(\xi^2 + \eta^2)\right] \tag{6.13}$$

is the spatial impulse response.

In order to achieve a Fraunhofer diffraction, we first let l_1 approach infinity. Thus, we see that

$$\lim_{l_1 \to \infty} \delta(\xi, \eta) * h_{l1}(\xi, \eta) = \lim_{l_1 \to \infty} C \iint_{-\infty}^{\infty} \delta(\xi, \eta) \exp\left\{\frac{ik}{2l_1}[(\alpha - x)^2 + (\beta - y)^2]\right\} d\xi\, d\eta$$

$$= C \iint_{-\infty}^{\infty} \delta(\xi, \eta) = C, \tag{6.14}$$

which is the plane wavefront. Thus, as $l_1 \to \infty$, Eq. 6.12 becomes

$$g(\alpha, \beta) = Cf(x, y) * h_{l2}(x, y), \tag{6.15}$$

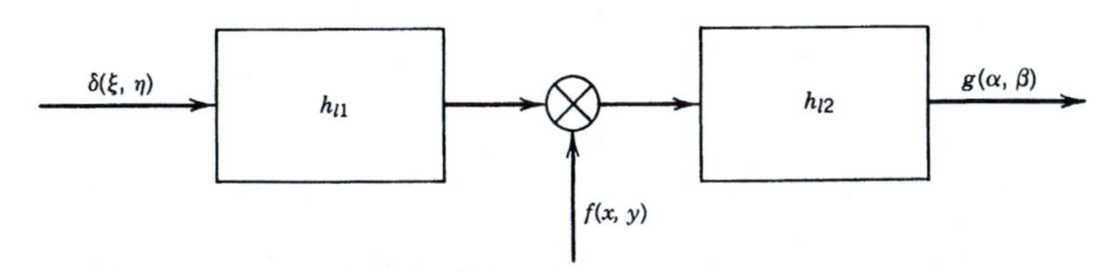

FIGURE 6.8 An analog system diagram of Figure 6.7.

which can be written as

$$g(\alpha, \beta) = C' \iint_{-\infty}^{\infty} f(x, y) \exp\left\{\frac{ik}{2l_2}[(\alpha - x)^2 + (\beta - y)^2]\right\} dx\, dy, \tag{6.16}$$

where C' is a proportionality complex constant. By expanding the quadratic exponent of Eq. 6.16, we have

$$g(\alpha, \beta) = C' \exp\left[\frac{ik}{2l_2}(\alpha^2 + \beta^2)\right] \iint_{-\infty}^{\infty} f(x, y) \exp\left[-\frac{ik}{l_2}(\alpha x + \beta y)\right] \cdot \exp\left[i\frac{k}{2l_2}(x^2 + y^2)\right] dx\, dy. \tag{6.17}$$

To achieve the Fraunhofer diffraction at (α, β), we should let l_2 be sufficiently large, as compared with the region of $f(x, y)$. Thus, Eq. 6.17 can be reduced to the following form:

$$g(\alpha, \beta) = C' \exp\left[\frac{ik}{2l_2}(\alpha^2 + \beta^2)\right] \iint f(x, y) \exp\left[-\frac{ik}{l_2}(\alpha x + \beta y)\right] dx\, dy. \tag{6.18}$$

If the quadratic phase variation caused by $\alpha^2 + \beta^2$ is assumed to be small, then

$$g(\alpha, \beta) \simeq C_1 \iint f(x, y) \exp\left[-\frac{ik}{l_2}(\alpha x + \beta y)\right] dx\, dy, \tag{6.19}$$

which is essentially the Fourier transform of $f(x, y)$.

In addition, the corresponding intensity distribution at the output plane can be written as

$$\begin{aligned} I(\alpha, \beta) &= g(\alpha, \beta) g^*(\alpha, \beta) \\ &= |g(\alpha, \beta)|^2 \\ &= K|F(p, q)|^2, \end{aligned} \tag{6.20}$$

which is proportional to the power spectral distribution of $f(x, y)$, where K is a proportionality constant, and $p = (2\pi/\lambda l_2)\alpha$ and $q = (2\pi/\lambda l_2)\beta$ are the angular spatial-frequency axes. Thus, we see that the Fraunhofer diffraction is indeed approaching the Fourier transformation.

Example 6.4

Assuming that a diffraction screen with a square aperture of dimension W is illuminated by a monochromatic plane wave, calculate the far-field Fraunhofer diffraction.

The transmission function of the diffraction screen can be written as

$$f(x, y) = \begin{cases} 1, & |x| \leq \dfrac{W}{2} \quad \text{and} \quad |y| \leq \dfrac{W}{2}, \\ 0, & \text{otherwise.} \end{cases}$$

According to the Fresnel–Kirchhoff theory, the complex light distribution behind the diffraction screen can be written as

$$g(\alpha, \beta) = \iint_{-\infty}^{\infty} f(x, y) h_l(\alpha - x, \beta - y)\, dx\, dy,$$

where l is the distance behind the screen and $h_l(x, y)$ is the spatial impulse response. Since we assumed that distance l is sufficiently large in comparison with the size of the diffraction aperture, the complex light field can be written as (see Eq. 6.19)

$$g(\alpha, \beta) = C \int_{-W/2}^{W/2} \int_{-W/2}^{W/2} \exp\left[-\frac{ik}{l} (\alpha x + \beta y) \right] dx\, dy,$$

where C is a proportionality complex constant. Making use of the separable nature of this equation, we have

$$\begin{aligned} g(\alpha, \beta) &= C \int_{-W/2}^{W/2} \exp\left(-\frac{ik}{l}\, dx \right) dx \int_{-W/2}^{W/2} \exp\left(-\frac{ik}{l}\, \beta y \right) dy \\ &= C \left[\frac{\sin\left(\dfrac{\pi W \alpha}{\lambda l} \right)}{\dfrac{\pi W \alpha}{\lambda l}} \right] \left[\frac{\sin\left(\dfrac{\pi W \beta}{\lambda l} \right)}{\dfrac{\pi W \beta}{\lambda l}} \right], \end{aligned}$$

which is the Fourier transform of the square aperture. ■

6.4 THE FRESNEL ZONE PLATE

A diffraction screen contains concentric narrow circular slits, as shown in Figure 6.9. The screen is normally illuminated by a monochromatic plane wave of wavelength λ. At a point P behind the diffraction screen, the optical disturbances produced by the individual circular slits are in phase. We are to determine the distances between the narrow slits.

In order for all the optical disturbances to be in phase at distance f behind the screen, the path lengths from each of the circular slits to point P have to be on the order of a full wavelength:

$$l_n - f = n\lambda, \qquad n = 0, 1, 2, \ldots, \infty. \tag{6.21}$$

The radii of the concentric circles can now be computed:

$$r_n = \sqrt{l_n^2 - f^2}, \qquad n = 0, 1, 2, \ldots, \infty. \tag{6.22}$$

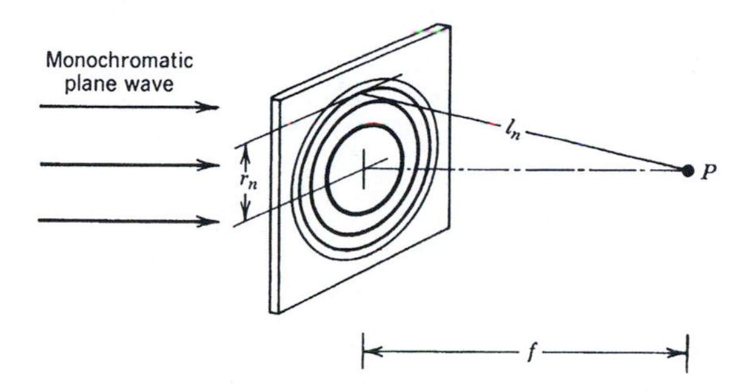

FIGURE 6.9 The focusing effect of a Fresnel zone plate.

By substituting Eq. 6.21 into Eq. 6.22, we get

$$\begin{aligned} r_n &= [(n\lambda + f)^2 - f^2]^{1/2} \\ &= [(n\lambda)^2 + 2fn\lambda]^{1/2}, \qquad n = 0, 1, 2, \ldots, \infty. \end{aligned} \tag{6.23}$$

Thus, we see that these concentric circular slits are capable of focusing a monochromatic plane wave into a very small region with very high intensity.

Suppose that the widths of the concentric slits are enlarged to include all the constructive light rays converging at focal point P. Then the slits would become half-period circular zones, also called *Fresnel zones.* In other words, by enlarging the widths of the slits, we change the optical path lengths of the diffracted light rays. If the variations of the optical paths are within the limit of half-wavelengths of the line-of-sight path from the light source, the diffracted light rays will be additively superimposed on each other at point P. Thus, we will see a very bright spot of light at P.

If the circular apertures are divided into Fresnel zones, as just described, and the alternate zones are covered with opaque material, we have what is called the *Fresnel zone plate*, as illustrated in Figure 6.10, where the central zone is shown open.

Notice that the focusing effect will be the same if we start with a closed zone at the center. Furthermore, if the transmittance of a Fresnel zone plate varies sinusoidally, it is known as a *Fresnel zone lens.*

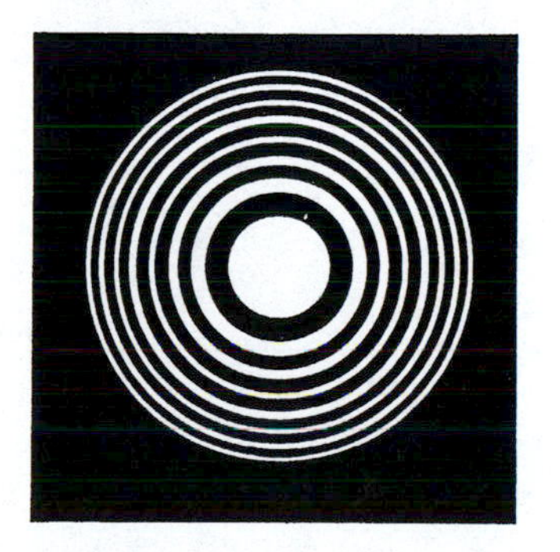

FIGURE 6.10 Fresnel zone plate with the center zone open.

Example 6.5

We are given an amplitude transmittance function of a one-dimensional Fresnel zone lens with infinite length,

$$T(x) = \frac{1}{2} + \frac{1}{2}\cos\left(\frac{\pi}{\lambda R}x^2\right),$$

where λ is an arbitrary wavelength and R is a positive constant.

If this Fresnel zone lens is illuminated by a monochromatic plane wave of wavelength λ, calculate the focal distance and the complex amplitude distribution of the convergent diffracted light rays.

Using the Fresnel–Kirchhoff theory, we can compute the complex light distribution behind the zone lens as follows,

$$g(\alpha) = \int_{-\infty}^{\infty} T(x) h_l(\alpha - x)\, dx,$$

where

$$h_l(x) = C \exp\left(\frac{i\pi x^2}{\lambda l}\right),$$

and l is the distance behind the zone lens. Since

$$\cos\left(\frac{\pi}{\lambda R}x^2\right) = \frac{1}{2}\left[\exp\left(-i\frac{\pi}{\lambda R}x^2\right) + \exp\left(i\frac{\pi}{\lambda R}x^2\right)\right],$$

we have

$$g(\alpha) = C\int_{-\infty}^{\infty}\left\{\frac{1}{2} + \frac{1}{4}\left[\exp\left(-i\frac{\pi}{\lambda R}x^2\right) + \exp\left(i\frac{\pi}{\lambda R}x^2\right)\right]\right\}$$
$$\cdot \exp\left[\frac{i\pi}{\lambda l}(\alpha - x)^2\right] dx,$$

whose terms can be integrated.

Let us now evaluate the second term, which is the convergent wave field,

$$g_2(\alpha) = C'\int_{-\infty}^{\infty} \exp\left(-\frac{i\pi}{\lambda R}x^2\right)\exp\left[\frac{i\pi}{\lambda l}(\alpha - x)^2\right] dx,$$

where C' is a proportionality complex constant. By expanding the quadratic factor, we can write this integral as

$$g_2(\alpha) = C'\exp\left(\frac{i\pi}{\lambda l}\alpha^2\right)\int_{-\infty}^{\infty}\exp\left[i\frac{\pi}{\lambda}\left(\frac{1}{l} - \frac{1}{R}\right)x^2\right]\exp\left[-\frac{i2\pi}{\lambda l}(\alpha x)\right] dx.$$

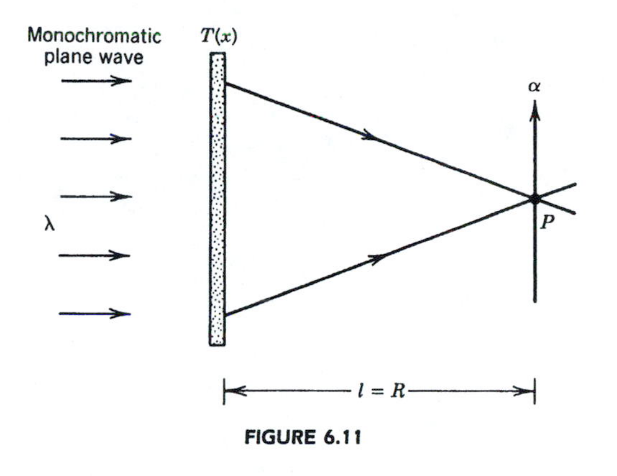

FIGURE 6.11

To eliminate the quadratic phase factor of x^2, we would let, $l = R$, which is the focal distance of the zone lens. Thus, at $l = R$, the integral reduces to

$$g_2(\alpha) = C' \exp\left(\frac{i\pi}{\lambda R}\alpha^2\right)\int_{-\infty}^{\infty} \exp\left[-\frac{i2\pi}{\lambda R}(\alpha x)\right] dx,$$

which is the Fourier transform of a unit constant function. This integral can therefore be reduced to the following result,

$$g_2(\alpha) = C''\delta(\alpha),$$

where C'' is a proportionality complex constant. From this result we see that the diffracted rays (i.e., those caused by the second term) from the Fresnel zone lens will be focused into a point P at a distance $l = R$ behind the zone lens, as sketched in Figure 6.11. ■

■ Example 6.6

If the Fresnel zone lens of Example 6.5 is illuminated by a monochromatic plane wave of wavelength λ_1, that is, $\lambda_1 \neq \lambda$, calculate the distance of the focused diffracted light rays and discuss the effect under white-light illumination. Since the zone lens is illuminated by a monochromatic plane wave with different wavelengths, the Fresnel–Kirchhoff integral should be written in terms of λ_1,

$$g(\alpha; \lambda_1) = \int_{-\infty}^{\infty} T(x; \lambda) h_l(\alpha - x; \lambda_1)\, dx,$$

where

$$h_l(x; \lambda_1) = C \exp\left(i\frac{\pi}{\lambda_1 l} x^2\right).$$

With substitutions for $T(x; \lambda)$ and $h_l(x; \lambda_1)$, the preceding equation becomes

$$g(\alpha; \lambda_1) = C \int_{-\infty}^{\infty} \left\{ \frac{1}{2} + \frac{1}{4} \left[\exp\left(-i \frac{\pi}{\lambda R} x^2 \right) + \exp\left(i \frac{\pi}{\lambda R} x^2 \right) \right] \right\} \cdot \exp\left[i \frac{\pi}{\lambda_1 l} (\alpha - x)^2 \right] dx.$$

The second integral is

$$g_2(\alpha; \lambda_1) = C' \int_{-\infty}^{\infty} \exp\left(-i \frac{\pi}{\lambda R} x^2 \right) \exp\left[\frac{i\pi}{\lambda_1 l} (\alpha - x)^2 \right] dx,$$

which can be written as

$$g_2(\alpha; \lambda_1) = C' \exp\left(i \frac{\pi}{\lambda_1 l} \alpha^2 \right) \int_{-\infty}^{\infty} \exp\left[i\pi \left(\frac{1}{\lambda_1 l} - \frac{1}{\lambda R} \right) x^2 \right] \exp\left[-i \frac{2\pi}{\lambda_1 l} (\alpha x) \right] dx.$$

To eliminate the quadratic phase factor of x^2, we see that

$$l = \frac{\lambda}{\lambda_1} R,$$

which is the focal distance of the covergent light rays,

$$g_2(\alpha; \lambda_1) = C'' \delta(\alpha; \lambda_1), \qquad l = \frac{\lambda}{\lambda_1} R.$$

Since the focal distance is inversely proportional to the illuminating wavelength λ_1, the focal point will be closer to the zone lens for illumination of longer wavelengths and farther from the zone lens for illumination of shorter wavelengths.

It is apparent that if the zone lens is illuminated by a white-light plane wave, the focal point will smear into rainbow colors. For example, the red color focal point will be located closer to the zone lens, and the violet color will be focused at

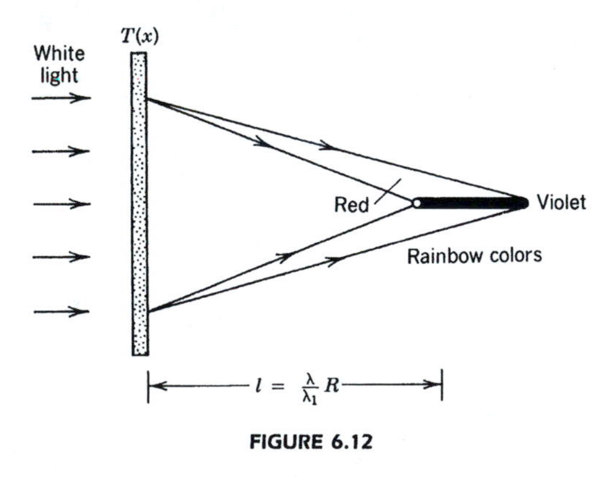

FIGURE 6.12

a greater distance, as illustrated in Figure 6.12. ■

6.5 PARTIAL COHERENCE

The widespread application of partially coherent light makes it appropriate to include a discussion of the principles of coherence in radiation. If the radiations from two point sources maintain a fixed phase relation between them, they are said to be *mutually coherent*. An extended source is coherent if all points of the source have fixed phase differences between them. Here we discuss some general aspects of partial coherence.

In the classical theory of electromagnetic radiation, as in the development of Maxwell's equations, it is usually assumed that the electric and magnetic fields are always measurable at any position. Thus, in these situations no accounting of partial coherence theory is needed. There are situations, however, for which this assumption of known fields cannot be made and for which it is often helpful to apply partial coherence theory. For example, if we want to determine the diffraction pattern caused by the radiation from several sources, we cannot obtain an exact result unless the degree of coherence from the separate sources is taken into account. In such a situation, however, it is desirable to obtain an ensemble average to represent the *statistically* most likely result from any such combination of sources. It may thus be more useful to provide a statistical description than to follow the dynamical behavior of a wave field in detail.

Our treatment of partial coherence uses such ensemble averages. Let us assume an electromagnetic wave field propagating in space, as depicted in Figure 6.13, where $u_1(t)$ and $u_2(t)$ denote the instantaneous wave disturbances at positions 1 and 2, respectively. We choose the second-order moment as the quantity to be averaged, which is why it is called the *mutual coherence function*. This function is defined in the following,

$$\Gamma_{12}(\tau) \triangleq \langle u_1(t+\tau)u_2^*(t)\rangle, \tag{6.24}$$

where the asterisk denotes the complex conjugate, and $\Gamma_{12}(\tau)$ is the mutual coherence function between these points for a time delay τ. The angle brackets $\langle\ \rangle$

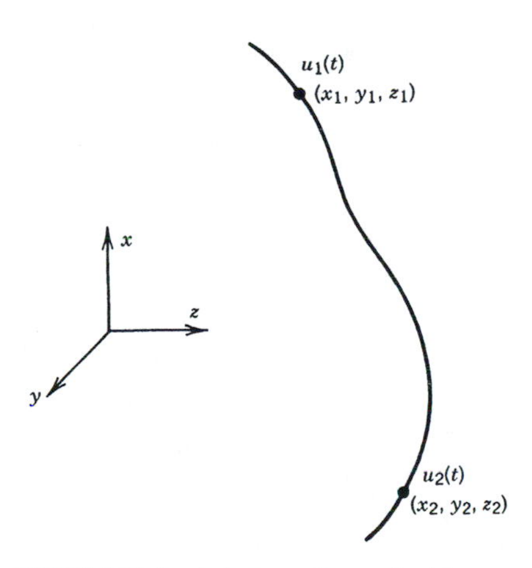

FIGURE 6.13 An electromagnetic wavefront in space.

denote a time average, which can be written as

$$\Gamma_{12}(\tau) = \lim_{T\to\infty} \frac{1}{T} \int_0^T u_1(t+\tau)u^*(t)\,dt. \tag{6.25}$$

It is apparent that the mutual coherence function defined by Eq. 6.25 is essentially the temporal *cross-correlation function* between the complex disturbances of $u_1(t)$ and $u_2(t)$.

The *normalized mutual coherence function* can therefore be defined as

$$\gamma_{12}(\tau) \triangleq \frac{\Gamma_{12}(\tau)}{[\Gamma_{11}(0)\Gamma_{22}(0)]^{1/2}}, \tag{6.26}$$

where $\Gamma_{11}(\tau)$ and $\Gamma_{22}(\tau)$ are the *self-coherence functions* of $u_1(t)$ and $u_2(t)$, respectively. The function $\gamma_{12}(\tau)$ may also be called the *complex degree of coherence* or the degree of correlation.

We now demonstrate that the normalized mutual coherence function $\gamma_{12}(\tau)$ can be measured by applying Young's experiment on interference. In Figure 6.14, Σ represents an extended source of light, which is assumed to be incoherent but nearly monochromatic; that is, its spectrum is of finite width. The light from this source falls upon a screen at a distance r_{10} from the source, and upon two small apertures (pinholes) in this screen, Q_1 and Q_2, separated by a distance d. On an observing screen located r_{20} away from the diffracting screen, an interference pattern is formed by the light passing through Q_1 and Q_2. Now let us suppose that the changing characteristics of the interference fringes are observed as the parameters of Figure 6.14 are changed. As a measurable quantity, let us adopt *Michelson's visibility* of the fringes, which is defined as

$$\nu \triangleq \frac{I_{\max} - I_{\min}}{I_{\max} + I_{\min}}, \tag{6.27}$$

where $I_{\max}$ and $I_{\min}$ are the maximum and minimum intensities of the fringes.

For the visibility to be measurable, the conditions of the experiment, such as the narrowness of the spectrum and the closeness of optical path lengths, must be such that they permit $I_{\max}$ and $I_{\min}$ to be clearly defined. Let us assume that these ideal conditions exist.

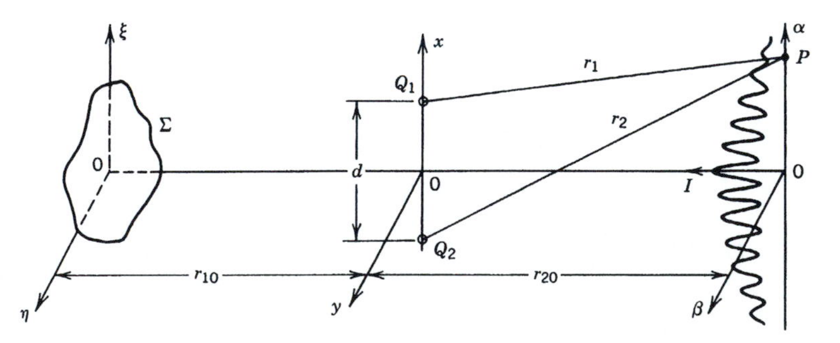

FIGURE 6.14 Young's experiment. The Σ is an extended but nearly monochromatic source.

As we begin, our parameters change. First, we find that the average visibility of the fringes increases as the size of the source Σ is made smaller. Next, as the distance between pinholes Q_1 and Q_2 is increased, while Σ is held constant (and circular in form), the visibility shifts in the manner shown in Figure 6.15. When Q_1 and Q_2 are very close together, the intensity between the fringes falls to zero, and the visibility becomes unity. As d is increased, the visibility falls rapidly and reaches zero as I_{max} and I_{min} become equal. An additional increase in d causes the fringes to reappear, although they are shifted on the screen by half a fringe; that is, the areas that were previously light are now dark, and vice versa. As the distance d between pinholes is made even greater, there are the repeated fluctuations in visibility shown in the figure. A curve similar to that of Figure 6.15 is obtained when the pinhole spacing is kept constant while the size of Σ is changed. These effects can be predicted from the Van Cittert–Zernike theorem (see Section 6.6.1). The visibility versus pinhole separation curve is sometimes used as a measure of *spatial coherence*, as discussed in Section 6.6.1. Screen separations r_{10} and r_{20} are both assumed to be large compared with the aperture spacing d and with the dimensions of the source. Beyond this limitation, changes in r_{10} or r_{20} shift the scale of effects, as is shown in Figure 6.15, but without altering their general character.

When the point of observation P is moved away from the center of the observing screen, visibility decreases as the path difference $\Delta r = r_2 - r_1$ increases, until it eventually becomes zero. The effect also depends on how nearly monochromatic the source is. The visibility of the fringes has been found to be appreciable only for a path difference of

$$\Delta r \simeq \frac{2\pi c}{\Delta\omega}, \tag{6.28}$$

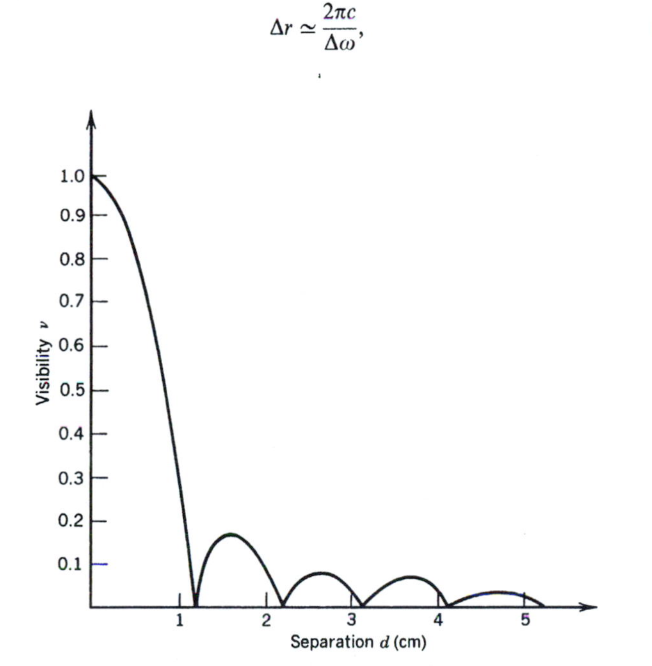

FIGURE 6.15 Visibility as a function of pinhole separation.

where c is the velocity of light and $\Delta\omega$ is the spectral width of the source. Equation 6.28 is often used to define the *coherence length* of the source, which is the distance that its light beam is longitudinally coherent.

The preceding discussion has indicated that it is not necessary to have completely coherent light to produce an interference pattern, but that under the right conditions such a pattern may be obtained from an incoherent source. This effect is called *partial coherence.*

It will be helpful to develop the preceding equations further. Thus, $u_1(t)$ and $u_2(t)$ of Eq. 6.24, the complex wave fields at points Q_1 and Q_2, are subject to the scalar wave equation in free space,

$$\nabla^2 u = \frac{1}{c^2}\frac{\partial^2 u}{\partial t^2}, \tag{6.29}$$

where c is the velocity of light. This is a linear equation, and the wave field at point P on the observing screen is a sum of the fields from Q_1 and Q_2,

$$u_p(t) = c_1 u_1\left(t - \frac{r_1}{c}\right) + c_2 u_2\left(t - \frac{r_2}{c}\right), \tag{6.30}$$

where c_1 and c_2 are the appropriate complex constants. The corresponding irradiance at P may be written as

$$I_p = \langle u_p(t)u_p^*(t)\rangle = I_1 + I_2 + 2\,\mathrm{Re}\left\langle c_1 u_1\left(t - \frac{r_1}{c}\right)c_2^* u_2^*\left(t - \frac{r_2}{c}\right)\right\rangle, \tag{6.31}$$

where I_1 and I_2 are proportional to the squares of the magnitudes of $u_1(t)$ and $u_2(t)$. We now define the variables

$$t_1 = \frac{r_1}{c} \quad \text{and} \quad t_2 = \frac{r_2}{c}. \tag{6.32}$$

Then Eq. 6.31 can be written as

$$I_p = I_1 + I_2 + 2c_1c_2^*\,\mathrm{Re}\langle u_1(t - t_1)u_2^*(t - t_2)\rangle. \tag{6.33}$$

Thus, we see that the averaged quantity in Eq. 6.33 is the cross-correlation of the two complex wave fields.

If we make $t_2 - t_1 = \tau$, Eq. 6.33 can be written as

$$I_p = I_1 + I_2 + 2c_1c_2^*\,\mathrm{Re}\langle u_1(t + \tau)u_2^*(t)\rangle,$$

and combining this with Eq. 6.24, we obtain

$$I_p = I_1 + I_2 + 2c_1c_2^*\,\mathrm{Re}[\Gamma_{12}(\tau)]. \tag{6.34}$$

The *self-coherence functions* (i.e., the autocorrelations) of the radiations from the two pinholes can therefore be defined as

$$\Gamma_{11}(0) = \langle u_1(t)u_1^*(t)\rangle \quad \text{and} \quad \Gamma_{22}(0) = \langle u_2(t)u_2^*(t)\rangle. \tag{6.35}$$

If we let the following relations hold,

$$|c_1|^2\Gamma_{11}(0) = I_1 \qquad \text{and} \qquad |c_2|^2\Gamma_{22}(0) = I_2,$$

the intensity at P can then be written in terms of the degree of complex coherence, as given in Eq. 6.26:

$$I_p = I_1 + I_2 + 2(I_1 I_2)^{1/2}\,\mathrm{Re}[\gamma_{12}(\tau)]. \tag{6.36}$$

Let us write $\gamma_{12}(\tau)$ in the form

$$\gamma_{12}(\tau) = |\gamma_{12}(\tau)|\exp[i\phi_{12}(\tau)], \tag{6.37}$$

and assume also that $I_1 = I_2 = I$, which can be called the *best condition.* Then Eq. 6.36 becomes

$$I_p = 2I[1 + |\gamma_{12}(\tau)|\cos\phi_{12}(\tau)]. \tag{6.38}$$

Thus we see that the maximum value of I_p is $2I[1 + |\gamma_{12}(\tau)|]$ and the minimum value is $2I[1 - |\gamma_{12}(\tau)|]$. By substituting these values into the visibility equation, Eq. 6.27, we have

$$v = |\gamma_{12}(\tau)|. \tag{6.39}$$

That is, under the best condition, the visibility of the fringes is a measure of the degree of coherence.

Example 6.7

Assume that two monochromatic plane waves of equal intensity are falling obliquely over an observation plane, as shown in Figure 6.16*a*. The fringe intensity distribution is as plotted in Figure 6.16*b*.

(a) Calculate the degree of coherence between these two beams of light.

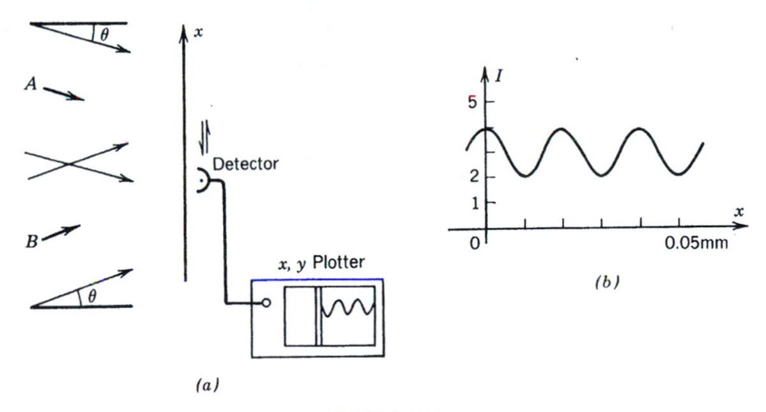

FIGURE 6.16

(b) Determine the spatial frequency of the fringe pattern and the oblique angle of the plane waves. The wavelength of illumination is assumed to be $\lambda = 0.633\ \mu m$.

(c) If the intensity ratio is equal to 2 (i.e., $I_A/I_B = 2$), calculate the visibility of the pattern of interference fringes.

Answers

(a) Referring to Figure 6.16*b*, we note that

$$I_{max} = 4,$$
$$I_{min} = 2.$$

From the relation given in Eq. 6.39 and the definition given by Eq. 6.27, we have

$$\gamma = v = \frac{I_{max} - I_{min}}{I_{max} + I_{min}} = \frac{4 - 2}{4 + 2} = \frac{1}{3}.$$

(b) The intensity distribution over x can be written as

$$I = I_A + I_B + 2\sqrt{I_A I_B}\,|\gamma| \cos(\phi_A - \phi_B),$$

where $\phi_A = -kx \sin\theta$ and $\phi_B = kx \sin\theta$. By substituting ϕ_A and ϕ_B, we have

$$I = I_A + I_B + 2\sqrt{I_A I_B}\,|\gamma| \cos\left(\frac{4\pi}{\lambda} x \sin\theta\right).$$

Thus, we see that the spatial frequency of the fringe pattern is

$$f_x = \frac{2}{\lambda} \sin\theta.$$

From the intensity traces of Figure 6.16*b*, the spatial frequency is computed as

$$f_x = \frac{5}{0.1} = 50 \text{ cycles/mm}.$$

The oblique angle of the plane waves can therefore be determined:

$$\sin\theta = \frac{f_x \lambda}{2} = \frac{50 \times 633 \times 10^{-6}}{2} = 0.01583$$

$$\theta = 0.01583 \text{ radian}.$$

(c) Using Eq. 6.27, we can write the visibility as

$$v = \frac{I_{max} - I_{min}}{I_{max} + I_{min}} = \frac{(I_A + I_B + 2\sqrt{I_A I_B}\,|\gamma|) - (I_A + I_B - 2\sqrt{I_A I_B}\,|\gamma|)}{(I_A + I_B + 2\sqrt{I_A I_B}\,|\gamma|) + (I_A + I_B - 2\sqrt{I_A I_B}\,|\gamma|)}$$

$$= \frac{2\sqrt{I_A I_B}\,|\gamma|}{I_A + I_B}$$

$$= \frac{2\sqrt{I_A/I_B}\,|\gamma|}{1 + I_A/I_B}.$$

By substituting $I_A/I_B = 2$, and $|\gamma| = \frac{1}{3}$, we have

$$v = \frac{(2\sqrt{2})(\frac{1}{3})}{1+2} = 0.314.$$ ■

6.6 SPATIAL AND TEMPORAL COHERENCES

The term *spatial coherence* is applied to effects that are due to the size—in space—of the source of radiation. If we consider a point source, and look at two points that are at equal light path distances from the source, the radiations reaching these points will be exactly the same. This mutual coherence will be equal to the self-coherence at either point. That is, if the points are Q_1 and Q_2, then

$$\Gamma_{12}(Q_1, Q_2, \tau) = \langle u(Q_1, t + \tau)u^*(Q_2, t)\rangle = \Gamma_{11}(\tau). \tag{6.40}$$

As the source is made larger, we can no longer claim an equality of mutual coherence and self-coherence. This lack of complete coherence is a *spatial* effect. *Temporal coherence* is an effect that is due to the finite spectral width of the source. The coherence is complete for strictly monochromatic radiation but becomes only partial as other wavelengths are added, giving a finite spectral width to the source. It is never possible to separate completely the two effects (i.e., spatial and temporal coherence), but it is well to name them and point out their significance, as we do in the following discussion.

6.6.1 Spatial Coherence

We utilize Young's experiment to determine the angular size of a spatially incoherent source (e.g., an extended source) as it relates to spatial coherence. Thus, we see that spatial coherence depends on the size of a light source. Let us now consider the paths of two light rays from a point on a linearly extended source ΔS, as they pass through the two narrow slits Q_1 and Q_2 depicted in Figure 6.17. If we let $r_{10} \gg d$, the intensity distribution at observation screen P_3 can, as a consequence

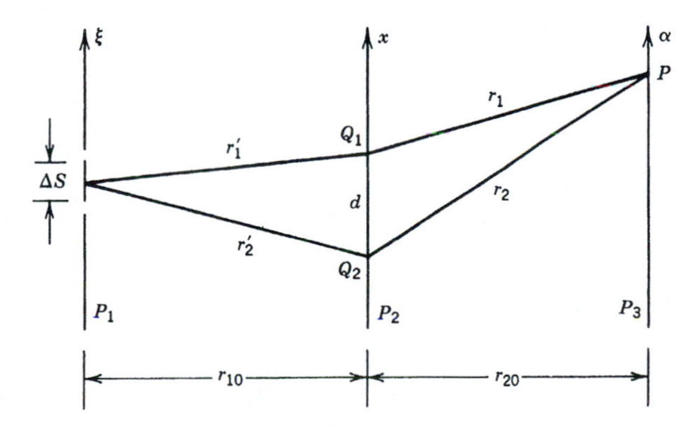

FIGURE 6.17 Young's experiment applied to the measurement of spatial coherence.

of Eq. 6.36, be written as

$$I_p(\alpha) = I_1 + I_2 + 2\sqrt{I_1 I_2}\cos\left[k\left(\frac{d}{r_{10}}\xi + \frac{d}{r_{20}}\alpha\right)\right]. \tag{6.41}$$

Here I_1 and I_2 are the corresponding intensity distributions that are produced by the light rays passing through Q_1 and Q_2, respectively, and $k = 2\pi/\lambda$ is the wave number.

Equation 6.41 describes a set of parallel fringes along the α axis that come from a single radiation point on the extended source, ΔS. Thus, we see that the other points on the extended source cause the superimposition of the fringes on the observation screen P_3. For a uniformly one-dimensional extended source of size ΔS, the resulting fringe pattern can be written as

$$\begin{aligned} I_p'(\alpha) &= I_1\,\Delta S + I_2\,\Delta S + 2\sqrt{I_1 I_2}\int_{-\Delta S/2}^{\Delta S/2}\cos\left[k\left(\frac{d}{r_{10}}\xi + \frac{d}{r_{20}}\alpha\right)\right]d\xi \\ &= \Delta S\left\{I_1 + I_2 + 2\sqrt{I_1 I_2}\,\mathrm{sinc}\left[\left(\frac{\pi d}{\lambda r_{10}}\right)\Delta S\right]\cos\left(k\frac{d}{r_{20}}\alpha\right)\right\}, \end{aligned} \tag{6.42}$$

where

$$\mathrm{sinc}\,\pi\chi \triangleq \frac{\sin\pi\chi}{\pi\chi}.$$

The sinc factor of Eq. 6.42 indicates that the fringes vanish at

$$d = \frac{\lambda r_{10}}{\Delta S}. \tag{6.43}$$

Similarly, we can write Eq. 6.43 in terms of the angular size of the source θ_s,

$$d = \frac{\lambda}{\theta_s}, \tag{6.44}$$

where $\theta_s \triangleq \Delta S/r_{10}$. We note that the sinc factor of Eq. 6.42 is primarily due to the incoherent addition of the point radiators over the extended source ΔS. Equation 6.42 presents a useful example for the discussion of coherent and incoherent illumination. For completely coherent illumination there are interference fringes, but for completely incoherent illumination the interference fringes vanish. Moreover, Eq. 6.42 shows that the interference pattern (i.e., the cosine factor) is weighted by a broad sinc factor, whereby the interference fringes go to zero and reappear as a function of d, ΔS, or $1/r_{10}$. If the extended source had been circular in shape, the sinc factor of Eq. 6.42 would have been

$$\frac{J_1(\pi\,\Delta S\,d/\lambda r_{10})}{\pi\,\Delta S\,d/\lambda r_{10}}, \tag{6.45}$$

where J_1 is the first-order Bessel function. The pinhole separation at which the fringes vanish would then be

$$d = 1.22\frac{\lambda}{\theta_s}, \tag{6.46}$$

where $\theta_s \triangleq \Delta S/r_{10}$ is the angular size of the circular source.

Since the region of spatial coherence is determined by the visibility of the interference fringes, the coherence region increases as the size of the source decreases or as the distance of the source increases. In other words, the degree of spatial coherence decreases as the distance of the propagated wave increases.

In addition, there is a relation between the intensity distribution of the source and the degree of spatial coherence. This relation is described in the *Van Cittert–Zernike theorem.* The theorem essentially states that the normalized complex degree of coherence $\gamma_{12}(0)$ between two points on a plane that come from an extended incoherent source at another plane is proportional to the normalized inverse Fourier transform of the intensity distribution of the source:

$$\gamma_{12}(0) = \frac{\int_{\Delta S} I(\xi) \exp[ik(\mathbf{x}_1 - \mathbf{x}_2)\xi]\, d\xi}{\int_{\Delta S} I(\xi)\, d\xi}, \tag{6.47}$$

Here the subscripts 1 and 2 denote the two points on the diffraction screen P_2; $\mathbf{x}_1$ and $\mathbf{x}_2$ are the corresponding position vectors; ξ is the position vector at the source plane, $k = 2\pi/\lambda$; $I(\xi)$ is the intensity distribution of the source; and the integration is over the source size. Thus, it is apparent that if the intensity distribution of the source is uniform over a circular disk, the mutual coherence function is described by a rotational symmetric Bessel function J_1. On the other hand, if the source intensity is uniform over a rectangular slit, the mutual coherence function is described by a two-dimensional sinc factor.

6.6.2 Temporal Coherence

It is possible for us to split a light wave into two paths, to delay one of them, and then to recombine them to form an interference fringe pattern. In this way, we can measure the temporal coherence of the light wave. The degree of temporal coherence is the measure of the cross-correlation of a wave field at one time with respect to another wave field at a later time. Thus, the definition of temporal coherence can also refer to *longitudinal coherence* as opposed to transverse (i.e., spatial) coherence. The maximum difference in the optical path lengths of two waves derived from a source is known as the *coherent length* of that source. Since spatial coherence is determined by the wave field in the transverse direction, and temporal coherence is measured along the longitudinal direction, spatial and temporal coherences would describe the degree of coherence of a wave field within a volume of space.

The *Michelson interferometer* is one of the instruments most commonly used to measure the temporal coherence of a light source (see Figure 6.18). Looking at this figure, we can see that the beam splitter BS divides the light beam into two paths, one of which goes to mirror M_1 and the other to mirror M_2. By varying the path length (i.e., τ) to one of the mirrors, we can observe the variation of the interference fringes at observation screen P. Thus, the measurement of the visibility of the interference fringes corresponds to the measurement of the degree of temporal coherence $\gamma_{11}(\tau)$ at a time delay τ, where subscript 11 signifies the interference derived from the same point in space. Therefore, when the difference in path lengths is zero, the measurement of visibility corresponds to $\gamma_{11}(0)$.

Since the nature of the source affects coherence, the spectral width of the source affects the temporal coherence of the source. The time interval Δt, during which

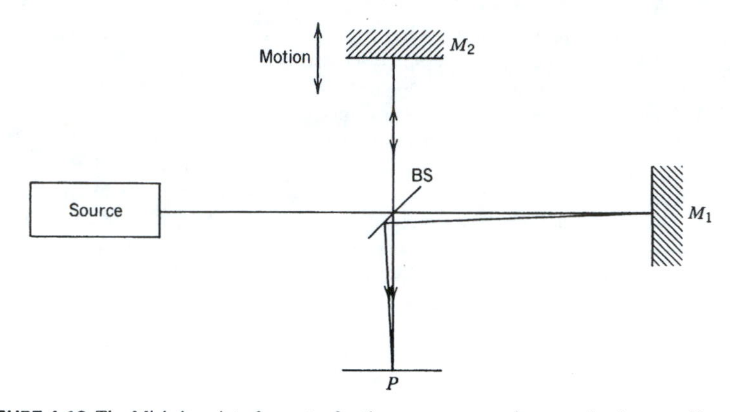

FIGURE 6.18 The Michelson interferometer for the measurement of temporal coherence. *Note:* BS, beam splitter; M, mirrors; P, observation screen.

the wave is coherent, can be approximated by

$$\Delta t \simeq \frac{2\pi}{\Delta\omega}, \tag{6.48}$$

where $\Delta\omega$ is the spectral bandwidth of the source. Thus, the coherence length of the light source can be defined, with Eq. 6.28 as

$$\Delta r = \Delta t c \simeq \frac{2\pi c}{\Delta\omega}, \tag{6.49}$$

where $c = 3 \times 10^8$ m/sec is the velocity of light.

By substituting the relations of

$$c = \frac{\omega\lambda}{2\pi} \quad \text{and} \quad \frac{\omega}{\Delta\omega} = \frac{\lambda}{\Delta\lambda},$$

we can also write Eq. 6.49 as

$$\Delta r \simeq \frac{\lambda^2}{\Delta\lambda}, \tag{6.50}$$

where λ is the center wavelength, and $\Delta\lambda$ is the spectral width of the light source.

Example 6.8

The intensity distribution of a one-dimensional monochromatic source is given by

$$I(\xi) = \begin{cases} 1, & |\xi| \leq W/2, \\ 0, & \text{otherwise.} \end{cases}$$

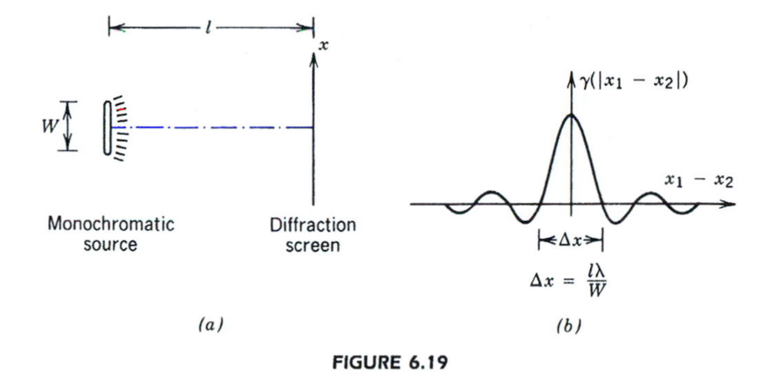

FIGURE 6.19

If the light source illuminates a diffraction screen (as shown in Figure 6.19*a*), calculate the complex degree of coherence (i.e., the normalized spatial coherence) at the screen.

Using the Van Cittert–Zernike theorem as given in Eq. 6.47, we can determine the complex degree of coherence:

$$\gamma_{12}(0) = \gamma_{12}(|x_1 - x_2|) = \frac{\int_{-W/2}^{W/2} \exp(ik|x_1 - x_2|\xi)\, d\xi}{\int_{-W/2}^{W/2} d\xi}$$

$$= \frac{\sin\left(\frac{\pi W}{l\lambda}|x_1 - x_2|\right)}{\frac{\pi W}{l\lambda}|x_1 - x_2|}$$

$$= \operatorname{sinc}\left(\frac{\pi W}{l\lambda}|x_1 - x_2|\right).$$

A sketch of the complex degree of coherence across the diffraction screen is given in Figure 6.19*b*. We see that the *coherence distance* can be written as $\Delta x = l\lambda/W$, which is equal to the width of the sinc factor. In other words, any two-points within distance Δx (i.e., $|x_2 - x_1| \leq \Delta x$) on the diffraction screen are highly coherent.

In order to gain a sense of magnitude, we let $W = 1$ mm, $l = 0.2$ m, and $\lambda = 600 \times 10^{-9}$ m. Then the coherence distance is

$$\Delta x = \frac{0.2 \times 600 \times 10^{-9}}{10^{-3}} = 0.12 \text{ mm}. \quad \blacksquare$$

■ Example 6.9

Refer to the depiction of Young's experiment in Figure 6.20, where Δs is a linear extended monochromatic source, $d = 1$ mm, $\lambda = 6330$ Å, and $L = 1$ m. Using these parameters, calculate the degrees of coherence between Q_1 and Q_2 for $\Delta s = 0.1$ mm and 0.5 mm, respectively.

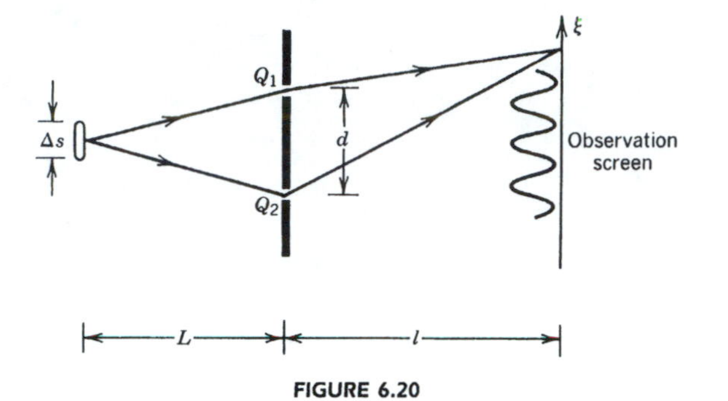

FIGURE 6.20

Using the Van Cittert–Zernike theorem given by Eq. 6.47 or as applied in Example 6.8, we obtain

$$\gamma_{12}(\Delta s) = \text{sinc}\left(\frac{\pi \, \Delta s}{\lambda L} \, d\right).$$

Substituting $d = 1$ mm, $\lambda = 6330$ Å, $L = 1$ m, and $\Delta s = 0.1$ mm, or

$$\frac{d}{\lambda L} = \frac{1}{6330 \times 10^{-7} \times 10^3} = 1.58 \text{ mm}^{-1},$$

we get the following degree of coherence,

$$\gamma_{12}(0.1 \text{ mm}) = \frac{\sin(\pi \times 0.1 \times 1.58)}{\pi \times 0.1 \times 1.58} = 0.959.$$

Similarly, for $\Delta s = 0.5$ mm, the degree of coherence is

$$\gamma_{12}(0.5 \text{ mm}) = \frac{\sin(\pi \times 0.5 \times 1.58)}{\pi \times 0.5 \times 1.58} = 0.247.$$ ■

6.7 COHERENCE MEASUREMENT

Although in the 1930s, Pieter H. Van Cittert and Frits Zernike predicted a profound relationship between the spatial coherence and the intensity distribution of a light source, it was Brian J. Thompson and Emil Wolf who demonstrated a two-beam interference technique to measure the degree of partial coherence. They showed that, under quasi-monochromatic illumination, the degree of spatial coherence depends on the size of the source and the distance between two arbitrary points. The degree of temporal coherence, however, depends on the spectral bandwidth of the light source. Thompson and Wolf also illustrated several coherence measurements that are very consistent with the Van Cittert–Zernike predictions.

In this section we illustrate some results obtained by interferometric techniques, as described in the previous section, for the measurement of spatial and temporal coherence.

Figure 6.21 shows the sequence of interference fringe patterns obtained with the dual-beam technique, which uses a circular extended source. This set of fringe

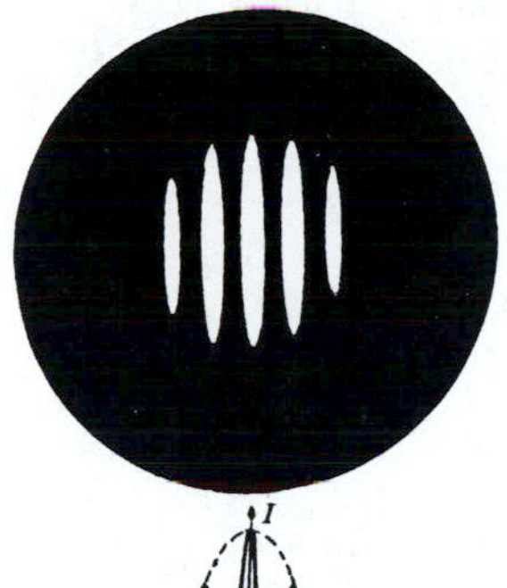

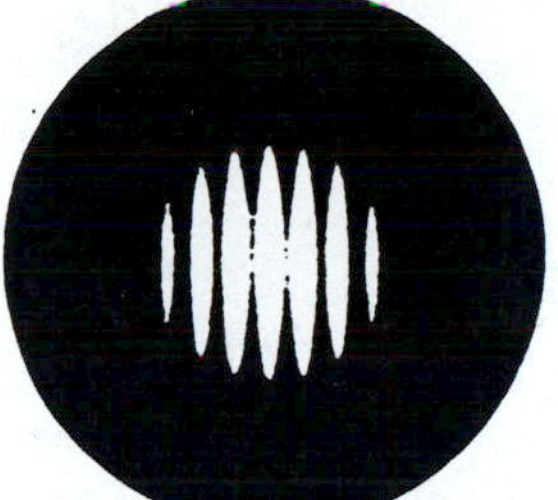

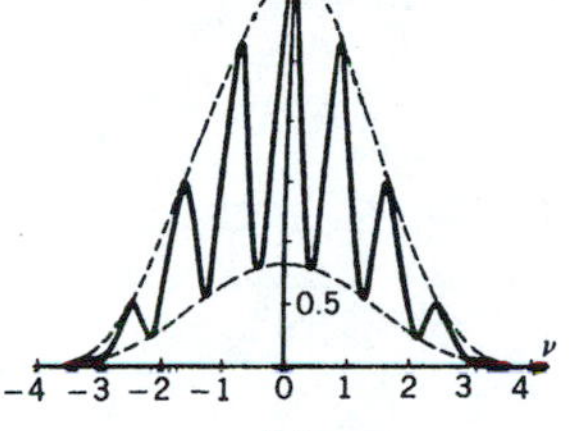

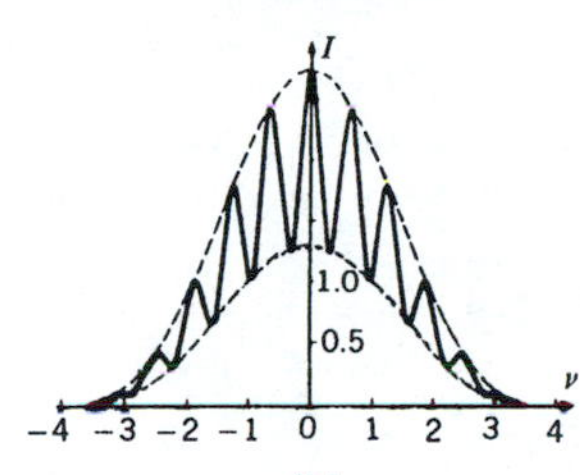

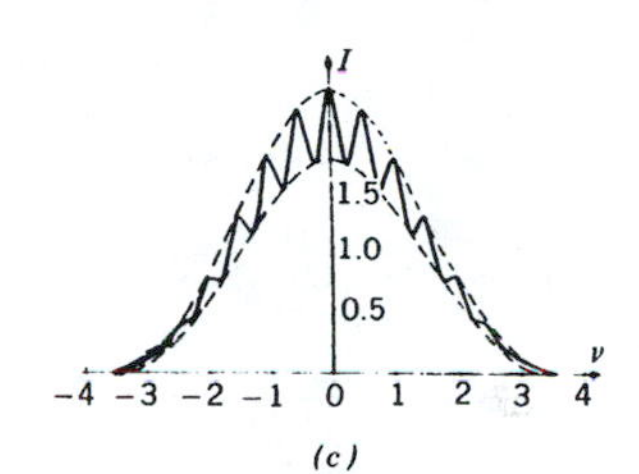

(a) (b) (c)

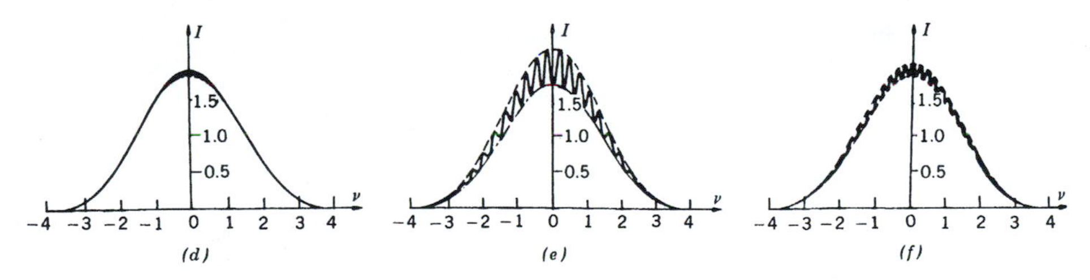

I
1.5
1.0
0.5
−4 −3 −2 −1 0 1 2 3 4
ν

(d) (e) (f)

FIGURE 6.21 Measurement of spatial coherence as a function of pinhole separation. *Note:* d, the separation between pinholes; γ, the degree of coherence.

(*a*) $d = 0.6$ cm; $|\gamma_{12}| = 0.593$. (*b*) $d = 0.8$ cm; $|\gamma_{12}| = 0.361$.
(*c*) $d = 1$ cm: $|\gamma_{12}| = 0.146$. (*d*) $d = 1.2$ cm; $|\gamma_{12}| = 0.015$.
(*e*) $d = 1.7$ cm; $|\gamma_{12}| = 0.123$. (*f*) $d = 2.3$ cm; $|\gamma_{12}| = 0.035$.
(By permission of Brian J. Thompson.)

patterns is obtained by increasing the separation d of the two pinholes. From this set of fringe patterns we see that the visibility (i.e., the degree of spatial coherence) decreases from Figure 6.21*a* to Figure 6.21*d*, then increases in Figure 6.21*e*, and decreases again in Figure 6.21*f*, in accord with the plotting in Figure 6.15. We also note that the phase of the fringe patterns is shifted a few times from Figures 6.21*a* through *f*, which is due to the bipolar nature of the first-order Bessel function, as described in Eq. 6.45. In other words, Figures 6.21*a*, *b*, and *c* give the fringe patterns recorded in the first lobe of Figure 6.15, Figures 6.21*d* and *e* give those in the second lobe, and Figure 6.21*f* gives the fringe patterns in the third lobe. It is also apparent from these six figures that the spatial frequency of the interference fringes is proportional to the separation d.

Figure 6.22 shows a set of interference fringe patterns produced when the separation d is held constant while the diameter of the circular source is increased. From this set of figures we see that the visibility decreases as the source size increases, and that the spatial frequency of the fringe patterns remains unchanged.

For the measurement of temporal coherence, however, we use the Michelson interferometer, as described earlier. Figure 6.23 shows a sequence of interference fringe patterns that we have obtained with this technique. In the Figures 6.23*a*, *b*,

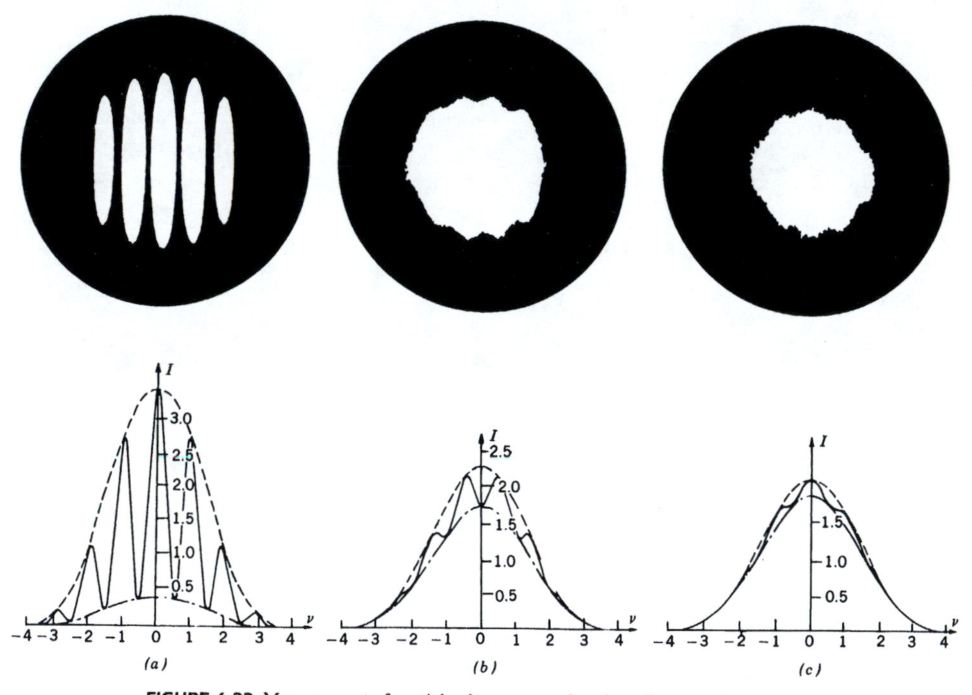

FIGURE 6.22 Measurement of spatial coherence as a function of source size. *Note:* ϕ, the phase shift; d, the separation between pinholes; γ, the degree of coherence.

(*a*) $\phi_{12} = 0$; $d = 0.5$ cm; $|\gamma_{12}| = 0.703$.
(*b*) $\phi_{12} = \pi$; $d = 0.5$ cm; $|\gamma_{12}| = 0.132$.
(*c*) $\phi_{12} = 0$; $d = 0.5$ cm; $|\gamma_{12}| = 0.062$.
(By permission of Brian J. Thompson.)

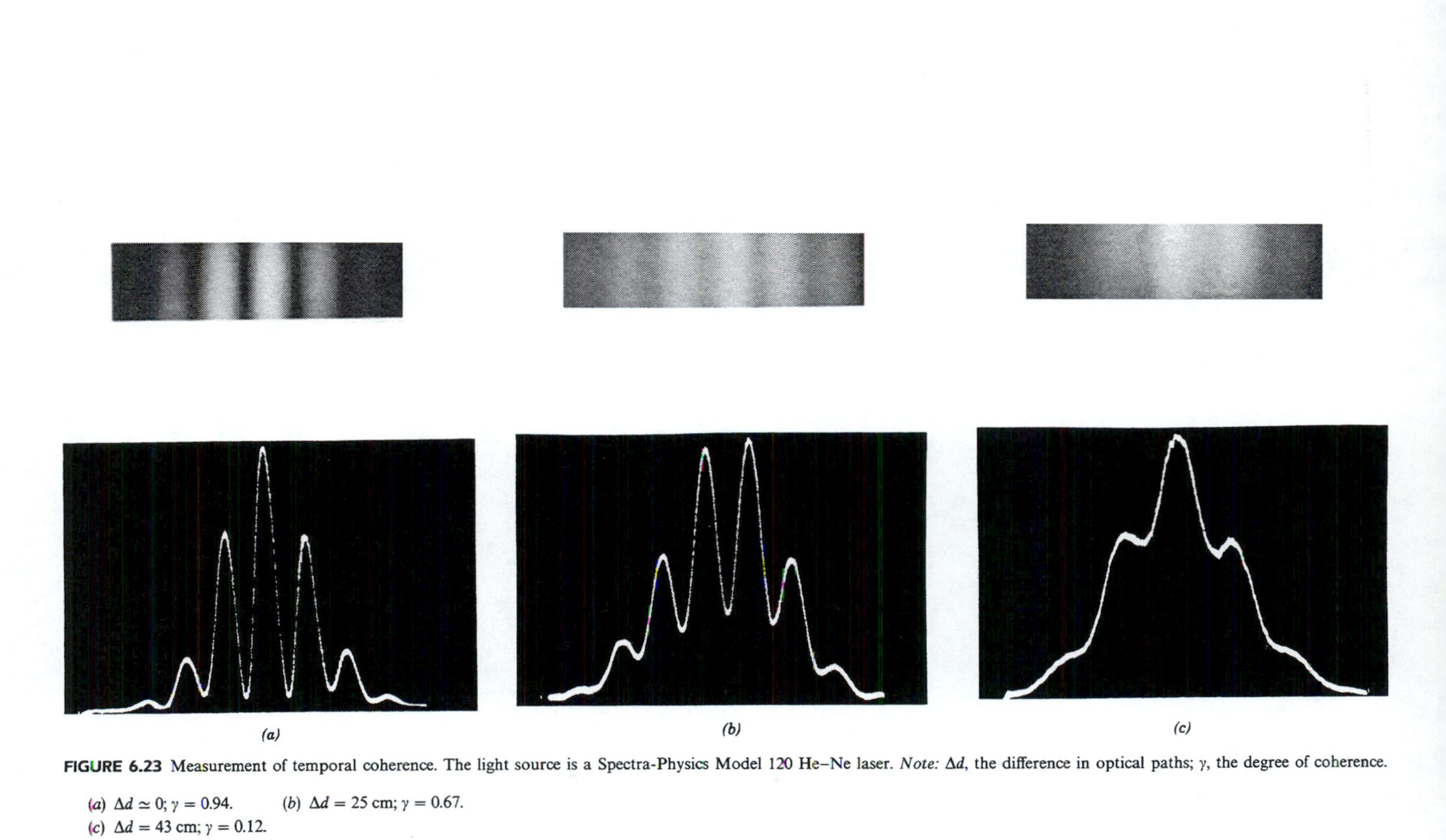

FIGURE 6.23 Measurement of temporal coherence. The light source is a Spectra-Physics Model 120 He–Ne laser. *Note:* Δd, the difference in optical paths; γ, the degree of coherence.

(*a*) $\Delta d \simeq 0$; $\gamma = 0.94$. (*b*) $\Delta d = 25$ cm; $\gamma = 0.67$.
(*c*) $\Delta d = 43$ cm; $\gamma = 0.12$.

and c we see that the visibility of the fringe patterns decreases as the difference in the path lengths (i.e., τ) of the two waves increases. Thus, we see that the coherence length of a temporal, partially coherent source can be measured with this technique.

Example 6.10

If the extended source of Example 6.9 is replaced by two point sources with a spacing b, calculate the degree of coherence for $b = 0.1$ mm and 0.5 mm, respectively. Since the two point sources can be described by two delta functions,

$$I(\xi) = \delta\left(\xi + \frac{b}{2}\right) + \delta\left(\xi - \frac{b}{2}\right),$$

by substituting into Eq. 6.47, we show that

$$\gamma_{12} = \frac{\int \left[\delta\left(\xi + \frac{b}{2}\right) + \delta\left(\xi - \frac{b}{2}\right)\right] e^{ik\xi d}\, d\xi}{\int \left[\delta\left(\xi + \frac{b}{2}\right) + \delta\left(\xi - \frac{b}{2}\right)\right] d\xi}$$

$$= \cos\left(\pi b \frac{d}{\lambda L}\right).$$

Thus, for the separation for $b = 0.1$ mm, the degree of coherence between Q_1 and Q_2 is

$$\gamma_{12}(0.1 \text{ mm}) = \cos(\pi \times 0.1 \times 1.58) = 0.879.$$

Similarly for $b = 0.5$ mm, we have

$$|\gamma_{12}(0.5 \text{ mm})| = |\cos(\pi \times 0.5 \times 1.58)| = 0.79.$$

Note that $0 \leq |\gamma_{12}| \leq 1$. ■

Example 6.11

Given a partially coherent source, a central wavelength λ that is assumed equal to 4880 Å, and a spectral width $\Delta\lambda$ of about 0.03 Å, calculate the coherence length of the light source.

Using Eq. 6.49, we can compute the coherence length:

$$\Delta\gamma = \frac{2\pi c}{\Delta\omega} = \frac{\lambda^2}{\Delta\lambda} = \frac{(4880 \times 10^{-8})^2}{0.03 \times 10^{-8}} = 7.9 \text{ cm}.$$

We note that

$$c = \frac{\omega\lambda}{2\pi} \quad \text{and} \quad \frac{\omega}{\Delta\omega} = \frac{\lambda}{\Delta\lambda}.$$ ■

REFERENCES

1. M. BORN and E. WOLF, *Principles of Optics*, second revised edition, Pergamon Press, New York, 1964.
2. J. M. STONE, *Radiation and Optics*, McGraw-Hill, New York, 1963.
3. M. J. BERAN and G. B. PARRENT, JR., *Theory of Partial Coherence*, Prentice-Hall, Englewood Cliffs, N.J., 1964.
4. F. T. S. YU, *Optical Information Processing*, Wiley-Interscience, New York, 1983.
5. P. H. VAN CITTERT, "Die Wahrscheinliche Schwingungs verteilung in einer von einer lichtquelle direkt Oden Mittels einer linse," *Physica*, Vol. *1*, 201 (1934).
6. F. ZERNIKE, "The Concept of Degree of Coherence and Its Application to Optical Problems," *Physica*, Vol. *5*, 785 (1938).
7. B. J. THOMPSON and E. WOLF, "Two-Beam Interference with Partially Coherent Light," *Journal of the Optical Society of America*, Vol. *47*, 895 (1957).
8. B. J. THOMPSON, "Illustration of the Phase Change in Two-Beam Interference with Partially Coherent Light," *Journal of the Optical Society of America*, Vol. *48*, 95 (1958).

PROBLEMS

6.1 Consider a diffraction screen that contains an open circular aperture which is 5 mm in diameter. If this screen is normally illuminated by a red light plane wave of $\lambda = 6.3 \times 10^{-5}$ cm, calculate the separation between the diffraction screen and the observation plane to achieve the far-field Fraunhofer diffraction.

6.2 In Section 6.3 we showed the Fourier transform property of Fraunhofer diffraction. Compute the complex light distribution of a circular aperture of diameter D under monochromatic plane-wave illumination. You may use the following identities,

$$J_0(z) = \frac{1}{2\pi} \int_0^{2\pi} \exp[-iz\cos(\theta - \phi)]\, d\theta,$$

and

$$zJ_1(z) = \int_0^z \alpha J_0(\alpha)\, d\alpha,$$

where J_0 and J_1 are the zero-order and first-order Bessel functions of the first kind, respectively.

6.3 Assume that the expression of the complex wave field of an open aperture caused by Fraunhofer diffraction is given as $g(x, y)$. If two identical apertures with separation d are located in the diffraction screen, as shown in Figure 6.24, calculate the resultant intensity distribution, in terms of $g(x, y)$, at the observation screen. (*Hint:* Use the Fourier transform property of Fraunhofer diffraction.)

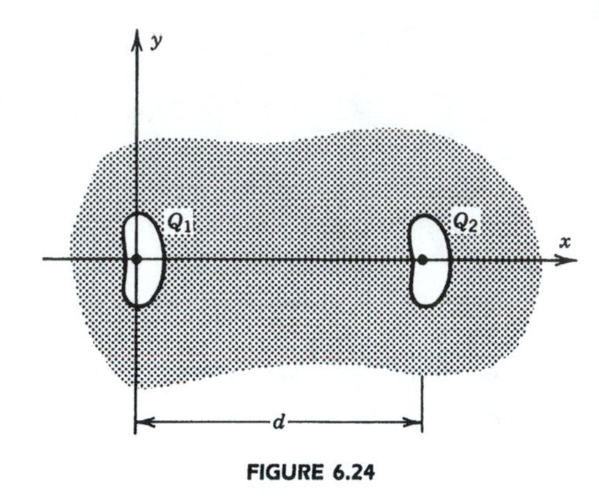

FIGURE 6.24

6.4 An image transparency, with an amplitude transmittance function $f(x, y)$, is normally illuminated by a monochromatic plane wave of λ, as shown in Figure 6.25.

(a) Draw an analog system diagram to evaluate the complex light field over the α, β plane.

(b) Calculate the corresponding light distribution, $g(\alpha, \beta)$.

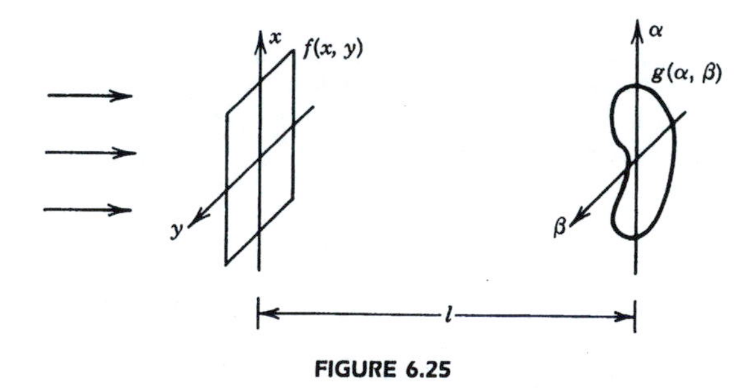

FIGURE 6.25

6.5 Use the image transparency of Problem 6.4, which is now illuminated by a monochromatic light field $u(x, y)$ instead of a plane wave.

(a) Draw an analog system diagram to represent this problem.

(b) Evaluate the corresponding complex light distribution over the α, β plane.

6.6 Use the result obtained in Problem 6.4b, that is, $g(\alpha, \beta)$.

(a) Draw an analog system diagram to evaluate the light field that emerges from the transparency.

(b) Show that the complex light field that emerges from the transparency is indeed equal to $f(x, y)$.

6.7 Using the optical imaging system shown in Figure 6.26,

(a) Draw an analog system diagram to represent this optical system.

(b) If a monochromatic point source is located at the front focal length of the optical system, calculate the complex light field at the output plane.

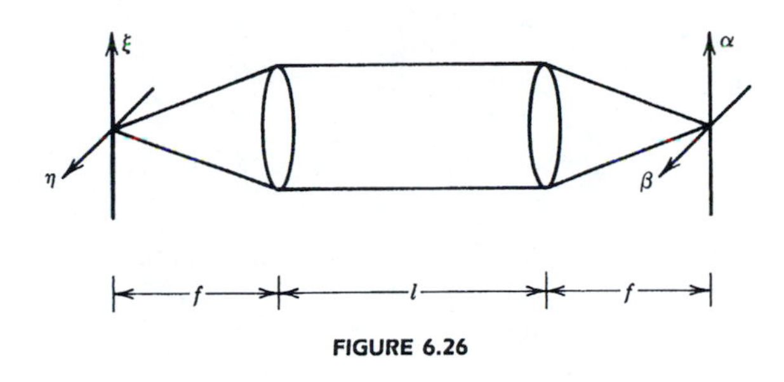

FIGURE 6.26

Remember that the phase transform of a positive lens is

$$T(x, y) = C \exp\left[-i \frac{k}{2f} (x^2 + y^2) \right].$$

6.8 A diffraction screen contains five narrow slits, as shown in Figure 6.27. If a monochromatic plane wave of $\lambda = 6000$ Å is normally incident on the screen, and if the optical disturbances produced by the individual slits, at a point P one meter behind the screen, are assumed to be in phase,

(a) Determine the separations between the narrow slits.

(b) Compute the corresponding intensity at P, in terms of the irradiances under incoherent illumination.

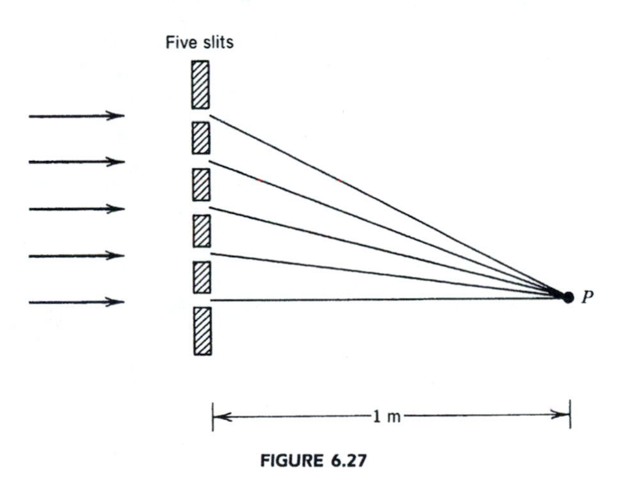

FIGURE 6.27

6.9 Referring to Eq. 6.23,

(a) Sketch the transmittance as a function of x for the Fresnel zone plate shown in Figure 6.10.

(b) Sketch the transmittance as a function of x^2.

(c) Expand the transmittance function of part b into a one-dimensional Fourier series.

6.10 The transmittance function of a Fresnel zone plate can be written in the following Fourier series expansion,

$$T(x, y) = A_0 + \sum_{n=1}^{\infty} A_n \cos\left[\frac{n\pi}{\lambda R}(x^2 + y^2)\right],$$

where A_0, A_n, and R are arbitrary constants. If this Fresnel zone plate is normally illuminated by a monochromatic plane wave of λ,

(a) Calculate the focal distances of the zone plate.

(b) Sketch the locations of the convergent focal points. Recall that $\cos\theta = \frac{1}{2}(e^{i\theta} + e^{-i\theta})$.

6.11 Assume that the transmittance function of a Fresnel zone lens is

$$T(x, y) = K_1 + K_2 \cos\left[\frac{2\pi}{\lambda}\left(\frac{x^2 + y^2}{2R} - x\sin\theta\right)\right].$$

If a monochromatic plane wave of λ is normally incident on this zone lens,

(a) Compute the focal distance of the zone lens.

(b) Sketch the location of the convergent focal point.

6.12 If the zone lens of Problem 6.11 is normally illuminated by a white-light plane wave,

(a) Show that the focal point would be smeared into rainbow colors.

(b) Sketch the location of the smeared focal point.

6.13 Two *mutually coherent* monochromatic light beams illuminate an observation screen, as shown in Figure 6.28. If the complex amplitude distributions produced by these two beams are represented by $U_1(x, y)$ and $U_2(x, y)$, compute the resultant intensity distribution $I(x, y)$.

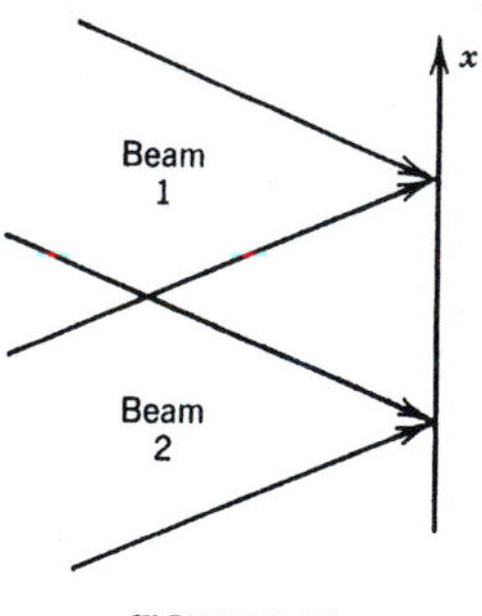

FIGURE 6.28

6.14 If two light beams such as those in Problem 6.13 are instead *mutually incoherent*,

(a) Calculate the resultant intensity distribution.

(b) Compare the results obtained in part a in Problem 6.13. State your observation.

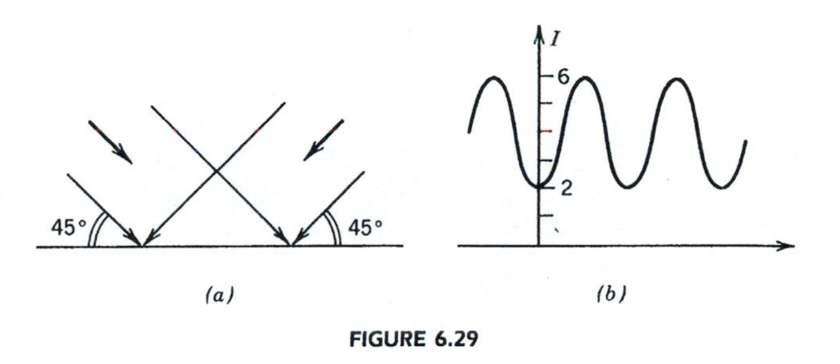

FIGURE 6.29

6.15 Two monochromatic plane waves of equal intensity illuminate an observation screen, as shown in Figure 6.29*a*. Use the fringe intensity distribution as plotted in Figure 6.29*b*.

(a) Calculate the degree of coherence between these two beams of light.

(b) If the wavelength of both of these partially coherent beams is $\lambda = 500$ nm, compute the spatial frequency of the fringes.

6.16 Refer to Young's experiment, which was discussed in Example 6.9. Assume that the light source is a point source (i.e., $\Delta s = 0$) that produces two spectral lines, $\lambda_1 = 5770$ Å and $\lambda_2 = 5790$ Å, of equal intensity.

(a) Calculate the visibility of the interference fringes at the observation plane.

(b) Sketch the corresponding interference fringes of part a.

6.17 Assuming that the spectral line width of an argon laser is $\Delta\lambda = 0.02$ Å and that its central wavelength is $\lambda = 4880$ Å, compute the coherent length of the light source.

7

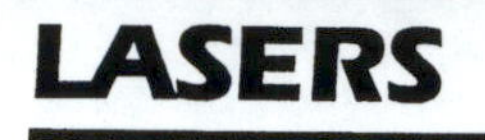

LASERS

Before the invention of the laser, advances in optical engineering were limited to instruments that did not require a light source with a high degree of coherence. This restriction precluded a vast spectrum of important applications. Indeed, most of the applications described in this book require the use of a laser source, and for those that can get by with a less coherent source, the use of the laser would provide superior results.

Why is laser light so important to optical engineering, and what makes a laser so different from other light sources? In Chapter 6 we discussed the temporal and spatial coherences that are needed for coherent optical processing and holography. Theoretically, it is possible to obtain coherent light from any light source. For example, if we have an extended polychromatic light source, such as an ordinary incandescent light bulb, we can obtain spatial coherence by letting only a small portion of the light pass through a pinhole aperture, as demonstrated in Figure 7.1. The smaller the pinhole aperture, the closer it is to being a point radiator, and the greater the degree of spatial coherence. Thus, we see that there is a direct trade-off between the degree of coherence and efficiency. Temporal coherence or monochromaticity can be achieved by filtering the light with a narrow-band interference filter, as shown in Figure 7.2. The narrower the bandwidth of the filter,

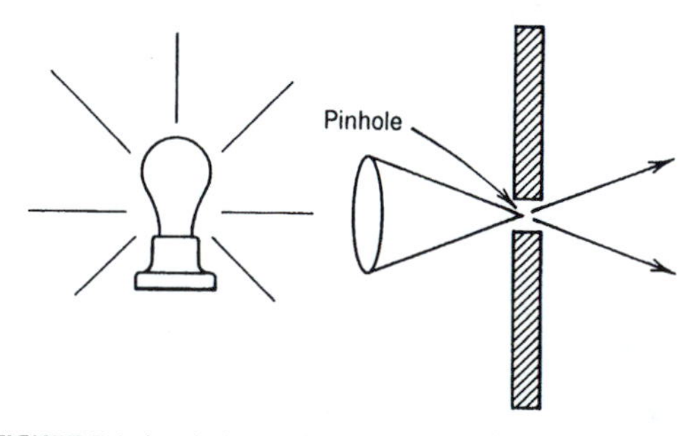

FIGURE 7.1 A polychromatic and spatially coherent light source.

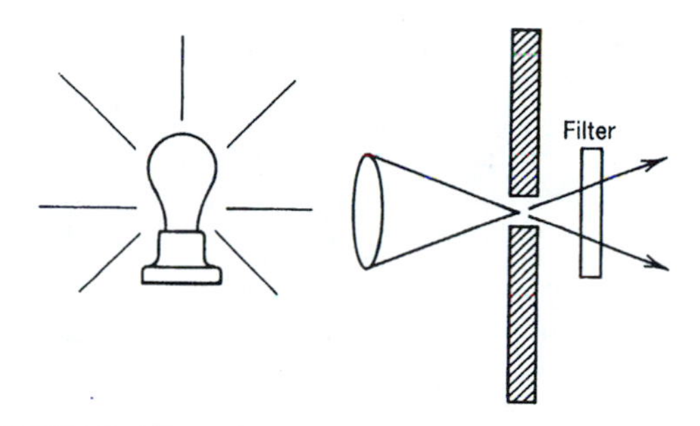

FIGURE 7.2 A "monochromatic" and spatially coherent light source.

the higher the degree of temporal coherence. Once again, however, there is a trade-off between coherence and efficiency. As we make the filter bandwidth narrower, the amount of light that passes through decreases. For the extreme and physically unrealizable situation in which the filter is infinitely narrow, the amount of light that passes through would approach zero.

Instead of using a light source containing a wide, continuous band of frequencies, we can improve the efficiency of the system by using a light source whose spectral content is concentrated around certain discrete frequencies. The most common light source of this type is the spectral arc lamp. The spectral content of a mercury arc lamp is shown in Figure 7.3. By using an interference filter to isolate one of the spectral lines, we can obtain a light source of good chromaticity that is fairly efficient. In order to achieve spatial coherence, however, we must still filter the

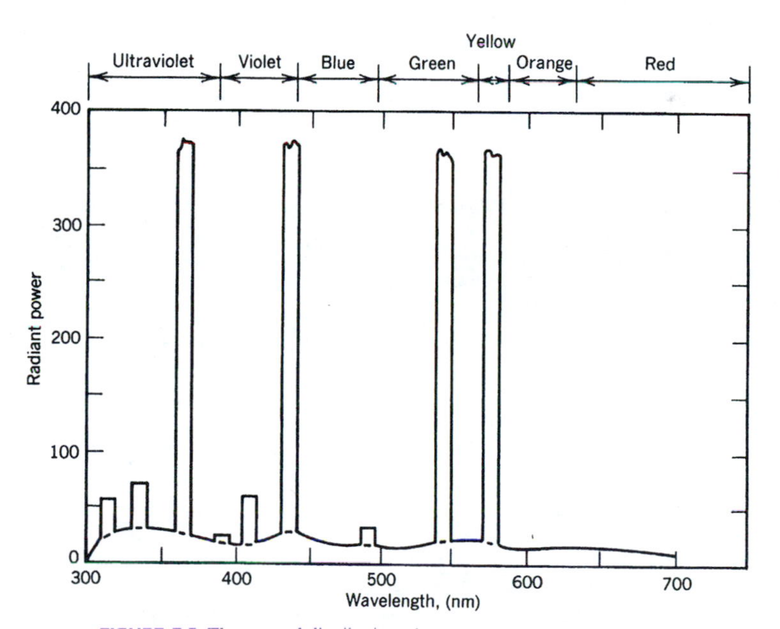

FIGURE 7.3 The spectral distribution of a commercial mercury arc lamp.

light of the arc lamp through a small pinhole, just as we did the incandescent light described earlier. Thus, in producing spatially coherent light, spectral lamps are still not very efficient. Moreover, the spectral lines are fairly wide. That is to say, the degree of monochromaticity, or temporal coherence, is not very good either, making the spectral lamp unsuitable for applications in which a very high degree of temporal coherence is required. We discuss this point more fully in a later section.

Example 7.1

As an example of how intense a laser is as a monochromatic light source, let us compare the radiation intensity (watts per square centimeter) from the sun on the surface of the earth with that associated with a small He–Ne laser.

The spectrum of the sun ranges from the ultraviolet to the infrared, which is a range of about 300 nm. The power per unit area, that is, the intensity, on the earth's surface is about 10 mW/cm^2 on the average. Therefore, the radiation intensity per unit wavelength is

$$\frac{10 \times 10^{-3}}{300} \simeq 3.3 \times 10^{-5}\ \mathrm{W/(cm^2 \cdot nm)}.$$

On the other hand, the output of the He–Ne laser is concentrated in a very narrow spectral range (<0.01/nm) in the red, and the beam diameter is about 0.1 cm. Consider a 1-mW He–Ne laser whose radiation intensity per unit wavelength is

$$\frac{1 \times 10^{-3}}{0.01 \times \pi(0.01)^2/4} \simeq 40\ \mathrm{W/(cm^2 \cdot nm)},$$

which is more than a million times that of the sun. ■

7.1 SPONTANEOUS AND STIMULATED EMISSIONS

Both arc lamps and lasers make use of the fact that atoms, ions, and molecules possess internal resonances of discrete frequencies. When the energy of an atom falls from one level to another (see Figure 7.4), it emits an amount of energy that is equal to the energy difference between the two levels. Consider a particular transition from a higher level 2 to a lower level 1, as shown in Figure 7.4. The energy emitted has a characteristic frequency ν_{21}, which is related to the energy difference between levels E_1 and E_2 by

$$\nu_{21} = \frac{E_2 - E_1}{h}, \tag{7.1}$$

where $h = 6.626 \times 10^{-34}$ J·sec is the universal constant called Planck's contant. There is, however, a big difference in the forms of emission from arc lamps and lasers. In arc lamps the atoms are lifted up to the higher energy level (level 2) by collision with the injected electrons, and then they drop back to the lower level

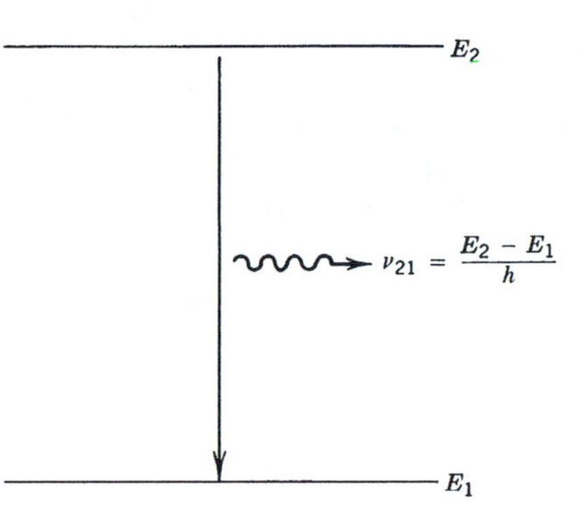

FIGURE 7.4 A particular transition in an atom or a molecule.

(level 1) spontaneously, emitting the corresponding amount of energy. This is called *spontaneous emission.* The form of emission associated with laser action, called *stimulated emission,* is fundamentally different from spontaneous emission. In spontaneous transitions the atoms can drop only from a higher energy level to a lower state, emitting a quantum of energy. Let N_2 be the number density of atoms in level 2. The rate of decay of level 2 is given by

$$\frac{dN_2}{dt} = -A_{21}N_2, \tag{7.2}$$

where A_{21} is the spontaneous transition rate to level 1 and is called the Einstein A coefficient. Suppose that initially ($t = 0$) the number density of level 2 is $N_2(0)$; then the number density at time t is given by

$$N_2(t) = N_2(0)e^{-At}. \tag{7.3}$$

Correspondingly, the intensity of light emitted from the atom is given by

$$I(t) = I(0)e^{-At}. \tag{7.4}$$

Except for some so-called quantum aspects, which we will not explore further, the emission of light from an excited atom is analogus to the radiation from a classical dipole. In fact, a rigorus quantum theory (see reference 6) shows that the Einstein A coefficient is related to the dipole moment of the atom, d, by

$$A = \frac{2n^3\omega^3d^2}{3hc^3\varepsilon}, \tag{7.5}$$

where n is the refractive index, ε is the dielectric constant, ω is the frequency, c is the velocity and h is Planck's constant. Classical radiation theory, which was discussed in Chapter 5, states that if the dipole moment is known, we can deduce that the initial radiated power, $p(0)$, is given by

$$p(0) = \frac{1}{3}\frac{\omega^4d^2}{c^3}. \tag{7.6}$$

In stimulated transitions, however, the atoms can drop from a higher to a lower level by emitting light, or they can absorb the light and jump back up. The number of transitions depends on the intensity of the electromagnetic field inducing the transition and the population density distribution of the energy levels. More quantitatively, when stimulated emission is percent, the level densities of atoms N_2 and N_1 are described by the equations

$$\frac{dN_2}{dt} = -\beta_{21}\rho(\nu)(N_2 - N_1) - AN_2 \tag{7.7}$$

and

$$\frac{dN_1}{dt} = -\beta_{12}\rho(\nu)(N_1 - N_2), \tag{7.8}$$

where $\rho(\nu)$ is the spectral energy density of the electromagnetic field ($\rho = In/c$) and has the unit of energy per unit volume per unit frequency, which is $\mathrm{J/(m^3 \cdot sec)}$. The presence of the term AN_2 on the right-hand side of Eq. 7.7 merely reflects the fact that level 2 also undergoes spontaneous emission to level 1. The coefficients β_{12} and β_{21} are known as the Einstein β coefficients for stimulated transitions $\beta_{21} = \beta_{12}$. In addition, β and A are related by

$$\beta_{21} = \frac{A_{21}c^3}{8\pi n^3 h\nu^3}. \tag{7.9}$$

It is sometimes more convenient to write $\beta_{12}\rho(\nu)$ and $\beta_{21}(\nu)$ in terms of an induced rate W_i [i.e., $W_i = \beta_{12}\rho(\nu) = \beta_{21}, \rho(\nu)$],

$$W_i = \frac{A_{21}c^3\rho(\nu)}{8\pi n^3 h\nu^3}. \tag{7.10a}$$

Note that W_i has the unit of per second, just as A_{21} does.

In many calculations it is more convenient to express the transition rate W_i in terms of the intensity, I_ν (in units of $\mathrm{W/m^2}$), of the monochromatic optical field at frequency ν and a spectral distribution function, $g(\nu)$ (in units of the inverse of frequency, i.e., in seconds). The exact form of the spectral distribution depends on the kind of mechanism causing the transition to broaden, as discussed in the next section. Therefore, Eq. 7.10a can have the alternative form

$$W_i = \frac{A_{21}c^2 I_\nu g(\nu)}{8\pi n^2 h\nu^3} \tag{7.10b}$$

for a monochromatic optical field interacting with a broadened two-level transition.

It is clear from Eqs. 7.7 and 7.8 that if the population density at level 2 is lower than that of level 1, more atoms would be induced to jump from level E_1 to E_2, absorbing a net amount of energy in the process. Usually, however, atoms are in thermal equilibrium, and level 1 is more populated than level 2. If we could invert the population density distribution , that is, make the population density at energy level E_2 greater than that of level E_1, the inversion would result in a net output of energy from the medium. We will return to this consideration in more detail in the next section.

The most significant differences between spontaneous and stimulated emission are that in spontaneous emission each atom drops back to the lower energy level independently, and the radiations emitted by the atoms are not in phase with one another. With stimulated emission the emitted field is always in phase with the coherent electromagnetic field inducing the transition. Because all the atoms are stimulated by the same coherent field, their radiations are coherent or in phase with one another and with the radiation of the inducing field.

7.2 LINE BROADENING

Atoms, especially in gaseous media, move randomly at different velocities and in different directions. The frequencies of the radiation emitted by these moving atoms are Doppler-shifted, which means that the observed frequency of the radiation emitted by an atom moving at a velocity v along the observed direction of propagation would become $\omega = \omega_0 + (v/c)\omega_0$, where ω_0 is the frequency of the field radiated by stationery atoms. Thus, even though all the atoms of the laser medium possess the same internal resonance, the observed frequencies of the radiation emitted are different because of the Doppler effect. The movements of the atoms follow certain statistical patterns, and the frequency spectrum of the radiation emitted by all the atoms has a certain characteristic line shape and width. This spectral line shape is often referred to as the Doppler-broadened line shape. The line shape indicates the power distribution of the radiated field at various frequencies, which in turn is determined by the number of atoms moving at the corresponding velocities and in the corresponding directions. Emission can be stimulated at any frequency within this line shape, and the power of the stimulated emission would be determined by the line shape. Thus, if the laser medium is stimulated by radiation at a frequency corresponding to the center of the Doppler-broadened line shape, the power of the stimulated emission would be the maximum obtained from the medium. If the radiation is of a frequency that is outside the line shape, no emission will be stimulated. For this reason, this spectral line shape is also referred to as the gain profile of the medium. In general, the gain profile follows the line shape of the materials.

The line shape, or the frequency spectrum of a particular transition, depends on the dynamics (i.e., time dependence) of the underlying mechanism. In general, the line shape can be obtained by simply performing a Fourier analysis of the dynamics. For example, in spontaneous emission, the dynamics is described by Eq. 7.3. A Fourier transform of the exponential functions gives a Lorentzian curve with width $\Delta\nu \simeq 1/2\pi\tau$. The frequency distribution takes the following form:

$$g(\nu) = \frac{\Delta\nu/2\pi}{(\nu - \nu_0)^2 + (\Delta\nu/2)^2} \tag{7.11}$$

Note that at $\nu = \nu_0$ and $g(\nu_0) = \Delta\nu^{-1}$.

Since all the atoms undergo the same spontaneous emission dynamics, the effect is classified as a *homogeneous broadening mechanism*. But the broadening of the line shape through the Doppler effect is classified as an *inhomogeneous broadening mechanism*, because different atoms "see" different Doppler effects. The derivation of the Doppler broadened line shape is more complicated than that of natural spontaneous emission broadening (see reference 4). Doppler broadening is described by a Gaussian function,

$$g(\nu) = \left(\frac{4\ln 2}{\pi}\right)^{1/2} \frac{1}{\Delta\nu_D} \exp\left[-4\ln 2\left(\frac{\nu - \nu_0}{\Delta\nu_D}\right)^2\right] \tag{7.12}$$

where $\Delta\nu_D = (2\pi K_B T/M)^{1/2}$, with K_B the Boltzmann's constant, T the absolute temperature, and M the mass of the atom.

Example 7.2

The energy levels of most atoms and molecules are expressed in terms of electron volts (eV). Calculate the wavelength and frequency of the light emitted in a atomic transition involving two energy levels $E_2 = 4$ eV and $E_1 = 1$ eV.

Recall that one electron volt is the energy needed to raise the potential of an electron by one volt. Since the charge of the electron is 1.6×10^{-19} coulomb, 1 eV corresponds to

$$1.6 \times 10^{-19}\ \mathrm{C \cdot V} = 1.6 \times 10^{-19}\ \mathrm{J}.$$

Therefore, $E_2 - E_1 = 3\ \mathrm{eV} = 4.8 \times 10^{-19}$ J.

The frequency of the emission ν_{21} is thus given by

$$\nu_{21} = \frac{E_2 - E_1}{h} = \frac{4.8 \times 10^{-19}\ \mathrm{J}}{8.6 \times 10^{-34}\ \mathrm{J \cdot sec.}}$$
$$= 7.2 \times 10^{14}/\mathrm{sec.}$$

The wavelength λ is therefore

$$\lambda = \frac{c}{\nu} = \frac{3 \times 10^{10}\ \mathrm{cm/sec}}{7.2 \times 10^{14}/\mathrm{sec}} = 0.41 \times 10^{-4}\ \mathrm{cm} = 410\ \mathrm{nm} = 4100\ \text{Å}$$ ■

Example 7.3

Given that the spontaneous emission rate A of a particular laser transition ($\lambda = 514.5$ nm) is 10^{-9} sec, calculate the Einstein β coefficient. What optical spectral intensity is required for the stimulated emission rate to equal the spontaneous rate?

From Eq. 7.9, we have

$$\beta_{21} = \frac{A_{21}c^3}{8\pi n^3 h\nu^3} = \frac{A_{21}}{8\pi n^3 h}\lambda^3 = \frac{10^{-9}(574.5 \times 10^{-9})^3}{8 \times \pi \times 1 \times (6.62 \times 10^{-34})}$$
$$= 5.42 \times 10^6\ \mathrm{m^3/(W \cdot sec^3)}.$$

From Eq. 7.10, if $W_i = A_{21}$, we have

$$1 = \frac{c^2 I}{8\pi n^2 h\nu^3} g(\nu).$$

Since $g(\nu) \simeq 1/\Delta\nu = \pi\tau = 10^{-9}\pi$ sec,

$$I = \frac{8\pi n^2 h\nu^3}{c^2 g(\nu)} = 8.4 \times 10^3 \text{ W/m}^2.$$

Note that I is the optical intensity. ■

7.3 PRINCIPLES AND REQUIREMENTS OF LASER ACTION

The arrangement of a laser is schematically depicted in Figure 7.5. The laser consists of a laser medium placed between two highly reflective mirrors which together act as an optical resonator. The atoms making up the lasing medium could be in the form of a gas, a solid, a crystal, a flowing dye solution, or a plasma, and the type of excitation mechanism employed varies accordingly. Similarly, the energy levels involved in the transition and stimulated emissions of light also varies. Nevertheless, the lasers can be roughly classified into two types.

1. *A Three-Level System.* As shown in Figure 7.6, the atom is excited from the ground state to level 3, which rapidly decays to level 2. The transition involved in the lasing process is from 2 to 1. The system is classified as a three-level one if level 1 is close to the ground state. By close we mean that

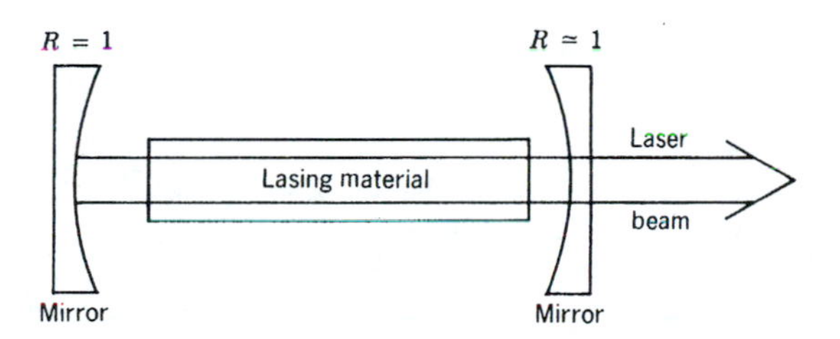

FIGURE 7.5 Schematic of a typical laser.

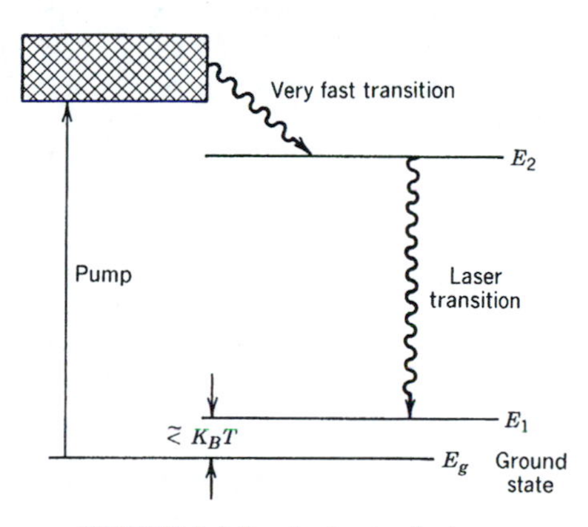

FIGURE 7.6 A three-level system for lasers.

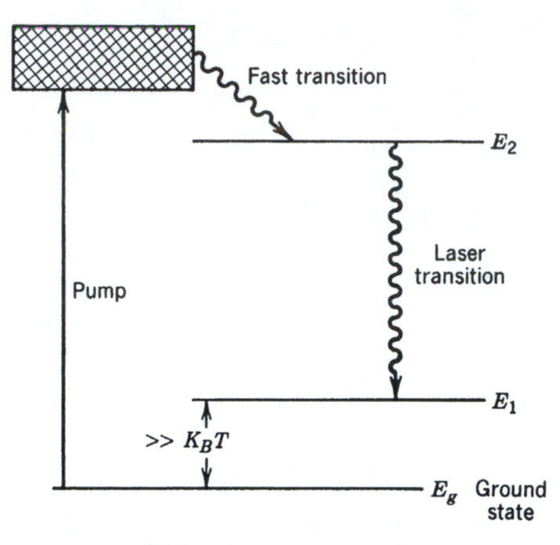

FIGURE 7.7 A four-level system for lasers.

the energy separation from level 1 to the ground is less than or comparable to the thermal energy K_BT. Here K_B is Boltzmann's constant, and T is the absolute temperature. Constant $K_B = 1.38 \times 10^{-23}$ Joule/°K, and at room temperature ($T = 300$ °K), $K_BT = 0.026$ eV. As a result of this closeness to the ground state, level 1 will be well populated. For a population inversion to take place between levels 2 and 1, a substantial number of atoms must be excited to level 2.

2. *A Four-Level System.* The level scheme for a four-level system is depicted in Figure 7.7. It is similar to the three-level system except now level 1 is energetically much removed from the ground state. Consequently, the atoms remain at the ground state, so the population of level 1 is practically zero. Any number of atoms excited to level 2 from the ground will therefore immediately create a population inversion between levels 2 and 1. In general, therefore, more pumping is needed to achieve population inversion in a three-level system than in a four-level system.

Now let us consider what happens when a population inversion has been achieved in the lasing medium. For the sake of simplicity, we will assume that the frequency spread of the transition is $\Delta\nu$ and that the transition frequency between levels 2 and 1 is ν. Consider a monochromatic plane wave of frequency ν traversing the medium as depicted in Figure 7.8. According to Eqs. 7.7 through 7.10, there will be an induced transition from level 2 of the amount $(N_2 - N_1)W_i/\Delta\nu$ per unit time per unit volume. The net electromagnetic *power* generated per unit volume

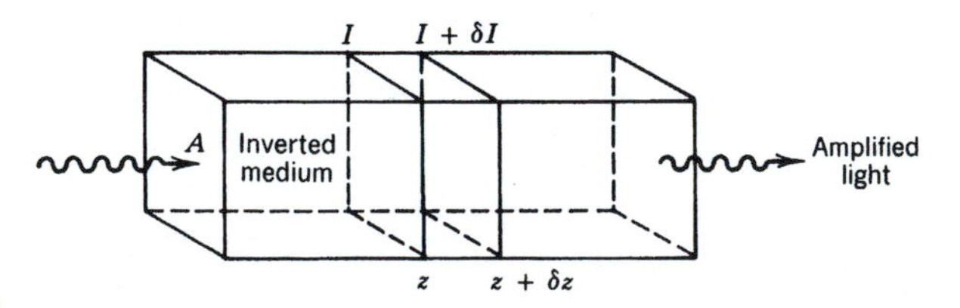

FIGURE 7.8 Light amplification in an inverted medium.

(P/volume) is therefore $(N_2 - N_1)W_i h\nu$. This radiation is added coherently to the incident wave. If we let the cross-sectional area of the material be A, we then have, between z and $z + dz$, a volume $\delta V = A\,\delta z$. The gain in the radiation power in this volume is

$$\frac{\delta P}{\delta V} = (N_2 - N_1)W_i h\nu,$$

or

$$\frac{\delta P}{A\,\delta z} = (N_2 - N_1)W_i h\nu. \tag{7.13}$$

Since $\delta P/A = \delta I$, and, in the limit of small δz and δI,

$$\frac{\delta I}{\delta z} \rightarrow \frac{dI}{dz},$$

we have

$$\frac{dI_\nu}{dz} = (N_2 - N_1)W_i h\nu \tag{7.14}$$

as the equation describing the rate of change of the optical intensity with the extent of its propagation into the medium. Using Eq. 7.10b, we get

$$\begin{aligned}\frac{dI_\nu}{dz} &= (N_2 - N_1)\frac{c^2 A_{21}}{8\pi n^2 \nu^2\,\Delta\nu} I_\nu g(\nu) \\ &= gI,\end{aligned} \tag{7.15}$$

where

$$g = \frac{(N_2 - N_1)c^2 A_{21} g(\nu)}{8\pi n^2 \nu^2}$$

is the gain.

Equation 7.15 does not account for the fact that the electromagnetic field also suffers losses. These losses come, for example, from spontaneous emissions from level 2, which is randomly radiated. Other possible losses come from scatterings caused by impurities, imperfections, and various other mechanisms. We can roughly lump all these losses together and represent them by $-\alpha I$. Taking both gain and loss into account, the equation describing the intensity as a function of its propagation through the medium is

$$\frac{dI}{dz} = (g - \alpha)I = \gamma I. \tag{7.16}$$

The solution of this equation is

$$I = I(0)e^{\gamma z}, \tag{7.17}$$

where $I(0)$ is the input intensity at the entrance face of the medium.

The preceding exercise shows how light amplification is achieved when there is a population inversion in the medium. The light intensity will grow exponentially whenever the gain exceeds the loss. *A laser is obtained when this amplification process is repeated as the light is bouncing between the mirrors. Hence, an initially small amount of light traveling in the axial direction of the cavity can gain several orders of magnitude and become a very intense collimated light beam. Part of this beam is transmitted through the front mirror of the laser cavity.*

Of course, the amplification cannot last forever, for every time a stimulated transition occurs, the population inversion decreases correspondingly. Obviously, amplification ceases when the gain no longer exceeds the loss. Therefore, the pumping process must be repeated. As a result, we have either a continuous-wave laser (under steady, continuous pumping) or a pulsed laser (with the pumping process pulsed, repetitive, or single-shot).

So far we have talked about the pumping mechanism without going into the details of the process. Pumping is nevertheless a very important part of making a laser, since it is the mechanism by which the externally supplied energy is converted into the energy of the excited lasing medium, which in turn generates the light (laser). The efficiency of the pumping mechanism varies, depending on the type, as well as on design of the optical cavity. Most solid-state lasers employ a flash lamp together with a highly reflective cavity, which directs the light from the flash lamp onto the laser rod. Since the output of the flash lamp is quite a bit wider than the laser line, considerable energy is wasted. The pumping mechanism in a gaseous laser (e.g., He–Ne or argon ion) consists of the electron collisions that occur in the plasma discharge. Optical pumping of one laser by second laser is another widely employed method.

Example 7.4

Let us calculate the exponential gain factor experienced by a monochromatic beam of light (at resonance with the atomic line center at $\nu = \nu_0$ or $\Delta\nu = 0$) propagating in an inverted medium that has the following characteristics:

spontaneous emission rate $A = 10^6$/sec
inversion $N_2 - N_1 \simeq 10^{13}$ atoms/cm^3
$n^2 \simeq 3$
$\nu \simeq 5 \times 10^{14}$ Hz
$\Delta\nu \simeq 2 \times 10^{11}$ Hz.

From Eq. 7.15 the exponential gain factor is

$$g = \frac{(N_2 - N_1)c^2 A_{21}}{8\pi n^2 \nu^2 \Delta\nu} \simeq \frac{10^{15} \times (3 \times 10^{10})^2 \times 10^6}{8\pi \times 3 \times (5 \times 10^{14})^2 \times 10^{11}}$$

$$\simeq 0.05 \text{ cm}^{-1}.$$

This calculation is of course based on the assumption that the incident light is in resonance with the atomic transition, and that the gain factor for frequencies slightly off the line center will be smaller. Still, this calculation allows us to make a quick estimate of the magnitude of the gain in comparison with losses. It is obvious from the preceding exercise that the gain factor is highly dependent on

$N_2 - N_1$, as well as on $\Delta\nu$. If $\Delta\nu$ is large, fewer of the atoms are stimulated by the incoming light, which is monochromatic and at line center. ■

■ Example 7.5

Calculate the population inversion needed in an He–Ne gas discharge to achieve a gain factor exceeding 5 percent per unit length. The characteristics of the discharge are

$$\lambda = 6.328 \times 10^{-5} \text{ cm}$$

$$A = 10^7$$

$$g(\nu_0) \simeq \Delta\nu \simeq 10^9.$$

From Eq. 7.15, we have

$$g = 0.05 = \frac{(N_2 - N_1)c^2 A_{21}}{8\pi n^2 \nu^2 \Delta\nu}$$

$$N_2 - N_1 = \frac{0.05 \times 8\pi \times n^2 \nu^2 \Delta\nu}{c^2 A_{21}}$$

$$\simeq 10^{11} \text{ atoms/cm}^3.$$ ■

■ Example 7.6

Consider the situation in which the number of excited atoms is insufficient, so that $N_1 > N_2$. These atoms can be stimulated to make more transitions upward than downward, and we will have resonant absorption of the light. For the same He–Ne discharge discussed in Example 7.5, when $N_1 - N_2 = 10^{11}$ atoms/cm^3, calculate the percentage intensity that the light has lost after traveling a distance of 10 cm.

Since $N_1 - N_2 = 10^{11}$, then $N_2 - N_1 = -10^{11}$. We therefore have a *negative* gain factor, that is, an absorption loss of

$$g = \frac{10^{11} c^2 A_{21}}{8\pi n^2 \nu^2 \Delta\nu} = \frac{10^{11} \times 10^7 \times (6.328 \times 10^{-5})^2}{8\pi \times 3 \times 10^9}$$

$$= -0.053 \text{ cm}^{-1}.$$

After the light travels 10 cm into the gas, its intensity is

$$I(z) = I(0)e^{-0.053 \times 10},$$

so the

$$\text{Percentage of loss} = \frac{I(0) - I(z)}{I(0)} \times 100 \text{ percent}$$

$$= (1 - e^{-0.53}) \times 100 \text{ percent}$$

$$= 54 \text{ percent}$$ ■

7.4 TYPES OF LASERS AND CHARACTERISTICS

By now several thousand laser transitions in various materials have been identified. Since it would be futile to review each and every one of them, we will briefly describe some commonly used lasers and refer the reader to readily available commercial pamphlets from laser vendors. By far the most commonly used laser is the He–Ne laser, with its familiar red beam. The laser medium is a mixture of helium and neon gases. An electrical discharge, in the form of direct current or radio-frequency current, is used to excite the medium to a higher energy level. The pumping action takes place in a complex and indirect manner. First, the He atoms are excited by the discharge to the 2^1s and 2^3s (two of the excited levels) levels. These two energy levels happen to be very close to the 3*s* and 2*s* levels of the Ne atoms. When the excited He atoms collide with the Ne atoms, energy is exchanged, pumping the Ne atoms to the respective energy levels. The atoms at the Ne 3*s* level eventually drop down to the 2*p* level as a result of stimulated emission, and emit light at a wavelength of 6328 Å. The atoms at the 2*s* level, on the other hand, drop to the 2*p* level by emitting light at 1150 Å. However, the atoms at the 3*s* level may instead drop down to the 3*p* level, by emitting light at 3390 Å, as shown in Figure 7.9. In fact, stimulated emission tends to occur between the 3*s* and 3*p* levels. Thus, laser action would normally take place at 3390 Å instead of at the desired 6328-Å line. This problem can be overcome by attenuating the 3390-Å radiation in a way that would not affect the 6328-Å emission, which means utilizing glass elements that strongly absorb the 3390 Å infrared radiations.

The argon laser is another very popular laser system. It can provide a coherent beam whose optical power is many times greater than that achievable with the He–Ne laser. And although only one of the laser lines emitted by the He–Ne laser is within the visible region, as many as ten lines, ranging from a light green to a deep blue, can be obtained from the Ar^+ laser. The argon laser can operate with pure argon gas or with a mixture of He and Ar gases. The gas medium is excited by highly energized electrons through an electrical discharge of either direct current or radio-frequency current. The argon atoms become ionized, and the ions are excited until they emit several lines of intense radiation. The transitions of the Ar^+ laser are shown in Figure 7.10.

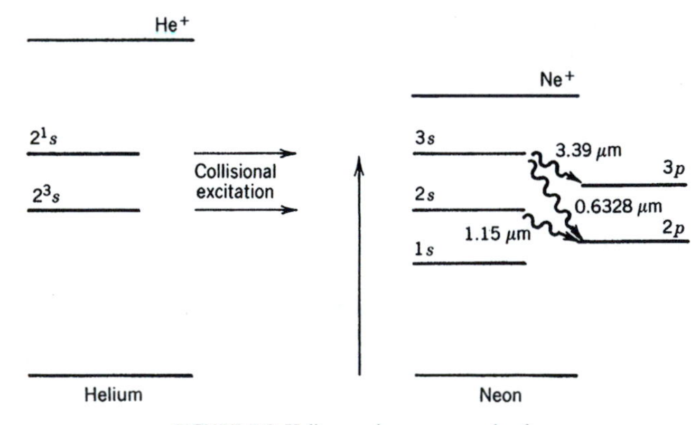

FIGURE 7.9 Helium and neon energy levels.

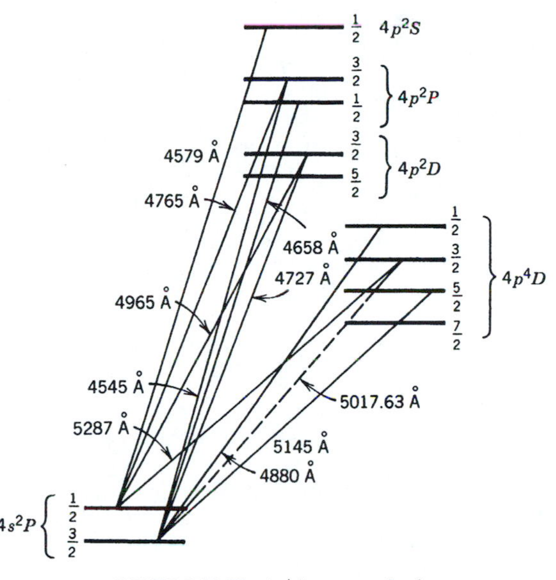

FIGURE 7.10 The Ar^+ ion energy levels.

The He–Ne and Ar^+ lasers are usually operated in the continuous-wave (CW) mode. A good example of a pulsed laser is the Nd–YAG laser, which is also often classified as a solid-state laser, because the lasing medium is in the form of a solid or crystalline rod. It is made up of yttrium–aluminum–garnet, with trivalent neodymium ions present as impurities. The lasing transition involved corresponds to a wavelength of 1.06 μm, which is in the infrared. Typically, the pulsed operation of these lasers yields rather high-power (> MW) pulses of nanosecond durations. The Nd–YAG lasers can also be continuously pumped to operate in the CW mode, which provides excellent high-power (> 10-W) infrared light sources.

A unique, and perhaps the most important type of laser in terms of optoelectronic applications, is the semiconductor laser. It is unique because of its dimension ($\simeq$mm × mm × μm), and its natural integration capabilities with microelectronic circuitry. Furthermore, because its light amplification by the process of stimulated emission is not exactly in the form we have been describing, it deserves some attention here.

A semiconductor laser uses the special properties of the transition region at the junction of a *p*-type semiconductor in contact with an *n*-type semiconductor. In solid-state materials, because of the extensive interactions of energy between atoms (or molecules), the energy levels form bands. The energy band diagrams for the *n*- and *p*-type semiconductors are depicted in Figures 7.11*a* and *b*, respectively. The energy gap between the valence and the conduction bands is designated by E_g and is measured in electron volts, eV_g. The Fermi level, E_f, is, roughly speaking, the level that divides the occupied from the unoccupied levels. The designation of *p*- and *n*-type semiconductors refers to the dopants, the impurities added usually in minute quantities to a pure substance to alter its properties. In the *p*-type semiconductor the dopants are such that there is an excess of conducting positively

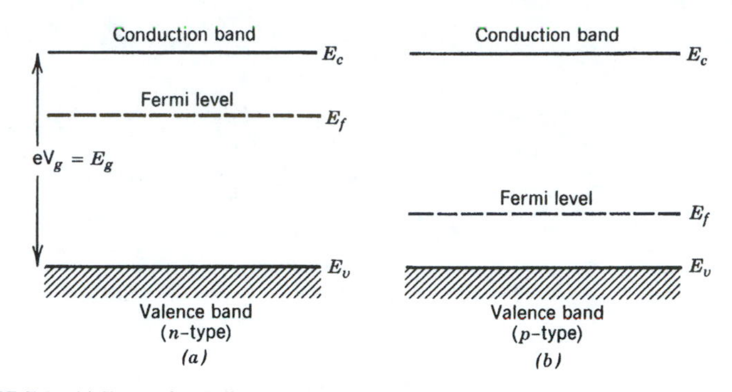

FIGURE 7.11 (a) Energy band diagram of an n-type semiconductor. (b) Energy band diagram of a p-type semiconductor.

charged sites, the so-called holes, whereas in the n-type semiconductor there is an excess of electrons.

In a p-n junction, as shown in Figure 7.12, the energy levels readjust in accordance with thermodynamics so that the E_f band is the same throughout the junction. The valence band E_v and the conduction band E_c of the p-type semiconductor are higher than the corresponding bands of the n-type semiconductor.

If we apply a positive voltage on the p side (forward biasing), the electrons (e^-) in the n side, having been attracted by the positive voltage, will cross into the junction region. There they recombine with the holes, which are attracted toward the n side. This process will continue as long as the external circuit is on, since the e^- and the holes that have recombined, and thus are depleted, are being continuously replenished.

When the electrons and the holes recombine, they emit energy in the form of radiation. The junction transition region in which this takes place is therefore a source of radiation and may be viewed as equivalent to the E_2 and E_1 transition levels we described earlier. The frequency of light emitted from the recombination can be expressed simply by

$$\nu = \frac{E_2 - E_1}{h} = \frac{eV_g}{h}. \tag{7.18}$$

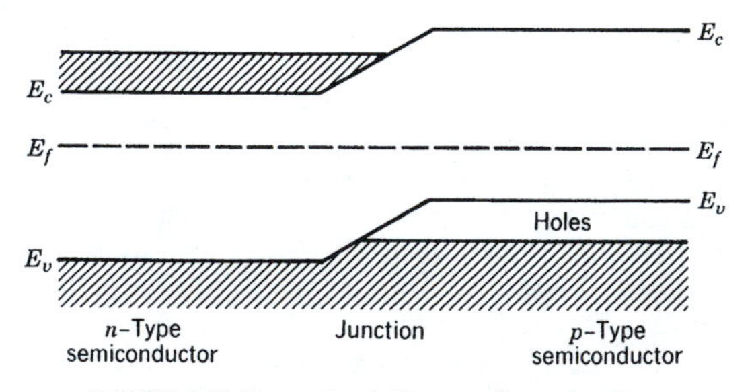

FIGURE 7.12 Energy band diagram of a p-n junction.

To obtain stimulated emission and amplification from this region, we need to create the equivalence of population inversion, which in the present case means we need to have in the region *simultaneously* a high density of electron and a high density of holes.

To achieve this, we use heavily doped *p-n* junctions. Figure 7.13*a* shows the resultant energy level, and Figure 7.13*b* shows what happens under a positive bias on the *p* side. There is a transition region with high concentration of e^- and hole. This region serves as an inverted medium, and it amplifies the radiation emitted within it through to the e^-–hole recombination.

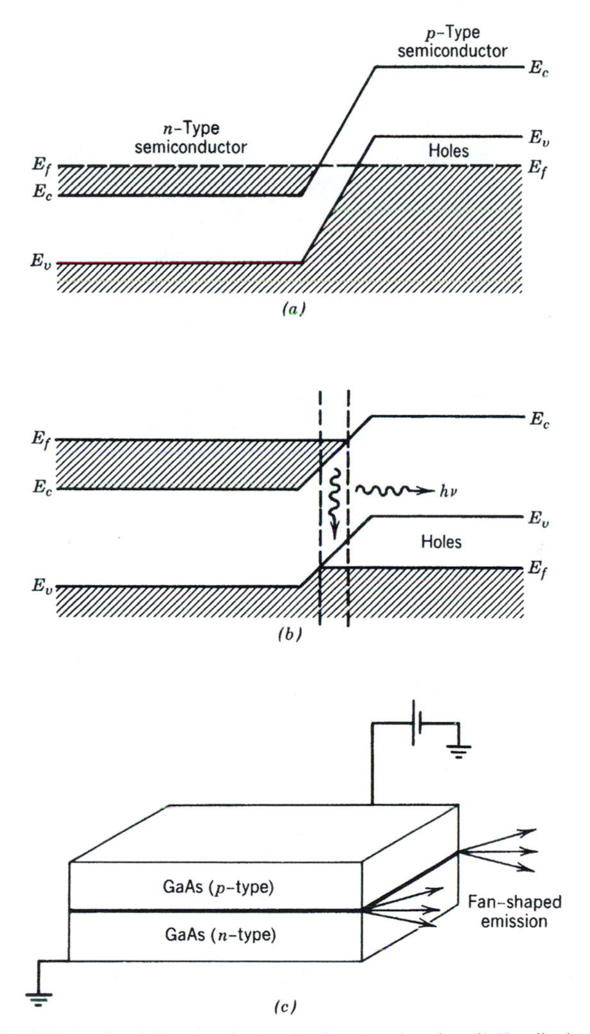

FIGURE 7.13 (*a*) Energy band diagram of a heavily doped *p-n* junction. (*b*) Heavily doped *p-n* junction under positive bias on the *p* side. (*c*) Schematic of a *p-n* junction laser made of the GaAs semiconductor.

The tiny, planar dimensions of the semiconductor laser cause a unique, but undesirable output pattern. For example, Figure 7.13*c* is a schematic of a *p-n* junction made of gallium arsinide (GaAs). The shaded area is the transition region where the lasing action takes place. This region is about one or two microns thick, whereas the dimensions of the GaAs semiconductors are much larger ($\leqq$mm). As a result, the emission is squeezed into a very thin plane from which it exits in a fan shape.

Example 7.7

A semiconductor laser is made up of a simple *p-n* junction. If the band gap of the material is 1.316 eV, what is the wavelength of the laser emission from the junction?

Given that $eV_g = 1.316 \text{ eV} = 1.316 \times 1.6 \times 10^{-19}$ J, we have

$$\frac{hc}{\lambda} = 2.11 \times 10^{-19} \text{ J}$$

or

$$\lambda = 837 \text{ nm}.$$

■

Example 7.8

The divergence of the beam of a semiconductor laser is very large, as we have just mentioned in the text. In this example we want to estimate the diffraction angle of the exiting laser beam.

Typically, the transition region has a width (t) of about 3 μm. Let the wavelength of the laser be $\lambda = 837$ nm. Light of this wavelength passing through a slit of width t (the other dimension is much larger than t) will be diffracted by an angle θ given by

$$\sin\theta \simeq \frac{\lambda}{t}.$$

Using $\lambda = 837$ nm and $t = 3$ μm, we get

$$\sin\theta \simeq \frac{837 \times 10^{-9}}{3 \times 10^{-6}} = 0.279$$

or

$$\theta = 16°.$$

■

7.5 LONGITUDINAL AND TRANSVERSE MODES

The optical oscillator formed by two mirrors behaves very much like a string stretched between two stationary points—it oscillates at only certain discrete resonant frequencies. The length of the resonant cavity is always an integer number of half-wavelengths of the resonant frequencies, that is, the resonant frequencies occur at

$$\nu_n = \frac{nc}{2l}, \tag{7.19}$$

when n is a positive integer, c is the velocity of light in a vacuum, and l is the length of the resonant cavity. A very large number of discrete frequencies (or longitudinal modes) are therefore allowed to oscillate within a given cavity length. Recall, however, that emission can be stimulated only within the Doppler-broadened line shape or gain profile of the laser medium. Obviously, no oscillation can take place at a frequency outside this bandwidth, since no radiation is emitted at that frequency. And since oscillations can occur only at certain discrete frequencies corresponding to the longitudinal modes of the cavity, stimulated emission can be sustained only at these frequencies. The resulting spectrum of the emitted light field is a product of these two effects, as illustrated in Figure 7.14.

The emissions at each of the longitudinal modes still suffer from all the homogeneous line-broadening effects, such as natural broadening, caused by the finite lifetime, and collisions. In most instances, however, the line widths resulting from

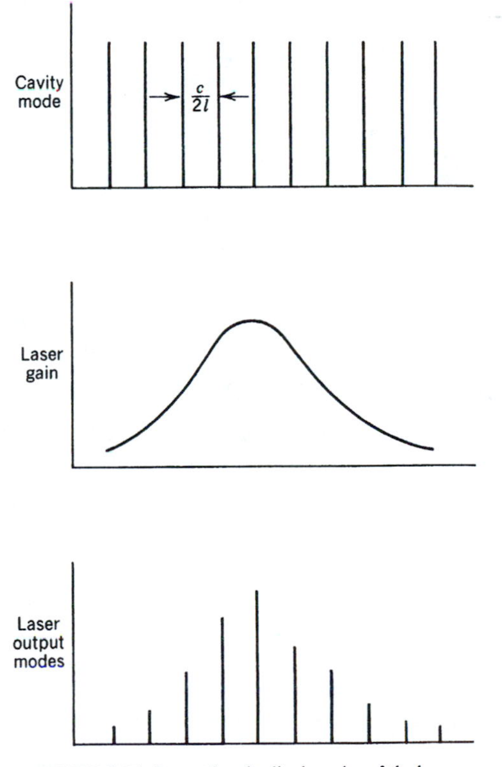

FIGURE 7.14 Output longitudinal modes of the laser.

these homogeneous broadening effects are significantly narrower than those produced by inhomogeneous Doppler broadening. Thus, by isolating the emission of a single longitudinal mode, we can obtain an extremely high degree of monochromaticity. Various dispersive devices and etalons can be combined to achieve a single-mode laser. The current state-of-the-art high-resolution technique is capable of producing a single-frequency laser with a line width on the order of one kilohertz, which is about 10^{12} times smaller than the center optical frequency ($\simeq 10^{15}$). In this instance the laser is for all intents and purposes a monochromatic light source.

The configuration of the optical cavity also determines the transverse modes of the laser output, that is, how the output beam shape looks depends on the type of cavity used. This is simply a natural consequence of electromagnetic waves confined in a cavity. In general, the allowed modes in mirrored cavities are designated as TEM_{xy} (transverse electromagnetic) modes where x and y are integers. The most common form is TEM_{00}, wherein the beam is circular with a radial intensity profile given by a Gaussian function. The beam is called a *Gaussian beam* (see Example 7.10). The light beam of a laser running on several transverse modes would resemble a plane wave passing through a circular aperture. As the beam propagates, far-field diffraction changes the profile of the beam and causes it to diverge. In addition, except for the TEM_{00} mode, the phases of all the transverse modes fluctuate across the beam diameter. If the laser is allowed to oscillate in several transverse modes, we will not have a uniphase wavefront (plane wave). It is therefore desirable to operate the laser with only the lowest-order mode, the TEM_{00} mode or Gaussian beam, as just described. This transverse mode has a Gaussian amplitude profile with the amplitude trailing off gradually at the edge. Such an amplitude profile will not suffer the edge diffraction effect and remains uniphase with very little divergence. We can look at this phenomenon another way. Remember that in Chapter 6 we showed that at the far field the diffraction pattern is equal to the Fourier transform of the amplitude distribution at the input plane. For the TEM_{00} mode, the amplitude distribution is a Gaussian function. The Fourier transform of a Gaussian function is again a Gaussian function. Thus, we see that the amplitude profile of the light beam does not change as it propagates if the laser oscillates only at the TEM_{00} mode. To achieve a single transverse-mode operation, we can place a small, circular aperture slightly larger than the spot size of the TEM_{00} mode in the middle of the laser cavity. The aperture will have no effect on the propagation of the TEM_{00} mode, but it will filter out or attenuate all other modes, so that only the TEM_{00} mode is able to oscillate. The same result can also be achieved by making the bore diameter of the laser tube just slightly larger than the spot size of the TEM_{00} mode.

7.5.1 Coherence Length

Earlier, we defined the line width of the emission spectrum as the frequency separation between the two points at which the intensities are half of the peak value. In describing the emission by laser devices, we commonly express the frequency spread in terms of the difference in wavelengths. The relation between wavelength (λ) and frequency (ν) can be written as

$$c = \nu\lambda, \tag{7.20}$$

where c is the velocity of light. The line width expressed in terms of wavelength $\Delta\lambda$ can therefore be obtained from the frequency bandwidth,

$$\Delta\lambda = \lambda - \frac{c}{c/\lambda + \Delta\nu}, \tag{7.21}$$

where

$$\Delta\nu = \frac{\Delta\lambda c}{\lambda^2 - \Delta\lambda\lambda} \tag{7.22}$$

$$\simeq \frac{\Delta\lambda c}{\lambda^2}. \tag{7.23}$$

We can also describe the degree of monochromaticity of a light beam in a third and very useful way—in terms of its coherence length. A measure of the coherence length is the distance traveled by the light beam before the waves with wavelengths λ and $\lambda + \Delta\lambda$ begin to differ by a half-length. Thus, if we let x be the number of cycles the waves have to oscillate before they differ by a half-wavelength, then

$$\lambda x + \frac{\lambda}{2} = (\lambda + \Delta\lambda)x \tag{7.24}$$

and

$$x = \frac{\lambda}{2\Delta\lambda}. \tag{7.25}$$

The coherence length l_c is therefore equal to

$$l_c = \lambda x = \frac{\lambda^2}{2\Delta\lambda} \simeq \frac{c}{2\Delta\nu}. \tag{7.26}$$

The coherence length is a very useful measure of temporal coherence because it tells how far apart two points along the light beam can be and still remain coherent with each other. In interferometry and holography, the coherence length is the maximum path difference that can be tolerated between the two interfering beams. For example, for the Michelson interferometer shown in Figure 7.15, the path difference $l_1 - l_2$ should be set less than the coherence length of the light source to assure coherence between the interfering beams.

Now let us compare the coherence length obtainable from an arc lamp source with that of a laser. Typically, the spectral line widths emitted by an arc lamp are in the neighborhood of 100 Å. If we take the 5461-Å line of a mercury arc lamp, then for a line width of 100 Å the coherence length is only 0.015 mm. By using a very narrow band interference filter to reduce the light intensity, we can decrease the spectral line width to 10 Å and improve the coherence length to 0.15 mm. However, if we let an He–Ne laser oscillate at all longitudinal modes, the total line width is effectively the Ne Doppler line width at 6328 Å, which is about 0.02 Å, and the corresponding coherence length is 100 mm, which is significantly better than that in arc lamps. If we single out only one of the longitudinal modes, the

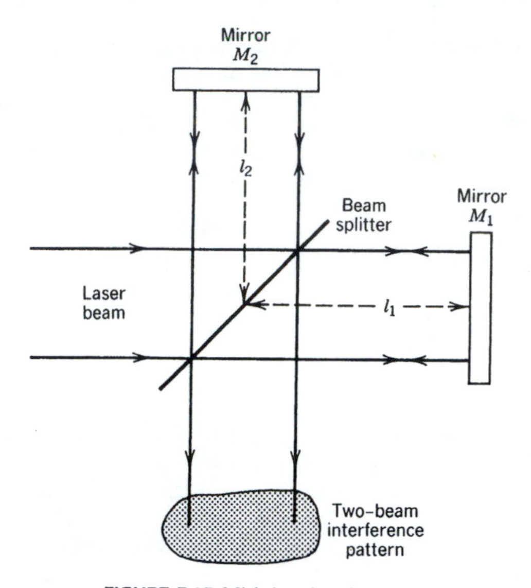

FIGURE 7.15 Michelson interferometer.

line width of the emission is less than 0.0002 Å, which would provide a coherence length of more than 10 m!

Example 7.9

The resonance frequency, $\nu_{21} = (E_2 - E_1)/h$, of the atoms within a laser cavity is not necessarily coincident with the cavity frequency given by $\nu_n = nc/2l$. Consequently, the output from the laser cavity lies somewhere between the cavity and the atomic frequencies. The laser frequency ν_l is given by (see reference 4, for example)

$$\nu_l = \nu_n - (\nu_n - \nu_{21}) \frac{c(1 - R)}{\Delta\nu\, 2\pi n l \sqrt{R}}, \tag{7.27}$$

where $\Delta\nu$ is the spectral width of the atomic transition and R is the reflectivity of the laser mirrors. ■

The importance of Eq. 7.27 is that

1. If the cavity is exactly tuned to the line center ν_{21} of the atomic transition, the frequency of the laser output is $\nu_n = \nu_{21}$.
2. If the cavity frequency ν_n is larger than ν_{21}, the laser output frequency ν_l will be smaller than ν_n by the amount given by the second term on the right-hand side of Eq. 7.27.
3. If ν_n is smaller than ν_{21}, the laser frequency will be slightly larger than ν_n by the same amount.

In situations 2 and 3 the output laser frequency is apparently "pulled" toward the atomic transition frequency—the so-called "frequency-pulling" effect.

Example 7.10

The exact form of the optical electric field of a TEM_{00} Gaussian laser beam is given by (see reference 2)

$$E(x, y, z) = E_0 \frac{\omega_0}{\omega(z)} \exp\left\{-i\left[kz - \tan^{-1}\left(\frac{z}{z_0}\right)\right]\right\} - r^2\left[\frac{1}{\omega^2(z)} + \frac{ik}{2R(z)}\right], \quad (7.28)$$

where E_0 is the initial amplitude (at $z = 0$ plane) of the electric field and ω_0 is the beam waist at the point where the cross section of the laser has the smallest diameter [where E drops to e^{-1} of its on-axis ($r = 0$) value]; $kz - \tan^{-1}(z/z_0)$ is a phase factor associated with propagation in the z direction where $k = 2\pi n/\lambda$; $R(z)$ is the radius of curvature of the beam; $\omega(z)$ is the beam waist at the z plane; and z_0 is the confocal parameter ($z_0 = \pi\omega_0^2 n/\lambda$).

The beam waist at z, $\omega(z)$, is related to $\omega_0(z)$ by

$$\omega^2(z) = \omega_0^2\left(1 + \frac{z^2}{z_0^2}\right), \quad (7.29)$$

and $R(z)$ is given by

$$R(z) = z\left(1 + \frac{z_0^2}{z^2}\right). \quad (7.30)$$

The propagation of a Gaussian beam is described by the three-dimensional Maxwell equations; Eqs. 7.28 through 7.30 are results obtained by solving for the fundamental TEM_{00} mode, assuming a cylindrical symmetry around z. They allow us to estimate the beam waist (the so-called "spot size") of the laser and its divergence at various propagation distances from the initial plane $z = 0$. ■

Example 7.11

Consider a beam of an Ar^+ ion laser ($\lambda = 514.5$ mm and $\omega_0 = 0.1$ mm) propagating in air. At a distance $z = 10$ m, what is the new beam waist.

From $z_0 = \pi\omega_0^2 n/\lambda$, we have

$$z_0 = \frac{\pi \times (0.1 \times 10^{-3}\text{ m})^2}{0.5145 \times 10^{-6}\text{ m}} \simeq 2\pi \times 10^{-2} \simeq 2\pi \times 10^{-2}\text{ m}.$$

Therefore,

$$\omega(z) = (0.1\text{ mm})\left(1 + \frac{10^2}{4\pi^2 \times 10^{-4}}\right)^{1/2}$$

$$= 159\text{ mm}.$$ ■

7.6 LASER OUTPUT MODULATIONS

As we discussed in the previous sections, the basic laser tends to oscillate in several longitudinal or transverse modes. In order to make the laser function in a single mode, transverse or longitudinal, we need to insert some intracavity devices such as a pinhole, which selects the transverse mode, or prisms or gratings, which select the appropriate output wavelength. In this section, we discuss some more specialized techniques for modulating the laser beam.

7.6.1 *Q*-Switching

One of the most frequently employed modulation techniques in a pulsed laser system is Q-switching, which literally means switching the Q, the quality factor, of the cavity. Figure 7.16 depicts a conventional Q-switch technique based on an electrooptical crystal (cf. Section 7.7) and polarizer combination placed within the laser cavity. After the pulse has excited the atoms (or molecules) to higher energy states, a high dc voltage pulse is applied to the electrooptical crystal, and this rotates the polarization of the light by 90°. The polarizer downstream will therefore block the passage of this light. When the dc voltage pulse is over, the electrooptical crystal will no longer change the polarization of the light, and it will propagate freely within the cavity. The success of the method depends on the period of time between the initial excitation pulse and the dc high-voltage pulse, and on the duration of the dc voltage pulse. These pulses should be adjusted so that the population inversion in the lasing medium reaches its peak value before the dc voltage is turned off. Essentially, the technique stops the system from lasing. Instead, the inversion keeps building up, which leads to a higher gain, and therefore a higher output, when the cavity is finally "opened" for lasing.

What is more important in terms of the power of the laser is that, in this kind of setup, the time it takes the cavity to decay determines the length of the laser pulse. That is, the energy stored within the cavity is expended during a period of time comparable to the time constant for cavity decay, τ_c, which is given by

$$\tau_c \simeq \frac{nl}{c(1 - R)}. \tag{7.31}$$

Typically, τ_c is on the order of nanoseconds (10^{-9} sec), and the laser pulse power (energy/time) is on the order of megawatts.

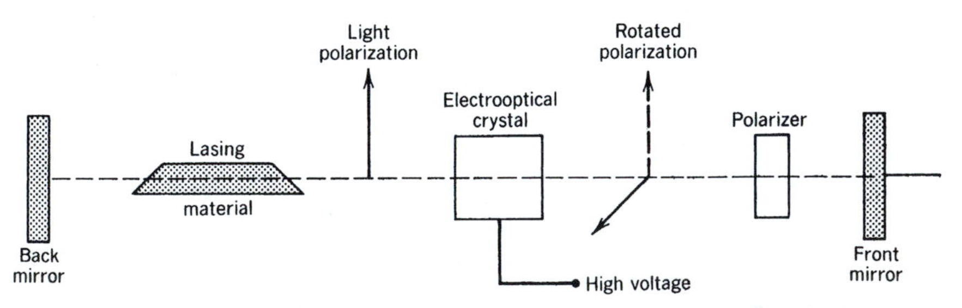

FIGURE 7.16 Q-switching of a laser with an electrooptical crystal.

The Q-switching method as described here is sometimes classified as an active Q-switching method. The passive Q-switching technique usually makes use of some intracavity absorptive materials, (such as dye solutions). The material absorbs up to the point at which its excited population is the same as its unexcited population, at which time it becomes transparent. Light is accordingly blocked or is free to propagate in the cavity.

7.6.2 Mode Locking

Because of its broad gain profile, a basic laser tends to oscillate in many longitudinal modes rather than in a single-cavity mode, as we discussed in Section 7.5. Without any control, these models oscillate with random phases, and the output, or intensity, of the laser is chaotic. In many ways this activity is analogous to the radiation from a number of randomly phased antennas. Mode locking is a technique that can be used to make these modes oscillate with the same phase. One method is to modulate the gain of the cavity periodically, with a period equal to the round-trip time τ of the light traveling between the mirrors ($\tau = 2nl/c$). This can be achieved by mounting an acoustooptical modulator on the rear mirror of the laser. The details of the theory are very complicated, but basically what happens is that, because of the periodic disturbances to the gain, the various modes build up and pick a fixed phase with respect to one another. The output consists of a continuous train of pulses with a period $\tau = 2nl/c$. The peak power of one of these pulses is N times the average power, where N is the number of modes locked together, and the individual pulse width τ_p is on the order of τ/N.

Because the line width of the atomic gain curve is $\Delta\nu$, the number of modes is roughly $\Delta\nu$ divided by the cavity frequency separation,

$$N \simeq \frac{\Delta\nu}{\nu_{\text{spacing}}} = \frac{\Delta\nu\, 2nl}{nc}.$$

Therefore,

$$\tau_p = \frac{\tau}{N} = \frac{2nlc}{c\, 2nl}\frac{1}{\Delta\nu} \simeq \frac{1}{\Delta\nu} \tag{7.32}$$

that is, the length of the mode-locked pulse is approximately the inverse of the gain line width. In some laser systems (e.g., dye lasers, whose active molecules are large organic molecules), the gain bandwidth is very large ($> 10^2\ \text{cm}^{-1}$). As a result, picosecond (10^{-12} sec) pulses are generated using the mode-locking technique.

Example 7.12

Estimate the Q-switched output laser pulsed from an Nd–YAG laser having the following characteristics:

$N_2 - N_1 = 9 \times 10^{15}\ cm^3$
length of rod $d = 10$ cm; refractive index $= 1.78$
cross-sectional area of beam $= 1\ \text{cm}^2$
mirror reflectivity $R = 0.8$
cavity length $l = 20$ cm.

The time it takes the cavity to decay is

$$\tau_c = \frac{nl}{c(1-R)} \simeq \frac{1.78 \times 20}{(3 \times 10^{10})(0.2)} \simeq 5 \times 10^{-9} \text{ sec},$$

and the total amount of stored energy E is

$$\begin{aligned} E &\simeq (N_2 - N_1) \times \text{crystal volume} \times \frac{hc}{\lambda} \\ &\simeq 9 \times 10^{19} \times 10 \times \frac{h(3 \times 10^8 \text{ m})}{1.064 \times 10^{-6} \text{ m}} \\ &\simeq 0.18 \text{ J}. \end{aligned}$$

The power of the Q-switched pulse is therefore

$$P \simeq \frac{E}{\tau_c} \simeq \frac{0.18}{5 \times 10^{-9}} \simeq 40 \text{ MW}.$$

■

■ Example 7.13

Calculate the mode-locked pulse length for a Nd–YAG laser, assuming that the laser cavity is 30 cm long and the line width of the laser transition is $\Delta\nu = 2 \times 10^{11}$ Hz.

From Eq. 7.32, the line width of the mode-locked pulse is

$$\begin{aligned} \tau_p = \frac{\tau}{N} \simeq \frac{1}{\Delta\nu} &= 0.5 \times 10^{-11} = 5 \times 10^{-12} \\ &= 5 \text{ psec}. \end{aligned}$$

We can also determine τ_p by calculating τ and N separately.

The time it takes the cavity to decay is

$$\tau_c = \frac{2l}{c} = \frac{2 \times 30}{3 \times 10^{10}} = 2 \times 10^{-9} \text{ sec}$$

The cavity frequency is therefore $\nu_c = c/2l = 0.5 \times 10^9$/sec; and the number of cavity modes in $\Delta\nu$ is

$$N = \frac{\Delta\nu}{\nu_c} = \frac{2 \times 10^{11}}{0.5 \times 10^9} = 400.$$

Therefore, the mode-locked pulse length is

$$\tau_p = \frac{\tau}{N} = \frac{2 \times 10^{-9}}{400} = 5 \times 10^{-12} \text{ sec}.$$

■

7.7 ELECTROOPTICAL MODULATIONS

After the laser output characteristics—gaussian beam transverse profile, pulsed or CW, *Q*-switched, mode-locked, single longitudinal mode, and so on—are determined by the appropriate *intracavity* optical elements, the exit beam can be modulated in many ways, depending on the applications desired. It would require a treatise to cover the wide variety of devices and applications making use of various electrooptical, acoustooptical, and optooptical modulations of the laser beams. To summarize, we can say that almost all these applications rely on amplitude modulation (AM) and frequency modulation (FM) of the laser beam.

Amplitude modulation of a laser beam by mechanical means is easy to accomplish. For example it is obvious that shutters, choppers, and rotating mirrors can turn the laser on and off and make it accessible for many common applications, such as bar code scanning, holography, and laser light shows. For more sophisticated applications, for example, those requiring very fast (gigahertz) on–off modulation of a laser, electrooptical or acoustooptical crystals are used. Figure 7.17 pictures a typical example of these high-speed modulation applications, the sending of audio signals (e.g., telephone messages from voices, computers, audio equipment, etc.). In the initial stage these signals are converted to electronic voltages which act on an electrooptical-crystal modulator. The modulator turns the laser (usually a semi-conductor laser) on and off accordingly, guiding the laser by the optical fiber to downstream electronic equipment for demodulation and reception.

Most electrooptical modulators function by changing their refractive index in accordance with the applied voltage. The voltage, applied across the crystals, produces an electric field which raises or lowers the refractive index, depending on the direction of the electric field with respect to the crystalline axes. The electrooptical crystals used in these modulation applications have three principal axes, x, y, and z (Figure 7.18). The corresponding refractive indices for the optical field polarized in these directions are given by

$$\begin{aligned} n_x &= n_o + r_{eo}E_z, \\ n_y &= n_o - r_{eo}E_z, \\ n_z &= n_e, \end{aligned} \tag{7.33}$$

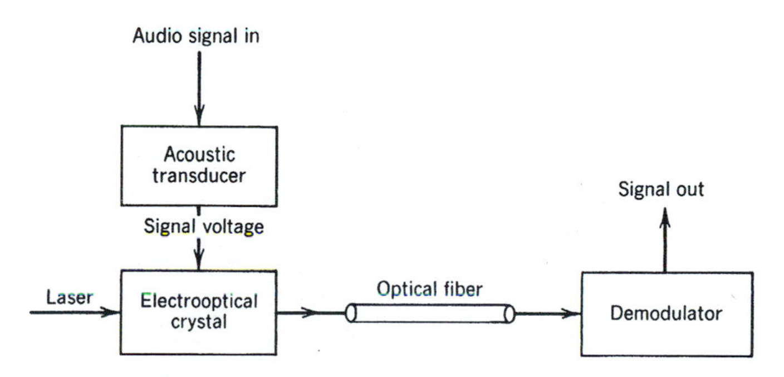

FIGURE 7.17 Schematic of a typical laser communication system.

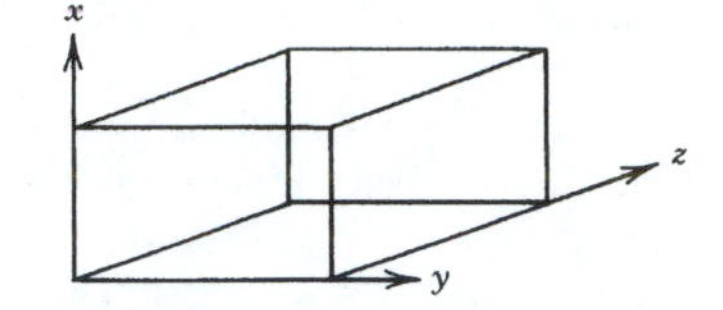

FIGURE 7.18 An anisotropic electrooptical crystal.

respectively, where E_z is the applied field (from the signal voltage), and r_{eo} is the electrooptical coefficient of the crystal. We should mention that besides the functional forms of n_x, n_y, and n_z given in Eq. 7.33, various other forms obviously exist, depending on which of the myriads of electrooptical crystals is used and to which of many fields the laser is applied. Our intention here, however, is to demonstrate the fundamental mechanism for *electrooptical amplitude modulation.*

Using an electrooptical crystal with the refractive index characteristics shown in Eq. 7.33, we can very easily devise an on–off modulation with a configuration similar to that shown in Figure 7.19. The laser is polarized in the $\hat{x}'$ direction, which makes an angle of 45° with the x and y axes of the crystal, and propagates along the z axis. The component of the laser field in the x axis will acquire a phase ϕ_x from

$$\phi_x = \frac{2\pi}{\lambda}(n_o + r_{eo}E_z)d, \tag{7.34}$$

where d is the thickness of the crystal in the z direction. On the other hand, the component of the laser field in the y axis acquires a phase ϕ_y from

$$\phi_y = \frac{2\pi}{\lambda}(n_o - r_{eo}E_z)d. \tag{7.35}$$

Therefore, the phase difference between the x and y components is given by

$$\Delta\phi = \phi_x - \phi_y = \frac{4\pi}{\lambda} r_{eo}E_z d. \tag{7.36}$$

If we control this phase difference with the applied field E_z, we can change the polarization of the existing beam accordingly.

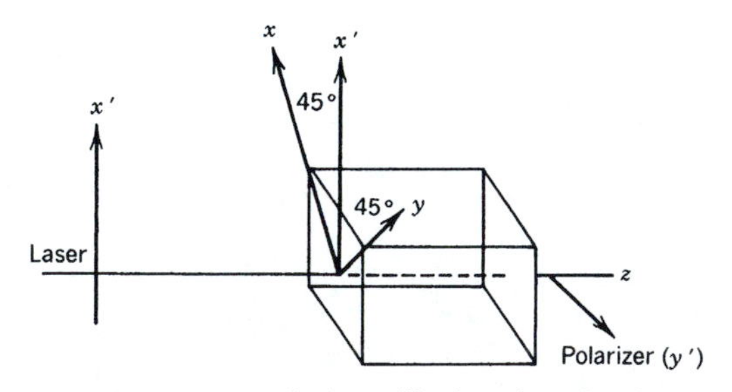

FIGURE 7.19 An amplitude-modification scheme for a laser.

If the phase difference $\Delta\phi$ is $\pi/2$, the existing beam will be circularly polarized. This means that only 50 percent of it will be transmitted through the polarizer in the y' direction. That is, we will get a modulation depth of 50 percent on the laser beam. On the other hand, if $\Delta\phi$ is π, the beam will be linearly polarized, but with the polarization rotated 90° from the original direction. In this instance the polarizer will transmit the beam totally in the y' direction, and we will have a 100 percent on–off modulator.

The modulation on the laser can be performed at a very high rate; a "bit" rate of tens of gigahertz is a practical reality. At such a high speed, the corresponding laser "pulses" are also very short, in tens of picoseconds. After traversing the optical fiber, these short pulses begin to experience significant broadening; the pulse becomes longer as a function of the propagation distance along the fiber. Dispersion mechanisms have come into play. A detailed discussion of this pulse broadening and the details of other problems associated with laser pulse propagation in optical fibers can be found in reference 7.

Frequency modulation (FM) of a laser beam can be achieved by simply imparting a time-varying phase shift to the laser. A simple setup is shown in Figure 7.20. The laser is polarized in the x direction and propagates along the z direction. The phase shift ϕ_x experience by the laser is given by

$$\phi_x = \frac{2\pi}{\lambda}(n_o + rE)d.$$

The optical electric field oscillation at the output end is therefore given by

$$E_{\text{out}} = A\cos\left[\omega t - \left(\frac{2\pi d}{\lambda}\right)(n_o + r_{eo}E)\right], \tag{7.37}$$

where A is the amplitude. If the applied electric field E on the crystal is of the form

$$E = E_m \sin\omega_m t, \tag{7.38}$$

we have

$$\begin{aligned} E_{\text{out}} &= A\cos\left(\omega t - r_{eo}E_m\frac{2\pi d}{\lambda}\sin\omega_m t - \frac{2\pi d n_o}{\lambda}\right) \\ &= A\cos\left(\omega t - \delta\phi\sin\omega_m t - \frac{2\pi d n_o}{\lambda}\right), \end{aligned} \tag{7.39}$$

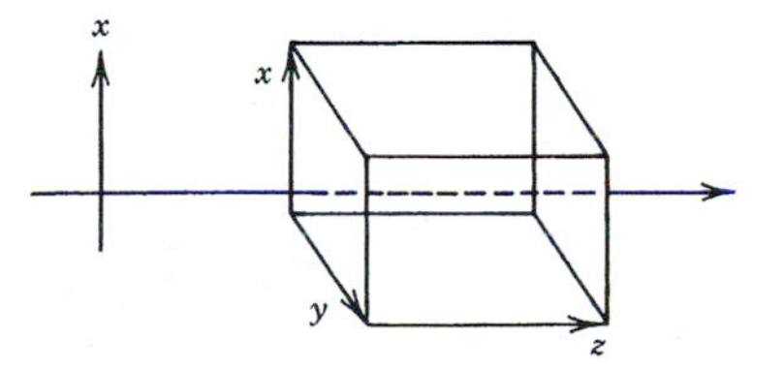

FIGURE 7.20 Phase modulation of a laser beam by an electrooptical crystal.

where

$$\delta\phi = \frac{2\pi d}{\lambda} r_{eo} E_m.$$

Using the Bessel function identities, we have

$$\cos(\delta\phi \sin \omega_m t) = J_0(\delta\phi) + 2J_2(\delta\phi)\cos 2\omega_m t + 2J_4(\delta\phi)\cos 4\omega_m t + J_6; J_8; \ldots, \quad (7.40)$$

and

$$\sin(\delta\phi \sin \omega_m t) = 2J_1(\delta\phi)\sin \omega_m t + 2J_3(\delta\phi)\sin 3\omega_m t + \cdots + J_5, J_7, \ldots, \quad (7.41)$$

We can easily see that the oscillation given in Eq. 7.39 contains frequency components $\omega \pm \omega_m$, $\omega \pm 2\omega_m$, $\omega \pm 3\omega_m, \ldots$, as a result of the phase modulation by the time-varying electric field on the electrooptical crystal.

REFERENCES

1. W. G. Driscoll and W. Vaughan, *Handbook of Optics*, McGraw-Hill, New York, 1978.
2. J. T. Verdeyen, *Laser Electronics*, Prentice-Hall, Englewood Cliffs, N.J., 1981.
3. D. C. O'Shea, W. R. Callan, and W. T. Rhodes, *Introduction to Lasers and Their Applications*, Addison-Wesley, Reading, Mass., 1977.
4. A. Yariv, *Introduction to Optical Electronics*, Holt, Rinehart and Winston, New York, 1976.
5. E. Hecht and A. Zajac, *Optics*, Addison-Wesley, Reading, Mass., 1979.
6. L. Allen and J. H. Eberly, *Optical Resonance and Two-Level Atoms*, Wiley-Interscience, New York 1975.
7. J. C. Palais, *Fiber Optic Communications*, Prentice-Hall, Englewood Cliffs, N.J., 1988.

PROBLEMS

The following instructions are for Problems 7.1–7.4.

A tube 10 cm long with a 1-cm^2 cross-sectional area is filled with 10^{10} two-level atoms with the transition energy as shown. A laser with a cross-sectional area equal to the tube, and a frequency close to the resonance frequency of the atoms is incident on this system. The broadening mechanism for the gaseous medium may be assumed to be Lorentzian:

$$g(\nu) \simeq \frac{2\Delta\nu/\pi}{4(\nu - \nu_0)^2 + \Delta\nu^2}$$

and the line width is $\Delta\nu = 1/2\pi\tau_{sp}$.

7.1 If all the atoms are initially in the lower state, and the incident laser intensity is 10^6 W/cm^2, what is the transmitted intensity? Assume that the laser is in resonance.

7.2 If the atoms are initially in the upper state, what incident laser intensity is needed to stimulate a downward transition rate equal to the spontaneous transition rate, assuming that the laser is on resonance?

7.3 If the transmitted intensity for Problem 7.1 is I_t and $\nu - \nu_0$ is four times the natural width, what is the transmitted intensity of the incident laser?

7.4 If you are designing a traveling wave amplifier with the type of atom shown in Figure 7.21, what density of the completely excited atom (i.e., the atom in level 2) is needed to obtain a gain of 30 for a resonant wave? A gain of 30 means that $I_{\text{transmitted}}/I_{\text{incident}} = 30$.

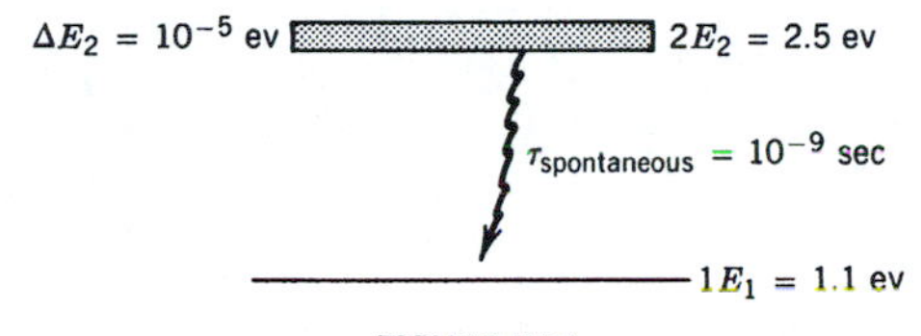

FIGURE 7.21

7.5 Derive the time constant for cavity decay, τ_c, given in Eq. 7.31.

7.6 Using Eqs. 7.39 and 7.41, obtain an explicit formula for the output of a laser in terms of all its frequency components.

7.7 An argon laser beam has a diameter of 1 mm just as it exits the front output mirror. The beam waist is located right on the rear mirror, one meter behind the front mirror.

(a) What is the radius of the beam waist ω_0?

(b) If this laser beam is shone on the moon, what will the size of the illuminated area be?

7.8 Estimate the number of photons emitted per second from a laser that puts out one watt of power. Assume a reasonable number for any of the quantities you need in your calculation.

7.9 The amplitude of the electric field strength of light emitted by an atom undergoing spontaneous transition may be written as

$$\varepsilon(t) = \varepsilon_0 e^{-t/\tau} \cos \omega_0 t,$$

where τ is the lifetime for the emission. Show that the emission spectrum is Lorentzian with a line width $\Delta\nu$ on the order of τ^{-1}. (*Note:* See page 195.) What is the line width for an atom whose emission has the lifetime

(a) $\tau = 10^{-9}$ sec (i.e., 1 nanosecond)?

(b) $\tau = 10^{-6}$ sec?

8 HOLOGRAPHY

The concept of holography, also known as wavefront reconstruction, was first introduced by Dennis Gabor in 1948. At that time he had encountered two major difficulties, namely that a high-intensity coherent source of light was not available and that virtual and real images could not be separated. Nevertheless, he set a basic foundation of holographic theory. In fact, the word "holography" was first coined by Gabor, by combining parts of two Greek words: *holos*, meaning "whole," and *graphein*, meaning "to write." Thus, *holography* means recording a complete message. In this chapter we discuss the fundamental concept of holography.

A *hologram* is a recording on a light-sensitive medium, such as a photographic plate, of interference patterns formed between two beams of coherent light coming from the same laser. In the process of holography the laser beam is first divided into two beams by a beam splitter and then broadened. One beam goes directly to the photographic plate. The second beam of light is directed onto the three-dimensional object under observation, each point of which reflects the part of the beam reaching it toward the plate. As each point of the object scatters the light, it acts as a point source of spherical waves. At the photographic plate the innumerable spherical waves from the object combine with the light wave of the first beam. Because they are from the same laser, the sets of light waves are coherent and are in a condition to interfere. They form interference fringes on the plane of the photographic plate and are thereby recorded. These interference fringes are a series of zone platelike rings, but these rings are also superimposed, making an incredibly complex pattern of lines and swirls. The developed negative of these interference fringes is the hologram.

In an ordinary photograph taken in noncoherent light, the film has recorded the intensity of light reaching it at every point. In holography the interference of the two beams allows the photographic plate to record both the intensity and the relative phase of the light waves at each point.

The complicated interference fringe pattern in the hologram would seem to bear no relation whatsoever to an image of the original object, but it does indeed contain all the information needed to reconstruct the wavefronts and reconstitute a three-dimensional image of great vividness and verisimilitude. For this reconstitution the hologram is illuminated by a parallel beam of light from the laser. Most of the light passes straight through, but the complex of fine fringes acts as an elaborate diffraction grating. Light is diffracted at a fairly wide angle. The diffracted rays form two images, a reconstituted virtual image of the object behind the transparency and a reconstituted real image on the eye side. Since the light rays pass through the point where the real image is, it can be photographed.

The vivid, three-dimensional virtual image of the hologram is for viewing. Observers can move to different positions and look around the image to the same extent that they would be able to were they looking directly at the real object. And the moving viewers observe parallax, the apparent displacement of nearer and more distant parts of the image.

8.1 ON-AXIS HOLOGRAPHY

Since the physical spaces involved in holographic construction and the reconstruction processes are generally assumed to be linear and spatially invariant, holographic processes can easily be evaluated by the simple object point approach. Thus, in holographic recording, we assume that an object point, to which a beam of coherent light is directed from a monochromatic light source, is located at a distance l_1 from the photographic plane. A coherent reference plane wave, derived from the same coherent source, is superimposed over the recording medium with the spherical beam diffracted from the object, as depicted in Figure 8.1. If the amplitude transmittance of the recording medium is assumed to be linearly proportional to the intensity distribution of the recording, an analog system diagram of the holographic construction, like the one in Figure 8.2, can be drawn.

The complex light field coming from the object and the plane wave derived from the coherent source, which is called the *reference beam*, are represented by

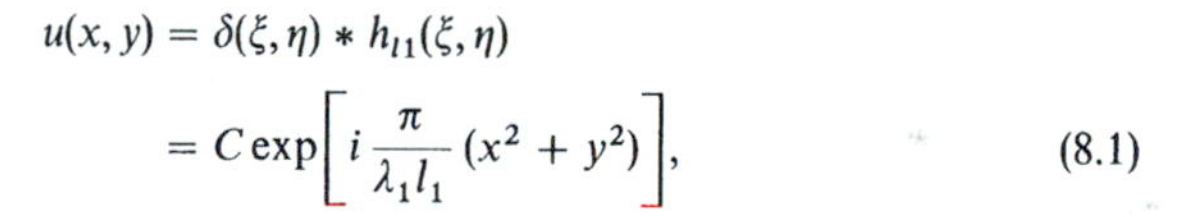

$$u(x, y) = \delta(\xi, \eta) * h_{l1}(\xi, \eta)$$
$$= C \exp\left[i \frac{\pi}{\lambda_1 l_1}(x^2 + y^2)\right], \tag{8.1}$$

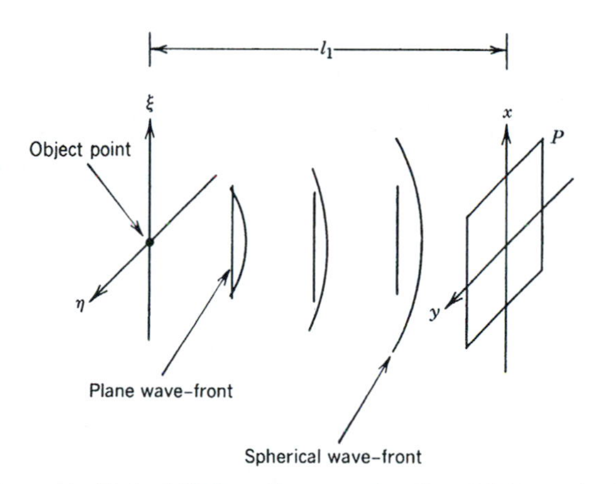

FIGURE 8.1 An on-axis object point holographic construction. *Note: P* designates the photographic plate.

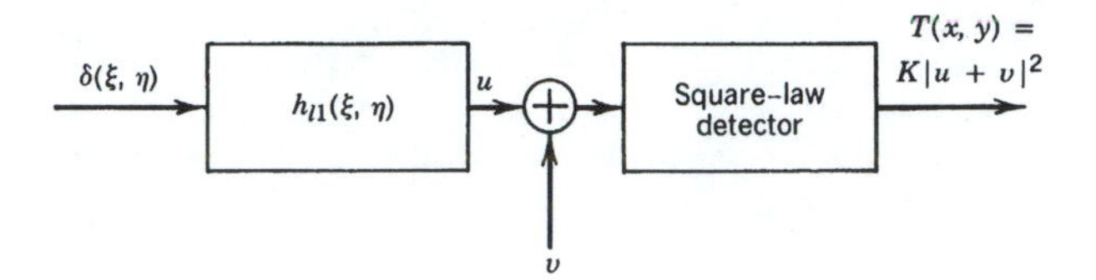

FIGURE 8.2 An analog system diagram of the holographic construction in Figure 8.1.

and

$$v(x, y) = C', \tag{8.2}$$

as they interfere over the recording plate. Here the asterisk denotes the convolution operation, and the C's are the proportionality complex constants. The resultant intensity distribution over the recording plate is

$$\begin{aligned} I(x, y; \lambda_1) &= |u(x, y) + v(x, y)|^2 \\ &= |C|^2 + |C'|^2 + 2|C||C'|\cos\left[\frac{\pi}{\lambda_1 l_1}(x^2 + y^2)\right]. \end{aligned} \tag{8.3}$$

Since the amplitude transmittance of the photographic plate is assumed to be linear in exposure, that is, the intensity multiplies the exposure time, we have

$$T(x, y; \lambda_1) = K_1 + K_2 \cos\left[\frac{\pi}{\lambda_1 l_1}(x^2 + y^2)\right], \tag{8.4}$$

which can be written as

$$T(x, y; \lambda_1) = K_1 + \frac{K_2}{2}\exp\left[i\frac{\pi}{\lambda_1 l_1}(x^2 + y^2)\right] + \frac{K_2}{2}\exp\left[-i\frac{\pi}{\lambda_1 l_1}(x^2 + y^2)\right], \tag{8.5}$$

where K_1 and K_2 are proportionality constants. We note further that the amplitude transmittance function of an object point hologram (i.e., Eq. 8.4) is identical to a *Fresnel zone lens*, which is composed of a glass plate and a negative and a positive lens (see Section 6.4).

In addition, if the object point hologram is normally illuminated by a monochromatic plane wave of λ_2, as shown in Figure 8.3, the holographic reconstruction

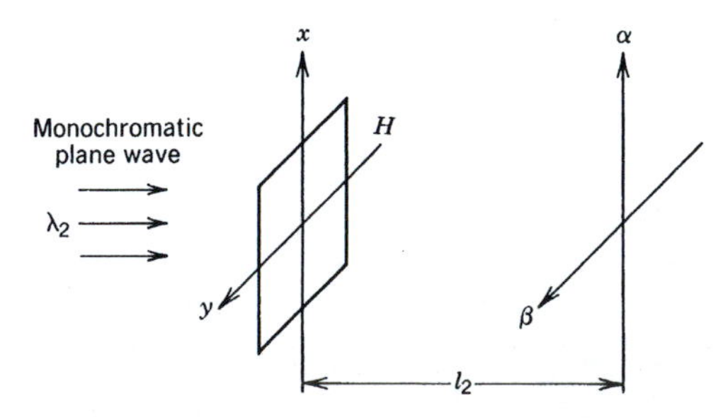

FIGURE 8.3 The geometry for hologram image reconstruction. *Note: H* designates the hologram.

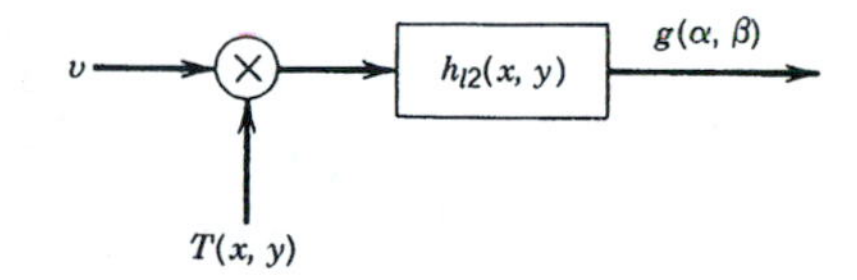

FIGURE 8.4 An analog system diagram of the hologram image reconstruction in Figure 8.3.

process can be evaluated, with the help of the analog system diagram in Figure 8.4. The complex light distribution over the α, β plane can be evaluated by

$$g(\alpha, \beta; \lambda_2) = CT(x, y; \lambda_1) * h_{l2}(x, y; \lambda_2)$$
$$= C \iint_S T(x, y; \lambda_1) h_{l2}(\alpha - x, \beta - y; \lambda_2)\, dx\, dy, \tag{8.6}$$

where C is a proportionality complex constant, S denotes the surface integration over the hologram, and

$$h_{l2}(x, y) = \exp\left[i \frac{\pi}{\lambda_2 l_2} (x^2 + y^2) \right] \tag{8.7}$$

is the spatial impulse response between the hologram and the α, β plane.

The terms in Eq. 8.6 can be evaluated by

$$g_0(\alpha, \beta; \lambda_2) = CK_1 \iint_S h_{l2}(\alpha - x, \beta - y; \lambda_2)\, dx\, dy, \tag{8.8}$$

$$g_v(\alpha, \beta; \lambda_2) = \frac{CK_2}{2} \iint_S \exp\left[i \frac{\pi}{\lambda_1 l_1} (x^2 + y^2) \right] h_{l2}(\alpha - x, \beta - y; \lambda_2)\, dx\, dy, \tag{8.9}$$

$$g_r(\alpha, \beta; \lambda_2) = \frac{CK_2}{2} \iint_S \exp\left[-i \frac{\pi}{\lambda_1 l_1} (x^2 + y^2) \right] h_{l2}(\alpha - x, \beta - y; \lambda_2)\, dx\, dy, \tag{8.10}$$

and

$$g(\alpha, \beta; \lambda_2) = g_0(\alpha, \beta; \lambda_2) + g_v(\alpha, \beta; \lambda_2) + g_r(\alpha, \beta; \lambda_2), \tag{8.11}$$

where the subscripts 0, v, and r denote the zero-order and the first-order virtual- and real-image diffractions. To simplify the analysis, we assume that the size of the hologram is infinitely extended and that the integral of Eq. 8.8 converges to a complex constant:

$$g_0(\alpha, \beta; \lambda_2) = C_1. \tag{8.12}$$

The integral of Eq. 8.9 can be evaluated so that

$$g_v(\alpha, \beta; \lambda_2) = C_2 \iint_{-\infty}^{\infty} \exp\left[i \frac{\pi}{\lambda_1 l_1} (x^2 + y^2) \right]$$
$$\cdot \exp\left\{ i \frac{\pi}{\lambda_2 l_2} [(\alpha - x)^2 + (\beta - y)^2] \right\} dx\, dy, \tag{8.13}$$

which can be written as

$$g_v(\alpha, \beta; \lambda_2) = C_2 \exp\left[i \frac{\pi}{\lambda_2 l_2}(\alpha^2 + \beta^2)\right] \iint_{-\infty}^{\infty} \exp\left[i\pi\left(\frac{1}{\lambda_1 l_1} + \frac{1}{\lambda_2 l_2}\right)(x^2 + y^2)\right]$$

$$\cdot \exp\left[-i \frac{2\pi}{\lambda_2 l_2}(\alpha x + \beta y)\right] dx\, dy. \qquad (8.14)$$

In view of the preceding equation, a hologram image of the object point can be reconstructed if and only if the quadratic phase factor can be eliminated:

$$\frac{1}{\lambda_1 l_1} + \frac{1}{\lambda_2 l_2} = 0.$$

Thus, the longitudinal location of the hologram image would be

$$l_2 = -\frac{\lambda_1}{\lambda_2} l_1, \qquad (8.15)$$

where λ_1 and λ_2 are the construction and reconstruction wavelengths, respectively, and l_1 is the separation between the object and the hologram. Since λ_1, λ_2, and l_1 are positive quantities, the minus sign in Eq. 8.15 means that the image is located at the front of the hologram. For $l_2 = -l_1\lambda_1/\lambda_2$ the virtual image is reconstructed on the optical axis:

$$g_v(\alpha', \beta'; \lambda_2)\Big|_{l_2 = -l_1\lambda_1/\lambda_2} = C_2 \exp\left[i \frac{\pi}{\lambda_2 l_2}(\alpha^2 + \beta^2)\right]$$

$$\times \iint_{-\infty}^{\infty} \exp\left[-i \frac{\pi}{l_2 \lambda_2}(\alpha x + \beta y)\right] dx\, dy$$

$$= C_2 \delta(\alpha', \beta'). \qquad (8.16)$$

We can also show that the real image is reconstructed behind the holographic aperture:

$$g_r(\alpha, \beta; \lambda_2)\Big|_{l_2 = l_1\lambda_1/\lambda_2} = C_3 \delta(\alpha, \beta). \qquad (8.17)$$

From the results of Eqs. 8.12, 8.16, and 8.17, we see that the holographic reconstruction process yields the zero-order, the virtual-image, and the real-image diffractions, as depicted in Figure 8.5. From this figure we see that the three orders of diffraction overlap, which causes spurious image distortion. We also notice that the overlapping diffractions cannot be separated, even under oblique illumination, as shown in Figure 8.6. For an oblique illumination the diffracted light field can be evaluated as follows:

$$g(\alpha, \beta; \lambda_2) = T(x, y; \lambda_1) \exp\left(i \frac{2\pi}{\lambda_2} x \sin\theta\right) * h_{l2}(x, y; \lambda_2)$$

$$= \iint_S T(x, y; \lambda_1) \exp\left(i \frac{2\pi}{\lambda_2} x \sin\theta\right) h_{l2}(\alpha - x, \beta - y; \lambda_2)\, dx\, dy. \qquad (8.18)$$

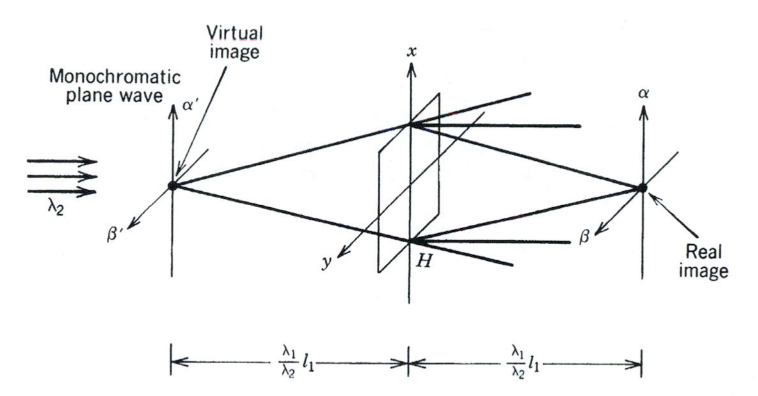

FIGURE 8.5 Hologram image reconstruction of an object point. *Note:* H designates the hologram.

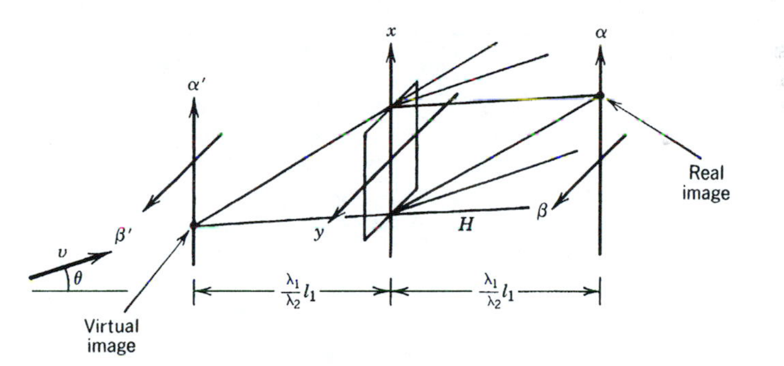

FIGURE 8.6 Oblique wavefront reconstruction of an on-axis object point hologram, with v designating the oblique plane wave.

Again, by substituting the amplitude transmittance function of Eq. 8.5 in Eq. 8.18, we obtain

$$g_o(\alpha, \beta; \lambda_2) = C_1 \exp\left(i \frac{2\pi}{\lambda_2} \alpha \sin\theta\right), \tag{8.19}$$

$$g_v(\alpha', \beta'; \lambda_2) = C_2 \delta\left(\alpha' + \frac{\lambda_1}{\lambda_2} l_1 \sin\theta, \beta'\right), \tag{8.20}$$

and

$$g_r(\alpha, \beta; \lambda_2) = C_3 \delta\left(\alpha - \frac{\lambda_1}{\lambda_2} l_1 \sin\theta, \beta\right). \tag{8.21}$$

The image reconstructions are illustrated in Figure 8.6.

Example 8.1

Assume that the on-axis object point hologram of Eq. 8.4 is constructed with a monochromatic red light of 600 nm and that the separation between the object and the holographic aperture is 50 cm. Calculate the locations of the hologram images if the hologram is illuminated by a normally incident plane wave of $\lambda = 500$ nm.

Let the construction wavelength be $\lambda_1 = 600$ nm, the separation be $l_1 = 0.5$ m, and the reconstruction wavelength be $\lambda_2 = 500$ nm. By substituting these parameters into Eq. 8.15, we have

$$l_2 = \frac{\lambda_1}{\lambda_2} l_1 = \frac{600}{500}(0.5) = 0.6 \text{ m}.$$

Thus, the virtual and the real hologram images are located 0.6 m from the front and behind the hologram, respectively. ■

Example 8.2

If the object point hologram of Example 8.1 is illuminated by a normally incident plane wave of white light, calculate the smearing length of the reconstructed image.

Since the spectral lines of visible light vary from 350 to 700 nm, a white-light source would have a spectral width of $\Delta\lambda = 700 - 350 = 350$ nm. By virtue of Eq. 8.15, the smearing length of real and virtual images can be calculated as

$$\Delta l = \frac{600}{350}(0.5) - \frac{600}{700}(0.5)$$
$$= 0.8571 - 0.4287 = 0.4284 \text{ m}.$$

Thus, we see that the hologram images smear into rainbow colors; the red image is located close to the holographic plate and the violet image is situated farther away. ■

8.2 OFF-AXIS HOLOGRAPHY

An oblique reference beam should be used in the construction process to separate the hologram image diffractions, as illustrated in Figure 8.7. An analog system diagram representing the off-axis (also called the carrier spatial frequency) holographic construction is depicted in Figure 8.8. Thus, the amplitude transmittance function of the constructed hologram is

$$T(x, y; \lambda_1) = K_1 + K_2 \cos\left[\frac{\pi}{2\lambda_1}\left(\frac{x^2 + y^2}{2l_1} - x \sin\theta\right)\right], \tag{8.22}$$

where the K's are proportionality constants.

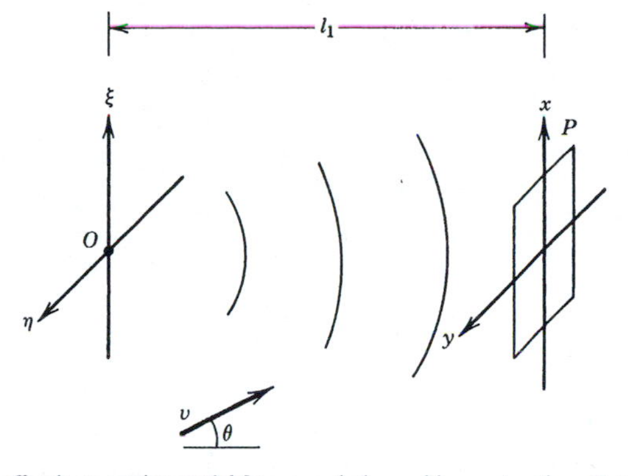

FIGURE 8.7 An off-axis or carrier spatial-frequency holographic construction. *Note: O*, object point; *P*, photographic plate; *v*, oblique plane wave.

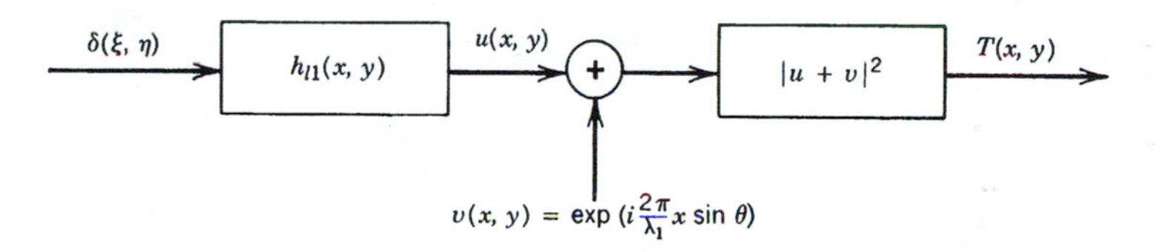

FIGURE 8.8 An analog system diagram of the holographic construction in Figure 8.7.

If this hologram is illuminated by the oblique conjugate plane wave of λ_2, as illustrated in Figure 8.9, the holographic image diffractions can be evaluated by

$$g(\alpha, \beta; \lambda_2) = \left[\exp\left(-i\frac{2\pi}{\lambda_2} x \sin\theta\right) T(x, y; \lambda_1)\right] * h_{l2}(x, y; \lambda_2), \tag{8.23}$$

where the asterisk denotes the convolution operation and $h_{l2}(x, y; \lambda_1)$ is the spatial impulse response.

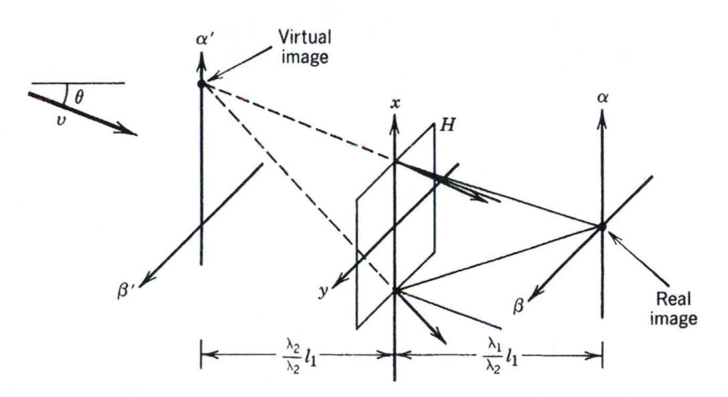

FIGURE 8.9 Reconstruction of the image in an off-axis hologram, with an oblique conjugate plane wave illumination.

By substituting the amplitude transmittance function of Eq. 8.22 in Eq. 8.23, we obtain

$$g_0(\alpha, \beta; \lambda_2) = C_1 \exp\left(-i \frac{2\pi}{\lambda_2} x \sin\theta\right), \tag{8.24}$$

$$g_v(\alpha', \beta'; \lambda_2) = C_2 \delta\left(\alpha' - 2\frac{\lambda_1}{\lambda_2} l_1 \sin\theta, \beta'\right), \tag{8.25}$$

and

$$g_r(\alpha, \beta; \lambda_2) = C_3 \delta(\alpha, \beta). \tag{8.26}$$

The corresponding reconstructions of the hologram virtual and real images are shown in Figure 8.9. Thus, we see that the real image can be separated from the other diffractions. Similarly, reconstructions of the hologram images with the same oblique reference plane wave, as shown in Figure 8.10, can be evaluated. The results are given by

$$g_0(\alpha, \beta; \lambda_2) = C_1 \exp\left(i \frac{2\pi}{\lambda_2} \alpha \sin\theta\right), \tag{8.27}$$

$$g_v(\alpha', \beta'; \lambda_2) = C_2 \delta(\alpha', \beta'), \tag{8.28}$$

and

$$g_r(\alpha, \beta; \lambda_2) = C_3 \delta\left(\alpha - 2\frac{\lambda_1}{\lambda_2} l_1 \sin\theta, \beta\right). \tag{8.29}$$

Thus, we see that the virtual image can be viewed through the holographic aperture without interference from the other diffractions.

Let us consider a hologram construction of an extended object, as shown in Figure 8.11. The complex light field distributed on the recording medium can be written as

$$u(x, y) = \iint_S O(\xi, \eta) h_{l1}(x - \xi, y - \eta; \lambda_1)\, d\xi\, d\eta,$$

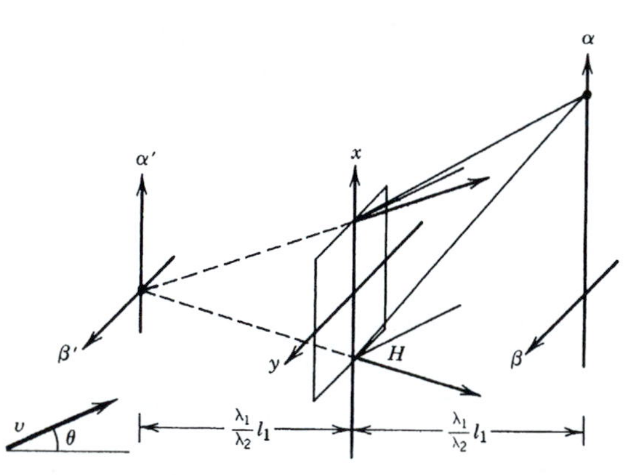

FIGURE 8.10 Hologram image reconstruction, with an oblique plane wave illumination.

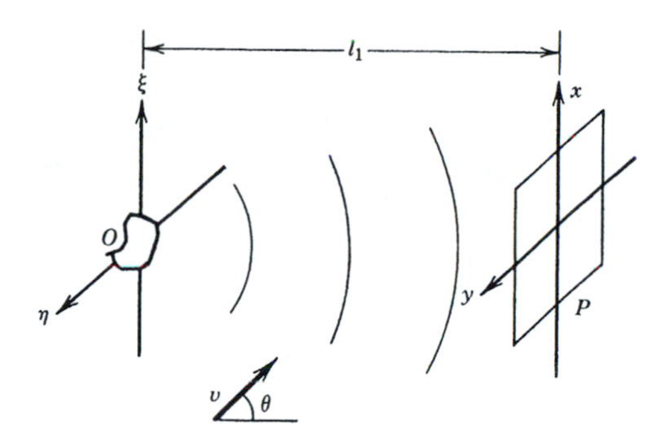

FIGURE 8.11 Holographic construction of an extended object. *Note: O*, object; *v*, oblique plane wave; *P*, photographic plate.

where $O(\xi, \eta)$ is the object function and S denotes the surface integration over (ξ, η). An analog system diagram representing the construction process appears in Figure 8.12. The amplitude transmittance function of the recorded hologram can be calculated as follows,

$$T(x, y; \lambda_1) = |u(x, y)|^2 + 1 + 2|u(x, y)| \cos[\phi(x, y) - k_1 x \sin\theta], \tag{8.30}$$

where we have ignored the proportionality constant for simplicity; here $k_1 = 2\pi/\lambda_1$ is called the wave number; $u(x, y) = |u(x, y)|e^{i\phi(x,y)}$ represents the object beam; and $\phi(x, y)$ is the phase distribution.

Needless to say, if the recorded hologram is illuminated by a conjugate plane wave of λ_2, as depicted in Figure 8.13, the real-image diffractions can be evaluated as

$$g_r(x, y; \lambda_2) = |u(x, y)| \exp\{i[\phi(x, y) - k_1 x \sin\theta]\} * h_{l2}(x, y; \lambda_2),$$

which can be shown equal to

$$g_r(x, y; \lambda_2) = O^*(\alpha, \beta; \lambda_2) \qquad \text{at} \quad l_2 = \frac{\lambda_1 l_1}{\lambda_2}, \tag{8.31}$$

where the asterisk denotes the complex conjugate. Similarly, the process of virtual-image reconstruction is

$$g_v(x, y; \lambda_2) = O\left(\alpha - 2\frac{\lambda_1}{\lambda_2} l_1 \sin\theta, \beta\right) \qquad \text{at} \quad l_2 = -\frac{\lambda_1 l_1}{\lambda_2}. \tag{8.32}$$

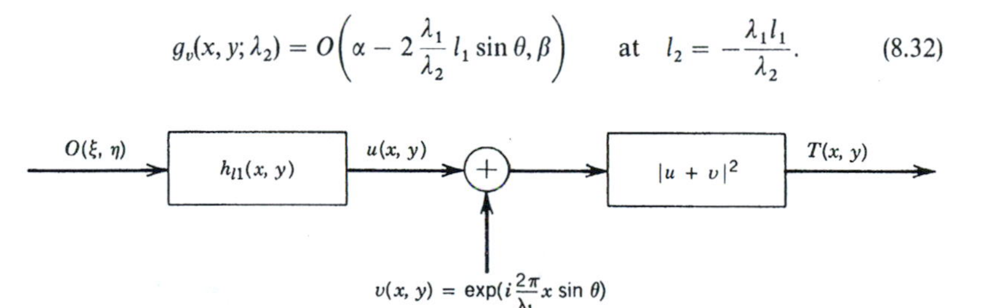

FIGURE 8.12 An analog system diagram of the holographic construction in Figure 8.11.

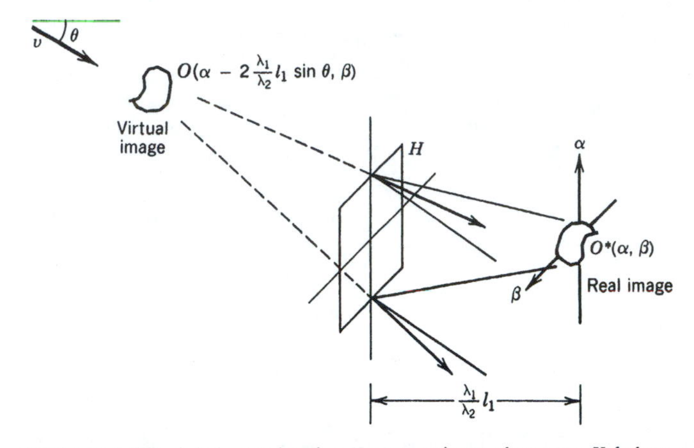

FIGURE 8.13 Off-axis hologram imaging. *Note:* v, conjugate plane wave; H, hologram.

This constructed virtual image is located at the front of the hologram. The image reconstructions are illustrated in Figure 8.13. Again we see that the real-image diffraction can be separated from the other diffractions. In addition, we notice that the real image is *pseudoscopic*, that is, the image is in reversed relief. The virtual image is *orthoscopic*, in proper relief.

The *diffraction efficiency* (D.E.) of a hologram can be defined as

$$\text{D.E.} = \frac{\text{output (hologram image intensity)}}{\text{input (incident intensity)}}. \tag{8.33}$$

In practice, this quantity is frequently used for determining the quality of a hologram.

Example 8.3

If the off-axis object point hologram of Eq. 8.22 is illuminated by a conjugate monochromatic point source, as shown in Figure 8.14,

(a) Draw an analog system diagram to represent the holographic reconstruction process.
(b) Evaluate the real- and virtual-image reconstructions.

Answers

(a) The location of the monochromatic point source can be represented by a delta function, $\delta(\alpha - r\sin\theta, \beta)$. The analog system diagram appears in Figure 8.15.
(b) The illuminating wavefront $u(x, y; \lambda_2)$ can be shown as follows:

$$u(x, y; \lambda_2) = \delta(\alpha - r\sin\theta, \beta) * h_l(\xi, \eta; \lambda_2)$$

$$= C\exp\left\{\frac{ik_2}{2l}\left[(x - r\sin\theta)^2 + y^2\right]\right\},$$

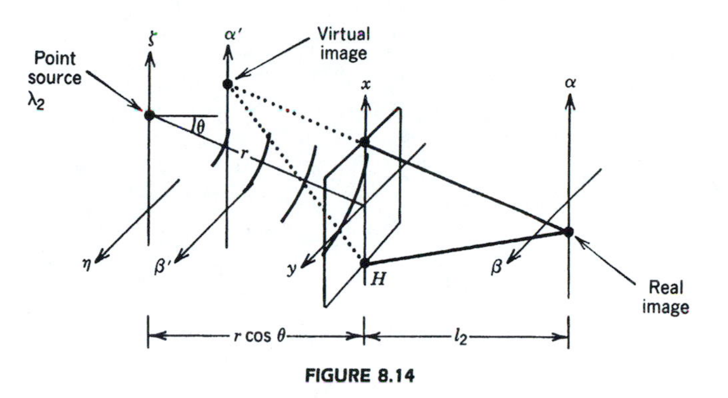

FIGURE 8.14

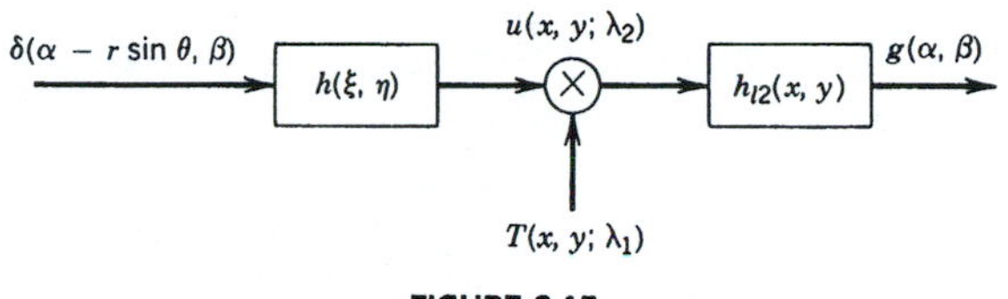

FIGURE 8.15

where

$$h_l(\xi, \eta; \lambda_2) = \exp\left[\frac{ik_2}{2l}(\xi^2 + \eta^2)\right], \qquad k_2 = \frac{2\pi}{\lambda_2}, \quad l = r\cos\theta.$$

Thus, the complex light distribution behind the hologram can be evaluated:

$$g(\alpha, \beta; \lambda_2) = u(x, y; \lambda_2)T(x, y; \lambda_1) * h_{l2}(x, y; \lambda_2).$$

For simplicity, we use a one-dimensional notation for the following analysis. The real-image reconstruction can be calculated as follows,

$$\begin{aligned}
g_r(\alpha; \lambda_2) &= u(x; \lambda_2)\exp\left[-ik_1\left(\frac{x^2}{2l_1} - x\sin\theta\right)\right] * h_{l2}(x; \lambda_2) \\
&= \int \exp\left[i\frac{k_2}{2l}(x - r\sin\theta)^2\right]\exp\left[-ik_1\left(\frac{x^2}{2l_1} - x\sin\theta\right)\right] \\
&\quad \cdot \exp\left[i\frac{k_2}{2l_2}(\alpha - x)^2\right]dx \\
&= C\exp\left(i\frac{k_2}{2l_2}\alpha^2\right)\int \exp\left[i\left(\frac{k_2}{2l} - \frac{k_1}{2l_1} + \frac{k_2}{2l_2}\right)x^2\right] \\
&\quad \cdot \exp\left\{i\left[\frac{k_2}{l}r\sin\theta - k_1\sin\theta + \frac{k_2}{l_2}\alpha\right]x\right\}dx,
\end{aligned}$$

where C is the proportionality constant.

If we let the quadratic phase factor equal zero, the longitudinal distance of the hologram image can be determined by

$$l_2 = \frac{l_1 l \lambda_1}{\lambda_2 l - l_1\lambda_1}, \tag{8.34}$$

where $l = r\cos\theta$. Thus, $g_r(\alpha; \lambda_2)$ can be reduced to the following form,

$$g_r(\alpha; \lambda_2) = C\exp\left(i\frac{k_2}{2l_2}\alpha^2\right)\int \exp\left\{i\frac{k_2}{2l_2}\left[\frac{l_2}{l}r\sin\theta - \frac{k_1 l_2}{k_2}\sin\theta + \alpha\right]x\right\}dx$$

$$= C_1\delta\left[\alpha + \left(\tan\theta - \frac{\lambda_2}{\lambda_1}\sin\theta\right)l_2\right], \tag{8.35}$$

where $l = r\cos\theta$. Similarly, the virtual-image reconstruction can be computed by

$$g_v(\alpha; \lambda_2) = \int \exp\left[i\frac{k_2}{2l}(x - r\sin\theta)^2\right]\exp\left[ik_1\left(\frac{x^2}{2l_1} - x\sin\theta\right)\right]$$

$$\cdot \exp\left[i\frac{k_2}{2l_2}(\alpha - x)^2\right]dx,$$

which can be shown as

$$g_v(\alpha; \lambda_2) = C\delta\left[\alpha + \left(\tan\theta + \frac{\lambda_2}{\lambda_1}\sin\theta\right)l_2\right], \tag{8.36}$$

where

$$l_2 = -\frac{l_1 l\lambda_1}{\lambda_2 l + l_1\lambda_1}, \tag{8.37}$$

and $l = r\cos\theta$.

The corresponding reconstructions of real and virtual images are depicted in Figure 8.14. ■

■ Example 8.4

We assumed that the hologram of Example 8.3 was constructed with $\lambda_1 = 500$ nm and $l_1 = 5$ cm. If, instead of a monochromatic source, the hologram is illuminated by a white-light point source at an oblique angle $\theta = 45°$ and with a radial distance $r = 0.5$ m away from H, as shown in Figure 8.14,

(a) Calculate the smearing length of the reconstructed hologram images.

(b) Sketch the smeared hologram images and identify the colors at the extreme ends. Assume that the white-light source has a continuous spectral line that varies from 350 to 700 nm.

Answers

(a) Using Eq. 8.34, we can compute the longitudinal locations of the violet and red real images as

$$l_v = \frac{(0.05)(0.5)\cos 45°(550)}{(350)(0.5)\cos 45° - (0.05)(500)}$$

$$= \frac{9.72}{123.73 - 25} = 0.0985 \text{ m}$$

and

$$l_r = \frac{9.72}{247.45 - 25} = 0.044 \text{ m}.$$

The longitudinal smearing length of the real image is therefore

$$\Delta l = l_v - l_r = 0.0985 - 0.044 = 0.0545 \text{ m}.$$

Equation 8.37 helps us to compute the violet and the red virtual images:

$$l'_v = -\frac{(0.05)(0.5)\cos 45°(550)}{(350)(0.5) + (0.05)(500)} = -0.065 \text{ m}$$

and

$$l'_r = -\frac{9.72}{247.45 + 25} = -0.036 \text{ m}.$$

Notice that the minus sign indicates that the image is reconstructed at the front of the hologram. Thus, the longitudinal smearing length of the virtual image is

$$\Delta l = 0.065 - 0.036 = 0.029 \text{ m}.$$

From this we see that, with divergent illumination, the smearing length for the real image appears longer than that for the virtual image.

(b) A sketch of the reconstructed hologram images appears in Figure 8.16. From this sketch we show that the red images appear closer to the holographic

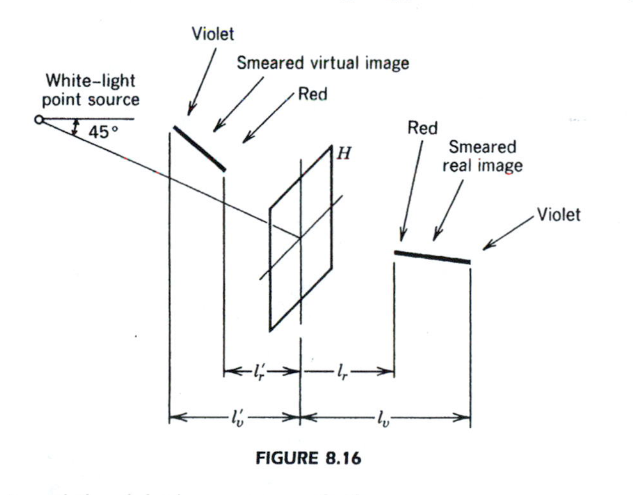

FIGURE 8.16

plate and the violet images appear farther away. ■

■ Example 8.5

Assuming that the modulation index of the object point hologram of Eq. 8.22 is 100 percent, determine the diffraction efficiency of the hologram.

Since the modulation index is 100 percent, Eq. 8.22 can be written as

$$T(x, y) = \frac{1}{2}\left\{1 + \frac{1}{2}\cos\left[\frac{\pi}{2\lambda_1}\left(\frac{x^2 + y^2}{2l_1} - x\sin\theta\right)\right]\right\}.$$

From Eqs. 8.25 and 8.26, the virtual- and real-image reconstructions are

$$g_v = (I_i)^{1/2}\left(\frac{1}{4}\right)\delta\left(\alpha' - 2\frac{\lambda_1}{\lambda_2}\, l_1 \sin\theta, \beta'\right)$$

and

$$g_r = (I_i)^{1/2}\left(\frac{1}{4}\right)\delta(\alpha, \beta),$$

where I_i is the input intensity.

Thus, the diffraction efficiency of the hologram is

$$\text{D.E.} = \frac{(\frac{1}{4})^2 I_i}{I_i} = \frac{1}{16} = 6.25\%. \qquad \blacksquare$$

8.3 HOLOGRAPHIC MAGNIFICATIONS

Since hologram images are generally three-dimensional in nature, in this section, we shall discuss the phenomenon of holographic magnifications.

To compute the *lateral magnification*, we again use the elementary object point holographic concept, as depicted in Figure 8.17. A simpler one-dimensional analog system diagram of Figure 8.17 is given in Figure 8.18.

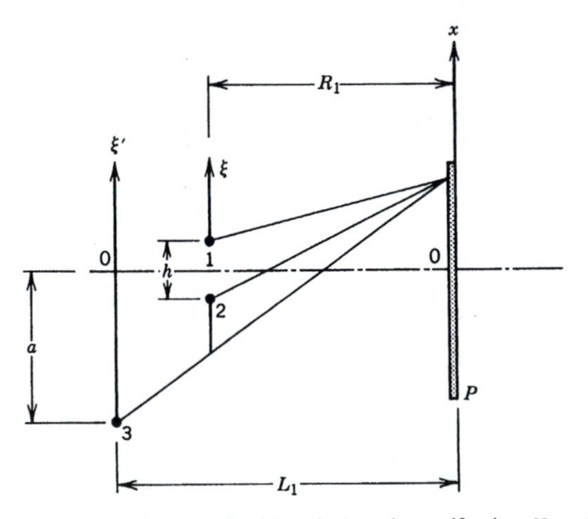

FIGURE 8.17 Recording geometry for determining the lateral magnification. *Note: P*, photographic plate.

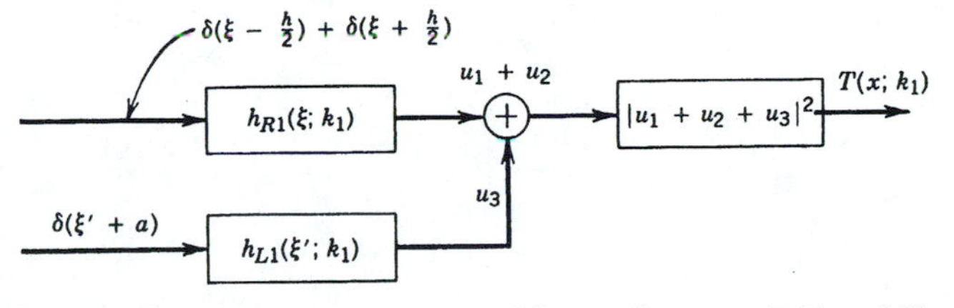

FIGURE 8.18. An analog system diagram of the recording process in Figure 8.17.

The complex light fields distributed over the holographic aperture can be written as

$$u_1(x;k_1) = \exp\left[\frac{ik_1}{2R_1}\left(x - \frac{h}{2}\right)^2\right], \tag{8.38}$$

$$u_2(x;k_1) = \exp\left[\frac{ik_1}{2R_1}\left(x + \frac{h}{2}\right)^2\right], \tag{8.39}$$

and

$$u_3(x;k_1) = \exp\left[\frac{ik_1}{2L_1}(x + \alpha)^2\right], \tag{8.40}$$

where $k_1 = 2\pi/\lambda_1$, and λ_1 designates the construction wavelength. Thus, the amplitude transmittance of the recorded hologram is

$$T(x;k_1) = K_0 + K_1 \cos\frac{k_1}{R_1}hx + K_2[e^{i(*)} + e^{-i(*)}] + K_3[e^{i(**)} + e^{-i(**)}], \tag{8.41}$$

where the K's are real proportionality constants, and

$$(*) = \frac{k_1}{2R_1}\left(x - \frac{h}{2}\right)^2 - \frac{k_1}{2L_1}(x + a)^2, \tag{8.42}$$

$$(**) = \frac{k_1}{2R_1}\left(x + \frac{h}{2}\right)^2 - \frac{k_1}{2L_1}(x + a)^2. \tag{8.43}$$

If the hologram is illuminated by a divergent wavefront of λ_2, as shown in Figure 8.19,

$$u_4(x;k_2) = \exp\left[\frac{ik}{2L_2}(x - b)^2\right], \tag{8.44}$$

then the complex light field behind the hologram can be calculated by

$$g(\alpha;k_2) = u_4(x;k_2)T(x;k_1) * h_l(x;k_2), \tag{8.45}$$

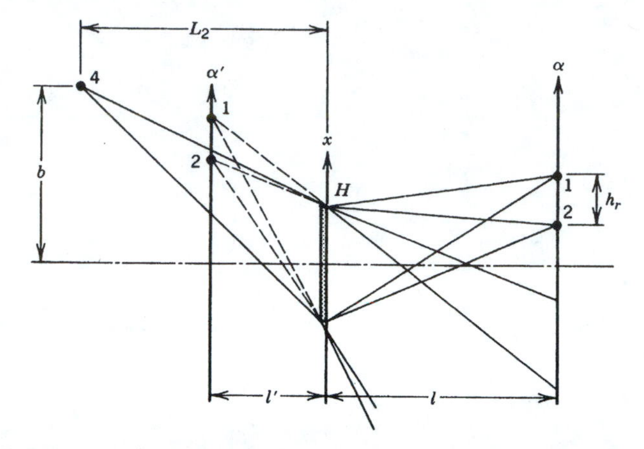

FIGURE 8.19 Reconstruction geometry for determining the lateral magnification. A monochromatic point source is located at 4. From it emanates the illuminating divergent wavefront of λ_2, *Note:* h_r, separation of the real images 1 and 2.

where the asterisk denotes the convolution integral, $k_2 = 2\pi/\lambda_2$, and

$$h_l(x; k_2) = \exp\left(\frac{ik_2}{2} x^2\right). \tag{8.46}$$

Substituting the real-image terms, which are the negative kernels of Eq. 8.41, we can evaluate the real-image diffractions:

$$\begin{aligned} g_r(x; k_2) = &\int \exp\left\{-i\frac{k_2}{2}\left(\frac{\lambda_2}{\lambda_1 R_1} - \frac{\lambda_2}{\lambda_1 L_1} - \frac{1}{L_2} - \frac{1}{l}\right)x^2 \right. \\ &\left. + \frac{2}{l}\left[\alpha + l\left(\frac{b}{L_2} - \frac{\lambda_2 h}{2\lambda_1 R_1} - \frac{\lambda_2 a}{\lambda_1 L_1}\right)\right]x\right\} dx \\ &+ \int \exp\left\{-i\frac{k_2}{2}\left(\frac{\lambda_2}{\lambda_1 R_1} - \frac{\lambda_2}{\lambda_1 L_1} - \frac{1}{L_2} - \frac{1}{l}\right)x^2 \right. \\ &\left. + \frac{2}{l}\left[\alpha + l\left(\frac{b}{L_2} + \frac{\lambda_2 h}{2\lambda_1 R_1} - \frac{\lambda_2 a}{\lambda_1 L_1}\right)\right]x\right\} dx. \end{aligned} \tag{8.47}$$

Eliminating the quadratic phase factors of Eq. 8.47, allows the real images to be reconstructed at

$$l = \frac{\lambda_1 R_1 L_1 L_2}{\lambda_2 L_1 L_2 - \lambda_2 R_1 L_2 - \lambda_1 R_1 L_1}, \tag{8.48}$$

and to take the form (see Figure 8.19)

$$g_r(x; k_2) = \delta\left[\alpha + l\left(\frac{b}{L_2} - \frac{\lambda_2 h}{2\lambda_1 R_1} - \frac{\lambda_2 a}{\lambda_1 L_1}\right)\right] + \delta\left[\alpha + l\left(\frac{b}{L_2} + \frac{\lambda_2 h}{2\lambda_1 R_1} - \frac{\lambda_2 a}{\lambda_1 L_1}\right)\right]. \tag{8.49}$$

Thus, the lateral magnification of the real-image can be written as

$$M_{\text{lat}}^r = \frac{h_r}{h} = \left(1 - \frac{\lambda_1 R_1}{\lambda_2 L_2} - \frac{R_1}{L_1}\right)^{-1}. \tag{8.50}$$

Similarly, the lateral magnification for the virtual image can be written as

$$M_{\text{lat}}^v = \frac{h_v}{h} = \left(1 + \frac{\lambda_1 R_1}{\lambda_2 L_2} - \frac{R_1}{L_1}\right)^{-1}. \tag{8.51}$$

From Eqs. 8.50 and 8.51, we note that

$$M_{\text{lat}}^r \geq M_{\text{lat}}^v$$

for *divergent* reference and *divergent* reconstruction beams. The equality holds if the reference and reconstruction beams are both plane waves.

■ Example 8.6

Referring to the geometry of Figures 8.17 and 8.19, evaluate the lateral magnifications for real and virtual images if both the reference and the reconstruction beams are oblique plane waves.

Letting L_1 and L_2 approach infinity in Eqs. 8.50 and 8.51, we have

$$M_{\text{lat}}^r = \lim_{\substack{L_1 \to \infty \\ L_2 \to \infty}} \left(1 - \frac{\lambda_1 R_1}{\lambda_2 L_2} - \frac{R_1}{L_1}\right)^{-1} = 1$$

and

$$M_{\text{lat}}^v = \lim_{\substack{L_1 \to \infty \\ L_2 \to \infty}} \left(1 + \frac{\lambda_1 R_1}{\lambda_2 L_2} - \frac{R_1}{L_1}\right)^{-1} = 1.$$

Thus, we see that $M_{\text{lat}}^r = M_{\text{lat}}^v$. ■

■ Example 8.7

We assume that a hologram is constructed by a plane reference beam of λ_1. If the hologram images are reconstructed with a *convergent* monochromatic wavefront of λ_2, as illustrated in Figure 8.20, determine the lateral magnifications.

Since the illuminating wavefront converges to a point at a distance behind the hologram, this distance L_2 is regarded as a *negative* quantity for the formulas of Eqs. 8.50 and 8.51:

$$M_{\text{lat}}^r = \left(1 + \frac{\lambda_1 R_1}{\lambda_2 L_2'}\right)^{-1}$$

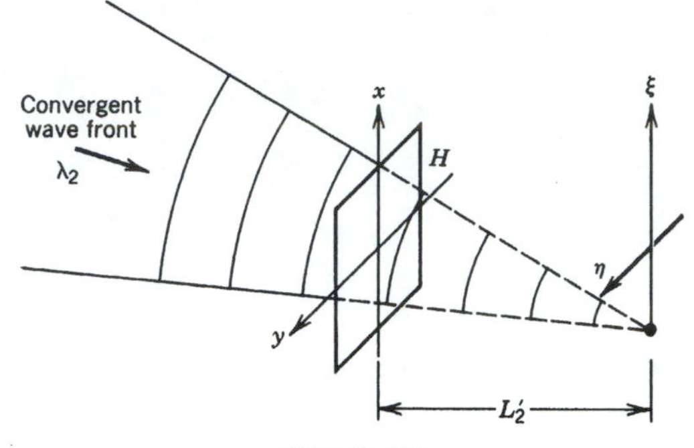

FIGURE 8.20

and

$$M_{\text{lat}}^{v} = \left(1 - \frac{\lambda_1 R_1}{\lambda_2 L_2'}\right)^{-1}.$$

Thus, we have

$$M_{\text{lat}}^{r} < M_{\text{lat}}^{v}$$

for the plane reference beam and the convergent construction beam. ■

8.4 REFLECTION HOLOGRAPHY

By a simple rearrangement of the holographic reconstruction process, it is possible to obtain hologram images using a simple white-light illumination. This reconstruction process is entirely dependent on reflection from the recorded hologram, rather than on transmission through the hologram. Since this technique utilizes a thick emulsion on the photographic plate, reflection holography is also known as thick-emulsion holography. The reflection holography that we discuss is similar in concept to Lippmann's color photography.[1] A reflection hologram is constructed by directing the object and reference beams in opposite directions into the photographic plate. The light waves traveling in opposite directions form standing waves. The interference fringes generated by these standing waves are recorded in the thick emulsion. Similar to Lippmann's photograph, the hologram image can be read out by illumination from a simple white-light source. If polychromatic coherent light is used in the construction process, a color hologram image will be read out by the white-light illumination. Thus, reflection holography is also known as white-light color holography; the concept was first suggested by Yuri Denisyuk in 1962.

[1] Lippmann's color photography depends on the interference fringes in standing electromagnetic waves generated when light is reflected by a mercury coating at the back of a special fine-grained photographic emulsion on the camera's photographic plate. Gabriel Lippmann was awarded the 1908 Nobel prize for physics for discovering this principle, first communicated in 1891.

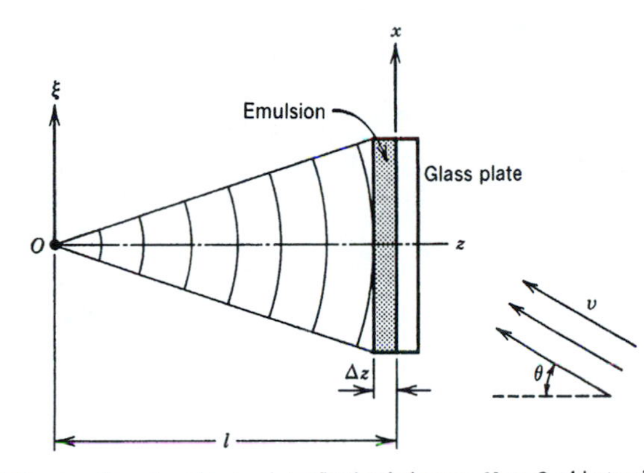

FIGURE 8.21 Construction of an object point reflection hologram. *Note: O*, object point; Δz, thickness of the photographic emulsion.

In the construction process of object point reflection holography, the object beam and the reference beam are combined from opposite directions within the recording medium, as shown in Figure 8.21. The complex light fields of the object and the reference beams as they come together and interfere within the recording plate can be expressed as

$$u(x; k_1) = \exp\left\{ik_1\left[z + \frac{x^2}{2(l+z)}\right]\right\} \tag{8.52}$$

and

$$u(x; k_1) = \exp[-ik_1(z - x\sin\theta)], \tag{8.53}$$

for $-\Delta z \leq z \leq 0$, where $k_1 = 2\pi/\lambda_1$.

If we assume that the *developed* photographic density is proportional to the intensity of the construction process, the density distribution of the encoded hologram is

$$D(x; k_1) = K_1 + K_2 \cos\left\{k_1\left[2z + \frac{x^2}{2(l+z)} - x\sin\theta\right]\right\}, \tag{8.54}$$

where the K's are proportionality constants. From this equation we see that there are thin holograms arranged in parallel layers within the emulsion, and that these thin holograms are spaced about half the recording wavelength apart, or $\lambda_1/2$. If we assume further that reflectance of these thin holograms is proportional to the photographic density, the reflectance function of the encoded hologram is

$$r(x; k_1) = K_1' + K_2' \cos\left\{k_1\left[2z + \frac{x^2}{2(l+z)} - x\sin\theta\right]\right\}, \tag{8.55}$$

for $-\Delta z \leq z \leq 0$, where the K's are proportionality constants.

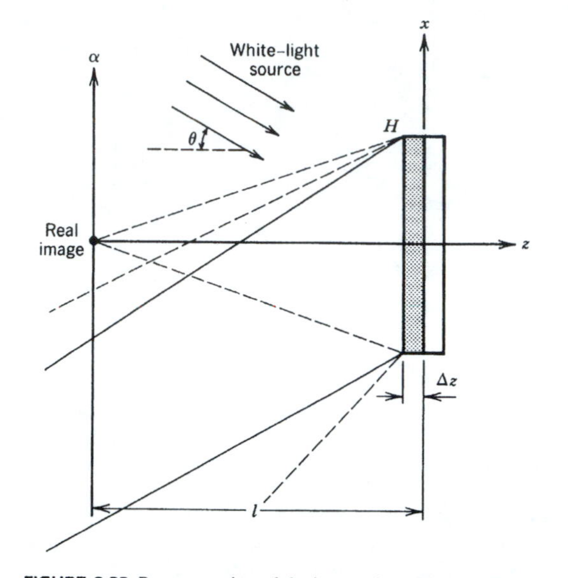

FIGURE 8.22 Reconstruction of the image of a reflection hologram.

If the recorded hologram is illuminated by a white-light source, as shown in Figure 8.22, the wavelength satisfying Bragg's law will be reflected. The complex light field of the selected wavelength can be computed by

$$g(x; k_1) = \exp[ik_1(z - x_1 \sin\theta)]r(x; k_1) * h_l(x; k_1), \tag{8.56}$$

with the asterisk denoting the convolution integral, and

$$h_l(x; k_1) = C \exp\left(\frac{ik_1}{2l} x^2\right).$$

Substituting Eq. 8.55 into Eq. 8.56, we have

$$\begin{aligned} g(x; k_1) &= C_1 \exp(-ik\alpha \sin\theta) \\ &+ C_2 \exp\left\{\frac{k_1}{4(l+z)} [\alpha - 2(l+z)\sin\theta]^2\right\} \\ &+ C_3 \delta(\alpha, \beta) \qquad \text{for} \quad -\Delta z \le z \le 0, \end{aligned} \tag{8.57}$$

where the C's are the complex constants. We note that the first term is the zero-order diffraction, and that the second and third terms are the divergent and the convergent hologram-image terms, respectively.

Example 8.8

Suppose that the object point hologram of Figure 8.21 is constructed with two coherent light sources of wavelengths λ_1 and λ_2.

(a) Draw an analog system diagram to represent this holographic construction with two wavelengths.

(b) What will the reflectance function of the encoded hologram be?

Answers

(a) Since the hologram construction is performed by two mutually coherent sources, the analog system diagram of the construction process is as shown in Figure 8.23.

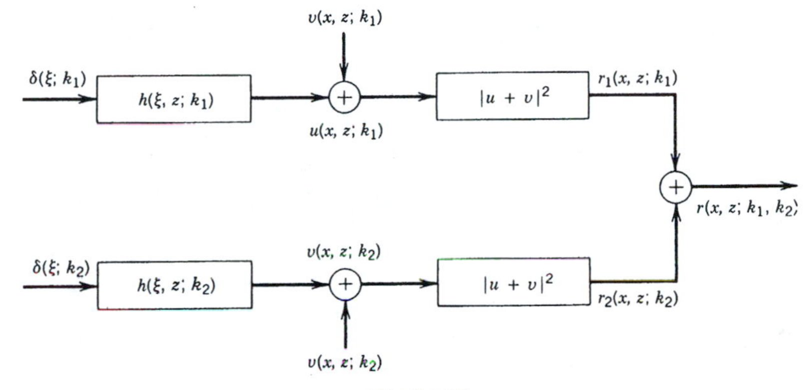

FIGURE 8.23

(b) Using Eq. 8.55, we can write the resulting reflectance function of the thick-emulsion hologram as

$$r(x, z; k_1, k_2) = r_1(x, z; k_1) + r_2(x, z; k_2)$$
$$= K_1 + K_2 \cos\left\{k_1\left[2z + \frac{x^2}{2(l+z)} - x\sin\theta\right]\right\}$$
$$+ K_3 \cos\left\{k_2\left[2z + \frac{x^2}{2(l+z)} - x\sin\theta\right]\right\},$$

where the K's are proportionality constants. ■

Example 8.9

Suppose that the reflection hologram of Eq. 8.55 is constructed with a coherent light source of wavelength $\lambda_1 = 500$ nm.

(a) Calculate the separation of the subholograms within the emulsion.

(b) Assuming that the photographic emulsion is about 10 μm thick, how many subholograms will be constructed within the holographic plate?

Answers

(a) Referring to Eq. 8.55, we see that the subholograms are constructed when

$$\cos\left(\frac{4\pi}{\lambda_1} z\right) = 1,$$

which can be shown as

$$\frac{4\pi}{\lambda_1} z = 2n\pi, \qquad n = 0, 1, 2, \ldots.$$

Thus, the separation of the subholograms is

$$z = \frac{\lambda_2}{2} = \frac{500}{2} = 250 \text{ nm}.$$

(b) Since the emulsion is about 10 μm thick, the number of subholograms would be

$$N = \frac{10}{0.25} = 40.$$

■

8.5 RAINBOW HOLOGRAPHY

We now discuss a technique for producing hologram images using a simple, inexpensive white-light source. This type of hologram is capable of producing brighter and more colorful hologram images than those previously discussed. Since these types of hologram images are observed through the transmitted light field, and because they produce rainbow color images, they are called *rainbow holograms.*

As indicated in Section 8.2, a real hologram image can be reconstructed with a conjugate coherent illumination. In fact, hologram images can be reconstructed using a very small holographic aperture, with only a minor degree of resolution loss. In other words, it is possible to reconstruct the entire hologram image when the aperture is reduced to a narrow slit, as shown in Figure 8.24. For convenience of discussion, we call hologram H_1 the primary hologram.

In rainbow holographic recording, we insert a fresh holographic plate H_2. To minimize color blurring in the rainbow hologram image, we recommend that this holographic plate be placed near the hologram image plane. Then, if the holographic plate is properly recorded in the linear region of the T–E curve, the resultant hologram H_2 will be a rainbow hologram.

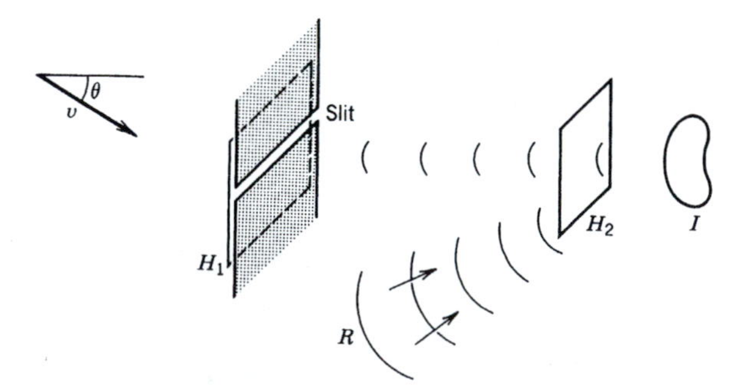

FIGURE 8.24 A two-step rainbow holographic construction. *Note:* v, conjugate coherent illumination; H_1, primary hologram; R, convergent reference beam; H_2, holographic plate; I, real image.

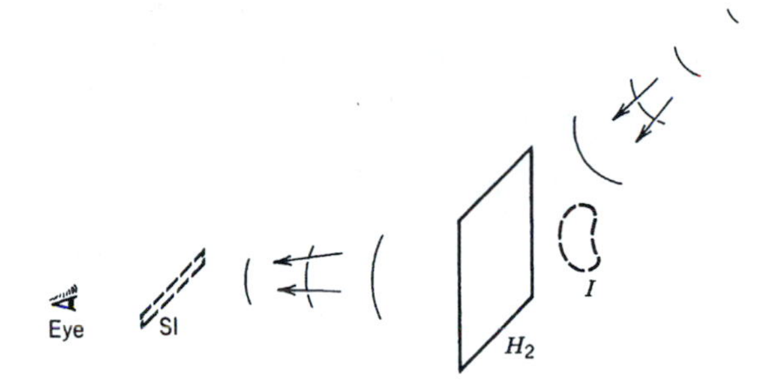

FIGURE 8.25 Holographic reconstruction with a coherent source. *Note:* SI, slit image; H_2, holographic plate; I, virtual hologram image; S, monochromatic point source.

In order to see the rainbow effect, we first reconstruct the hologram image from the rainbow hologram H_2. If we look through the real slit image, we expect to see a virtual hologram image behind holographic plate H_2. If holographic plate H_2 is inserted behind the hologram image I during the rainbow holographic construction, a real hologram image will be seen through the slit image because the reconstruction process shown in Figure 8.25 takes place. Since the real slit image is convergently reconstructed, we see a brighter image.

As we recall from our discussion of holographic magnification in Section 8.3, the location of the real slit image varies as a function of the reconstruction wavelength of the light source. In other words, when the reconstruction wavelength is longer, the slit width is wider and the slit image appears to be located higher and closer to the hologram aperture H_2. The same effect applies to the hologram image seen through the slit image. When the reconstruction wavelength is longer, the hologram image appears to be larger and closer to the holographic plate H_2.

Now, if hologram H_2 is illuminated by a conjugate divergent white-light source, as shown in Figure 8.26, the hologram slit images produced by the different wavelengths of the white-light source will separate into rainbow colors in the space of the real slit image. The hologram image of the object behind the hologram will take on the same rainbow effect. If we view this image transversely through the smeared slit image, we will see it in a succession of rainbow colors. In other words, if we

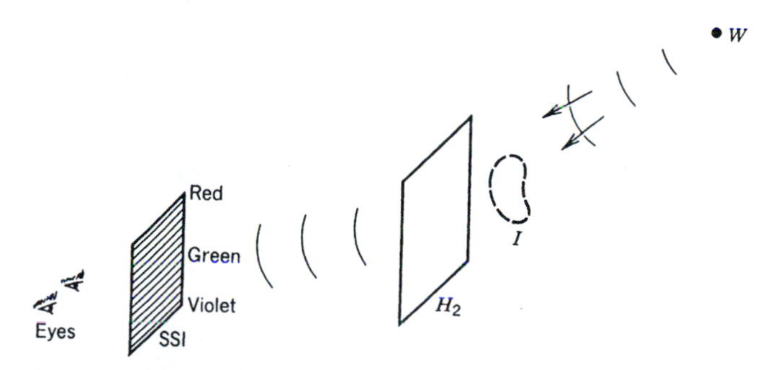

FIGURE 8.26 Rainbow holographic reconstruction with a white-light source. *Note:* SSI, smeared slit image; H_2, hologram; I, virtual hologram image; W, divergent white-light source.

view the image through the red-colored slit image, we will see a red-colored hologram image, and if we peer through the green-colored slit image, we will see a smaller green hologram image.

We are not able, however, to see the "over" or "under" of the object image by moving our eyes transversely up and down against the smeared slit image, as we would in conventional holography. Thus, one of the consequences of using this type of rainbow holography is that the vertical parallax is lost, although the full horizontal parallax is retained. This means that we still have a right-to-left view for binocular stereopsis and motion parallax, and that the sensation of a three-dimensional scene is preserved.

Example 8.10

Given a rainbow holographic construction, as shown in Figure 8.27.

(a) Draw an analog system diagram to represent the holographic construction.

(b) If the real hologram image, derived from the primary hologram, represents an object point, calculate the amplitude transmittance function of the rainbow hologram.

Answers

(a) A one-dimensional analog system diagram of the rainbow holographic construction is given in Figure 8.28.

(b) Using the analog system diagram, we have

$$u_1(x; k_1) = \exp\left[\frac{ik_1}{2l_1}\left(x - \frac{w}{2}\right)^2\right],$$

$$u_2(x; k_1) = \exp\left[\frac{ik_1}{2l_1}\left(x + \frac{w}{2}\right)^2\right],$$

$$u_3(x; k_1) = \exp\left(-i\frac{k}{2l_2}x^2\right),$$

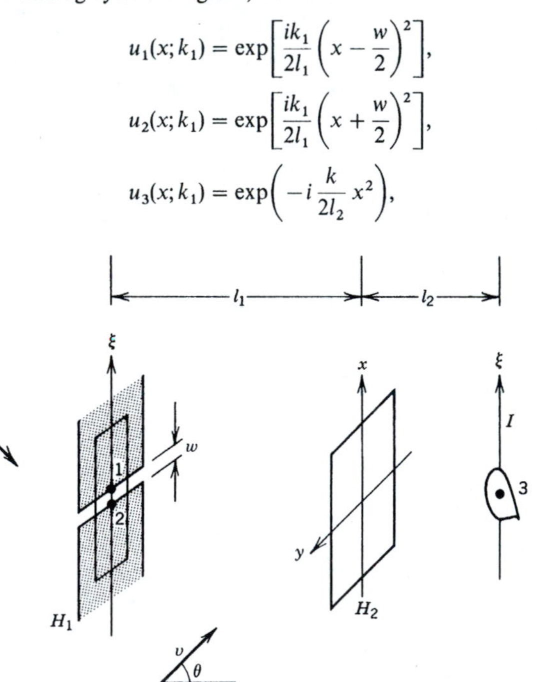

FIGURE 8.27

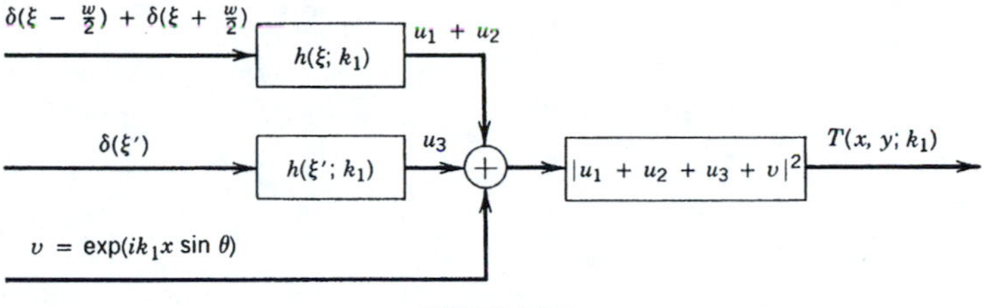

FIGURE 8.28

and

$$v(x; k_1) = \exp(ik_1 x \sin\theta).$$

Thus, the amplitude transmittal of the rainbow hologram, as an object point representation, is

$$\begin{aligned} T(x, k_1) &= |u_1 + u_2 + u_3 + v|^2 \\ &= K_1 + K_2 \cos\left(\frac{k_1}{l_1} xw\right) + K_3 \cos\left\{-\frac{k_1}{2}\left[\frac{x^2}{l_2} + \frac{(x - w/2)^2}{l_1}\right]\right\} \\ &\quad + K_4 \cos\left\{-\frac{k_1}{2}\left[\frac{x^2}{l_2} + \frac{(x + w/2)^2}{l_1}\right]\right\} \\ &\quad + K_5 \cos\left\{k_1\left[\frac{(x - w/2)^2}{2l_1} - x\sin\theta\right]\right\} \\ &\quad + K_6 \cos\left\{k_1\left[\frac{(x + w/2)^2}{2l_1} - x\sin\theta\right]\right\} \\ &\quad + K_7 \cos\left[-k_1\left(\frac{x^2}{2l_2} + x\sin\theta\right)\right]. \end{aligned}$$

■

■ Example 8.11

Let us assume that the parameters of the rainbow holographic construction in Figure 8.27 are $l_1 = 30$ cm, $l_2 = 2$ cm, $w = 2$ mm, $\lambda_1 = 600$ nm, and $\theta = 30°$. The rainbow hologram is illuminated by a broad-band plane wave of uniform spectral distribution, from $\lambda = 600$ nm to 400 nm, as shown in Figure 8.29.

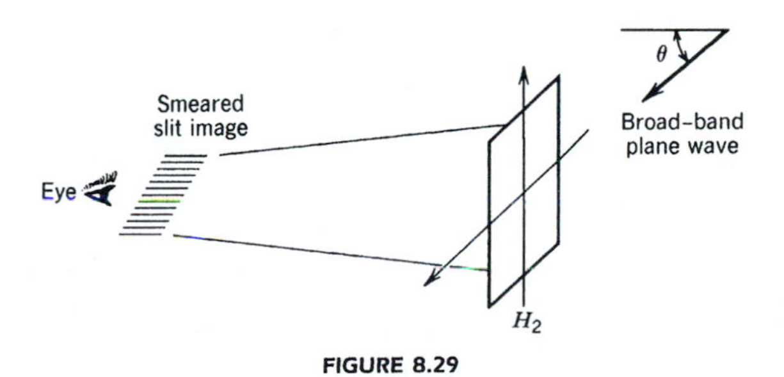

FIGURE 8.29

(a) Evaluate the length of the smeared slit image.

(b) If the rainbow hologram image is viewed through the smeared slit image by an unaided eye with a 2.5-mm pupil, calculate the wavelength spread over the pupil.

Answers

(a) The fifth and sixth terms of the amplitude transmittance function in Example 8.10 represent the construction of the real slit image, that is, object points 1 and 2 in Figure 8.27. The location of the real slit image can be computed, as

$$g(\alpha; k) = \exp(-ikx\sin\theta)\exp\left\{-ik_1\left[\frac{(x \pm w/2)^2}{2l_1} - x\sin\theta\right]\right\} * \exp\left(i\frac{k}{2l}x^2\right),$$

which can be written

$$g(\alpha; k) = C\exp\left(\frac{ik}{2l}\alpha^2\right)\int \exp\left[\frac{i}{2}\left(\frac{k}{l} - \frac{k_1}{l_1}\right)x^2\right]$$
$$\cdot \exp\left\{\frac{-ik}{l}\left[\alpha \pm \frac{k_1 l w}{2kl_1} - \frac{k_1 - k}{k} l\sin\theta\right]x\right\}dx.$$

When the quadratic phase factor is eliminated, the longitudinal distance of the slit image is $l = (\lambda_1/\lambda)l_1$. The transverse locations of the slit edge images can be found by

$$\alpha = \pm\frac{k_1 l w}{2kl_1} + \frac{k_1 - k}{k} l\sin\theta.$$

By substituting $l = (\lambda_1/\lambda)l_1$, we have

$$\alpha = \pm\frac{w}{2} + \frac{\lambda - \lambda_1}{\lambda} l_1 \sin\theta.$$

By substituting $\lambda = \lambda_1 = 600$ nm, we have $l = l_1$ and $\alpha_{12} = \pm w/2 = \pm 1$ mm. The slit image is formed at the same location as the slit. However, for $\lambda = 400$ nm, we have

$$l = \frac{600}{400} \times 30 = 40 \text{ cm},$$

$$\alpha_1 = 1 - \frac{600 - 400}{400}(300)(0.5) = -74 \text{ mm},$$

and

$$\alpha_2 = -1 - \frac{6 - 4}{4}(300)(0.5) = -76 \text{ mm}.$$

Thus, the length of the smeared slit image is

$$L = [(1 + 76)^2 + (400 - 300)^2]^{1/2} = 126.2 \text{ mm}.$$

(b) From part a, we see that the slit image spreads uniformly about 77 mm, as projected along the vertical axis. Thus, the wavelength spread over the pupil is

$$\frac{(200)(2.5)}{77} = 6.5 \text{ nm}. \quad \blacksquare$$

8.6 ONE-STEP RAINBOW HOLOGRAMS

In the previous section we discussed the general concept of the rainbow holographic process. This process requires two recording steps: first, use the conventional off-axis holographic technique to make a primary hologram from a real object; then record a rainbow hologram from the real hologram image of the primary hologram. Placing a narrow-slit aperture behind the primary hologram in the second step of the holographic process means that the reconstruction light source need not be coherent. However, a two-step holographic recording process is cumbersome and requires a separate optical setup for each step, a major undertaking for laboratories with limited resources for optical components.

In this section we illustrate one-step technique for producing rainbow holograms. This technique offers certain flexibilities in the construction of rainbow holograms, and the optical arrangement is simpler than that for the conventional two-step process.

As noted, making a rainbow hologram requires recording a real hologram image of the object through a narrow slit. If the rainbow hologram is illuminated by a monochromatic light source, a real hologram image of the object is produced, but the vertical parallax of the image is limited by the narrow slit. If the rainbow hologram is illuminated by a white-light source, however, the hologram image of the slit will disperse in rainbow colors. Therefore, the basic goal of rainbow holography is to form the image of the slit aperture between the hologram image of the object and the observer. Figures 8.30 and 8.31 show how a lens or lens system can in a one-step process simultaneously image both the object and the slit. In this manner a rainbow hologram can be made without a primary hologram.

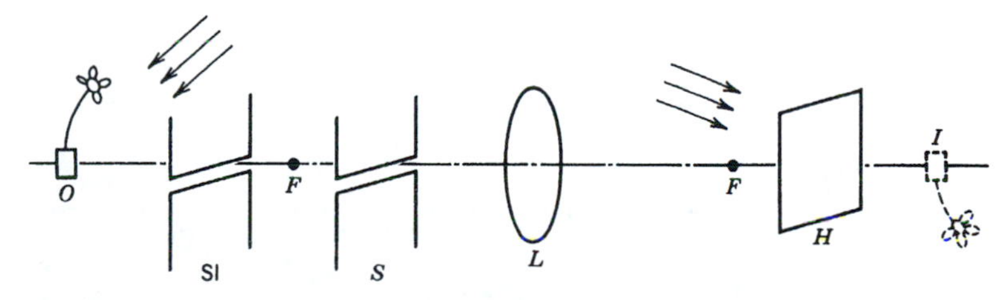

FIGURE 8.30 One-step rainbow holographic construction for pseudoscopic imaging. *Note:* SI, slit image; *F*, focal point of the lens; *S*, slit, *L*, imaging lens; *H*, holographic plate.

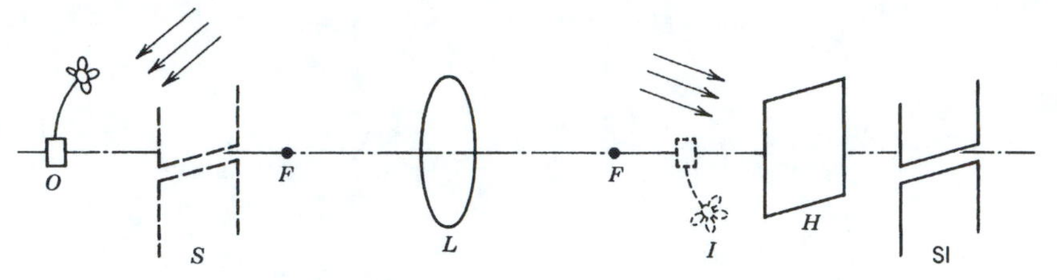

FIGURE 8.31 One-step rainbow holographic construction for orthoscopic imaging. *Note:* S, slit; F, focal point of the lens; L, imaging lens; H, holographic plate; SI, slit image.

Example 8.12

Consider the one-step rainbow holographic construction of Figure 8.32. If the narrow slit is located at $f/4$ away from the imaging lens,

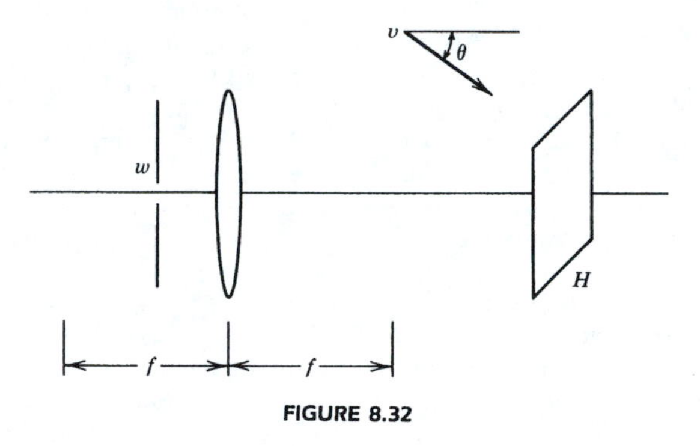

FIGURE 8.32

(a) Draw an analog system diagram to evaluate the slit image produced by the imaging lens.
(b) Determine the location of the slit image.
(c) Draw an equivalent optical setup to replace Figure 8.32.
(d) Draw an analog system diagram of the rainbow holographic construction.

Answers

(a) The analog system diagram is shown in Figure 8.33.

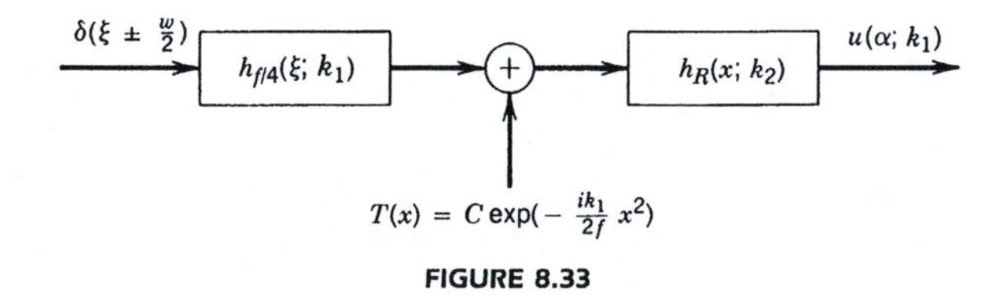

FIGURE 8.33

(b) The location of the slit image can be calculated by

$$u(\alpha; k_1) = \left\{\left[\delta\left(\xi \pm \frac{w}{2}\right) * h_{f/4}(\xi; k_1)\right] T(x; k_1)\right\} * h_R(x; k_1)$$
$$= C \exp\left(i\frac{k_1}{2R}\alpha^2\right) \int \exp\left[i\frac{k_1}{2}\left(\frac{3}{f} + \frac{1}{R}\right)x^2\right]$$
$$\exp\left\{-i\frac{k_1}{R}\left[\left(\alpha \pm \frac{2R}{f}w\right)x\right]\right\} dx\, dy.$$

Thus, we see that the slit image will be located at

$$R = -\frac{f}{3}$$

and

$$\alpha_{1,2} = \pm\tfrac{2}{3}w.$$

(c) An equivalent optical setup without the lens is shown in Figure 8.34.

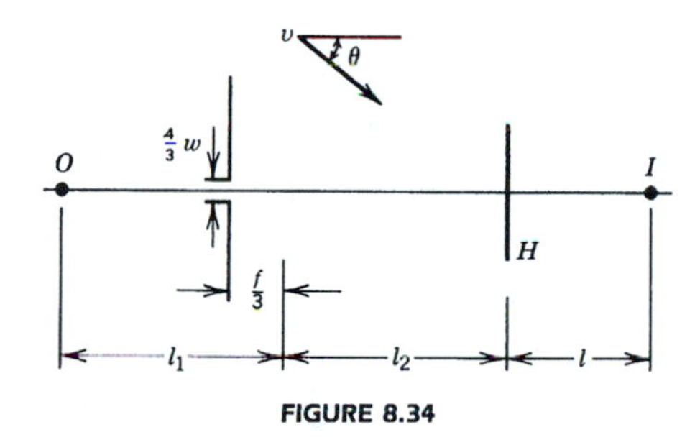

FIGURE 8.34

(d) An analog system diagram of the rainbow holographic construction is given

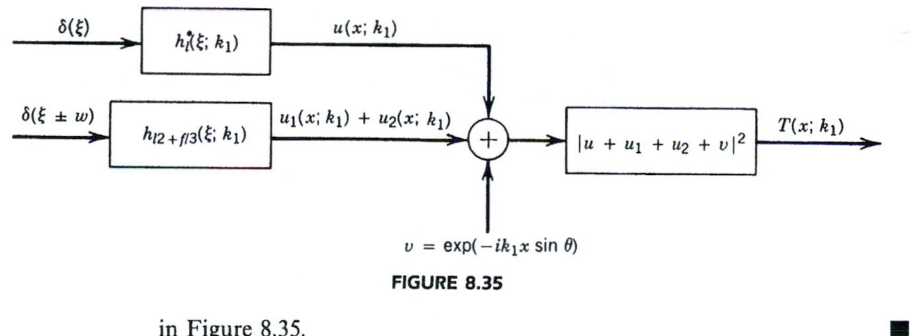

FIGURE 8.35

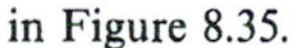

in Figure 8.35. ■

■ Example 8.13

Show that the rainbow hologram image produced by the one-step process in Example 8.12 is pseudoscopic.

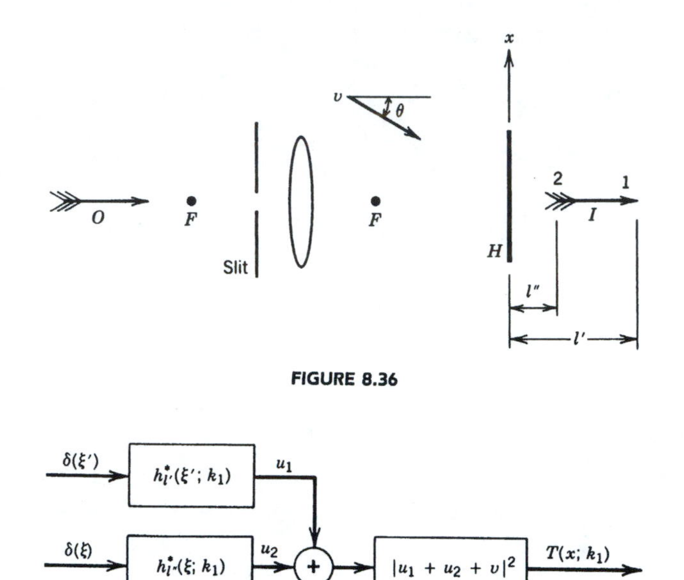

FIGURE 8.36

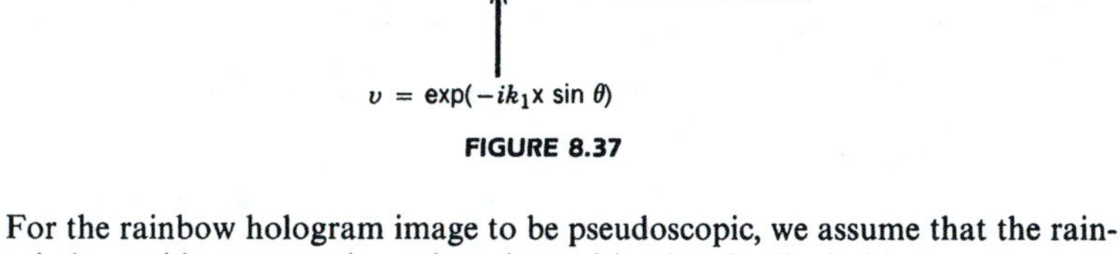

FIGURE 8.37

For the rainbow hologram image to be pseudoscopic, we assume that the rainbow holographic construction takes place with a longitudinal object, as shown in Figure 8.36. An analog system diagram of the construction process is given in Figure 8.37. Thus, the holographic amplitude transmittance function is

$$T(x; k_1) = K_1 + K_2 \cos\left[\frac{k_1}{2}\left(\frac{1}{l''} - \frac{1}{l'}\right)x^2\right] + K_3 \cos\left[k_1\left(\frac{x^2}{2l''} - x\sin\theta\right)\right]$$
$$+ K_4 \cos\left[k_1\left(\frac{x^2}{2l'} - x\sin\theta\right)\right].$$

If the hologram is illuminated by a conjugate plane wave of λ, as shown in Figure 8.38, the divergent or virtual-image diffractions can be computed by

$$g_2(\alpha; k) = \int \exp(ikx\sin\theta)\exp\left[ik_1\left(\frac{x^2}{2l''} - x\sin\theta\right)\right]$$
$$\cdot \exp\left[i\frac{k}{2l}(\alpha - x)^2\right]dx$$
$$= \exp\left(i\frac{k}{2l}\alpha^2\right)\int \exp\left[i\left(\frac{k_1}{2l''} + \frac{k}{2l}\right)x^2\right]\exp\left(i\frac{k}{l}\alpha x\right)dx,$$

$$g_2(\alpha; k)\Big|_{l = -(\lambda_1/\lambda)l''} = C\delta_2(\alpha),$$

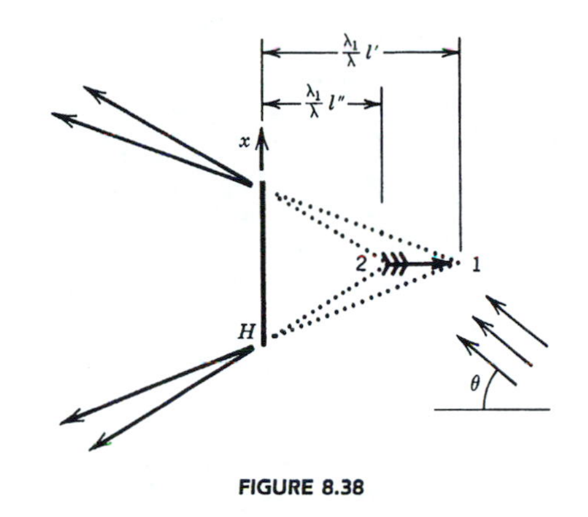

FIGURE 8.38

and

$$g_1(\alpha; k)\Big|_{l=-(\lambda_1/\lambda)l'} = C\delta_1(\alpha).$$

Since hologram image point 2 appears closer to the holographic plate than image point 1, we conclude that the rainbow hologram image is pseudoscopic. ■

8.7 COLOR HOLOGRAPHY

This chapter would be incomplete without a discussion of color holography. The aim of this section is to review briefly two frequently used techniques for generating color hologram images with white light.

The best-known color holographic process using white light is the reflection holography invented by Yuri Denisyuk. As we noted earlier, in 1962 he reported a technique in which the process of holography was combined with the form of color photography that had been developed by the French physicist Gabriel Lippmann in 1891. In other words, Denisyuk's work is one of the cornerstones of white-light holography, combining, as it does, the work of Lippmann and of Dennis Gabor by using coherent light for holographic construction and white light for hologram image reconstruction.

In this method a coherent polychromatic wave field, with the primary colors of light, passes through a recording plate, falls on a diffused color object, and then is reflected back to the recording plate, as shown in Figure 8.39. As in the Lippmann color photography process, interferometric fringes are formed throughout the depth of the emulsion covering the plate. The color hologram, which has the characteristics of a photograph produced by the Lippmann process, can be viewed with a white-light source of limited spatial extent, for example, an ordinary high-intensity desk lamp or a slide projector, as illustrated in Figure 8.40. Although such color hologram images have been widely demonstrated, reflection color holography does have several drawbacks which prevent widespread practical applications. Two of these drawbacks are that (1) an elaborate film-processing technique

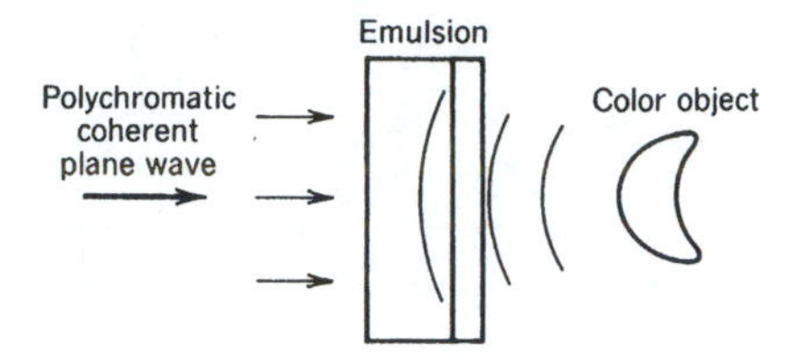

FIGURE 8.39 A reflection holographic construction.

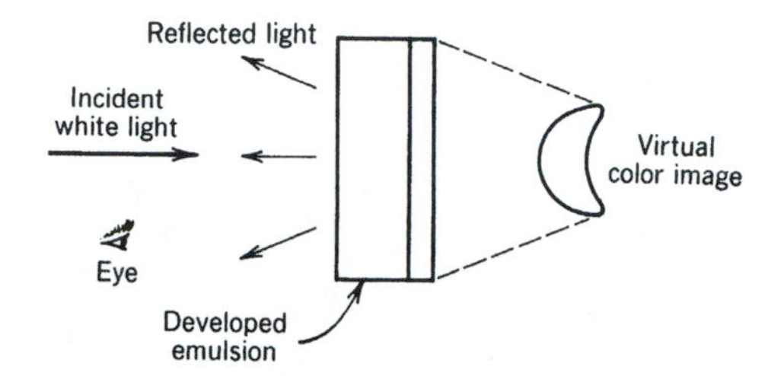

FIGURE 8.40 Reconstruction of a reflection hologram image using white-light illumination.

is required to prevent the emulsion from shrinking; and (2) the efficiency of the hologram image diffraction is rather low. Nevertheless, the reflection hologram image can be viewed by direct white-light illumination, and it is useful for decorative display purposes.

In 1969 another type of color holography was reported by Stephen Benton. The hologram produced has been called a white-light transmission hologram, but it is best known as a rainbow hologram (Section 8.5).

Benton's two-step technique for constructing a true-color rainbow hologram is rather cumbersome. First, three primary holograms have to be constructed using three primary color coherent sources. Then the projected real images of these three primary holograms are sequentially multiplexed onto a fourth hologram, again with three-color coherent read out. These three primary holograms must be aligned very carefully to make certain that their reconstructions fit exactly one on top of another. Thus, this technique is complicated and not easily implemented.

We shall now describe a one-step technique for generating a color rainbow hologram. The optical setup is illustrated in Figure 8.41. An He–Ne laser is used to provide the red light (6328 Å), and an argon laser provides the green and blue lights (5145 Å and 4765 Å). The illuminated object is imaged through an imagining lens to a plane just in front of the hologram. A narrow slit of about 1.5 mm is placed between the object and the focal plane of the imaging lens. A collimated reference beam ensures that the carrier spatial frequency is the same across the hologram. The intensities of the three lights are measured independently, and the exposure time for each is calculated. The hologram is first exposed to the red light of the He–Ne laser, then the green light (5145 Å) of the argon laser. The argon laser is then tuned to the blue line (4765 Å), and a third exposure is made.

A Kodak 649F plate is used because it has a relatively flat spectral response. As the plate is developed, a rainbow hologram is formed. When the hologram is viewed with a white-light point source, a very bright color image can be reconstructed. As in the two-step technique, a true-color hologram image can be observed when the

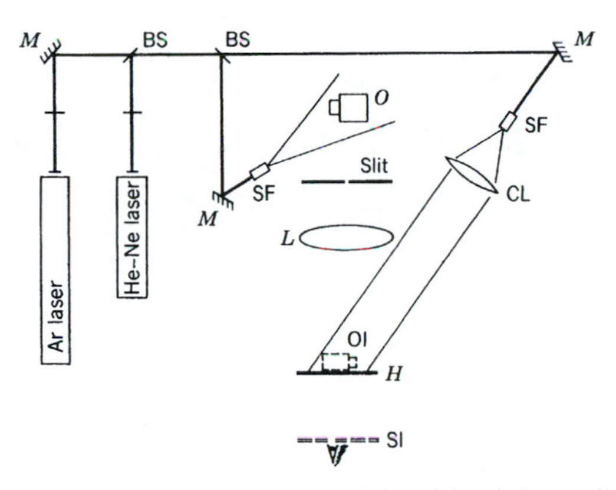

FIGURE 8.41 A one-step technique for constructing a color rainbow hologram. *Note: M*, mirror, BS, beam splitter; SF, spatial filter; *O*, object; *L*, imaging lens; CL, collimated lens; OI, object image; SI, slit image.

hologram is viewed in the correct plane. If the viewer moves off this plane, different shades of color can still be seen, but the color will be different from that of the original object.

REFERENCES

1. F. T. S. Yu, *Optical Information Processing*, Wiley-Interscience, New York, 1983, Chapters 10 and 11.
2. J. W. Goodman, *Introduction to Fourier Optics*, McGraw-Hill, New York, 1968, Chapter 8.
3. D. Gabor, "A New Microscope Principle," *Nature*, Vol. *161*, 777 (1948).
4. Y. N. Denisyuk "Photographic Reconstruction of the Optical Properties of an Object in Its Own Scattered Radiation Field," *Soviet Physics Doklady*, Vol. 7, 543 (1962).
5. S. A. Benton, "Hologram Reconstructions with Extended Light Sources," *Journal of the Optical Society of America*, Vol. *59*, 1545 (1969).

PROBLEMS

8.1 Let the on-axis object point hologram of Eq. 8.4 be normally illuminated by a *divergent* monochromatic point source, which is located at a distance $2l_1$ from the hologram.

(a) Draw an analog system diagram of the holographic construction process.

(b) Calculate the locations of the virtual and real hologram images, if the reconstruction wavelength is $0.8\lambda_1$.

8.2 If a *convergent* monochromatic wavefront, which converges at a distance $\frac{1}{2}l_1$ behind an on-axis object point hologram, is normally incident on the hologram.

(a) Draw a schematic diagram to represent the holographic reconstruction.
(b) Draw an analog system diagram of the reconstruction process.
(c) Evaluate the virtual and real images, assuming that the illuminating wave is the same as the construction wavelength.

8.3 Use the on-axis object point hologram of Eq. 8.4.
(a) Plot the amplitude transmittance function of the hologram along the x axis.
(b) If the hologram was constructed with wavelength $\lambda_1 = 500$ nm, at a distance of $l_1 = 20$ cm, and with a 10×10-cm^2 square recording aperture, compute the spatial frequency near the edge of the hologram.

8.4 Consider the on-axis hologram of Example 8.1. If the hologram is illuminated obliquely from a 45° angle by a collimated white light, calculate the smearing lengths of the hologram images. The spectral content of the white-light source is assumed to be uniformly distributed in the range from 350 to 700 nm.

8.5 For the on-axis object point hologram of Example 8.1, we assume that the separation between the object point and the holographic aperture is 0.2 m (i.e., $l_1 = 0.2$ m). The hologram is illuminated by a normally incident white light.
(a) Calculate the hologram image reconstruction.
(b) What is the smearing length of the reconstructed images?
(c) By comparing this result with the result obtained from Example 8.2, draw an explicit conclusion.

8.6 An on-axis object point hologram is constructed with a reference beam for which the object-to-reference beam (intensity) ratio is 1:3. If we assume that the hologram is recorded in the linear region of the T–E curve, what will the diffraction efficiency of the hologram be?

8.7 Consider the on-focus holographic construction shown in Figure 8.42.
(a) Draw an analog system diagram of the holographic construction.
(b) Show that the hologram image can be viewed by a simple white-light illumination.
(c) What color would you expect the hologram image to be? Explain briefly.

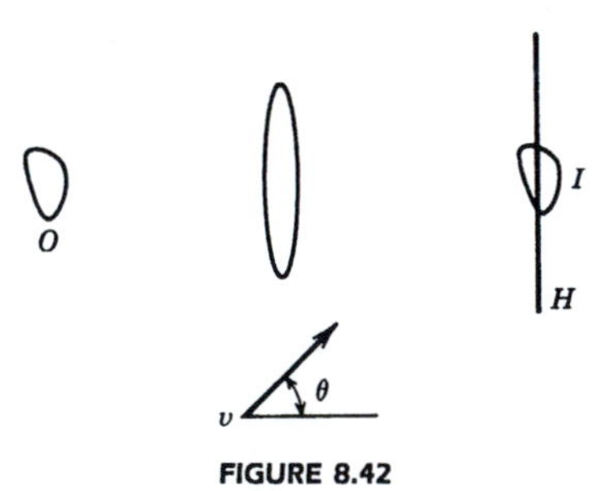

FIGURE 8.42

8.8 An off-axis object point hologram is constructed with a convergent reference beam as depicted in Figure 8.43.

(a) Draw an analog system diagram of the holographic construction.
(b) Evaluate the amplitude transmittance function of the recorded hologram.

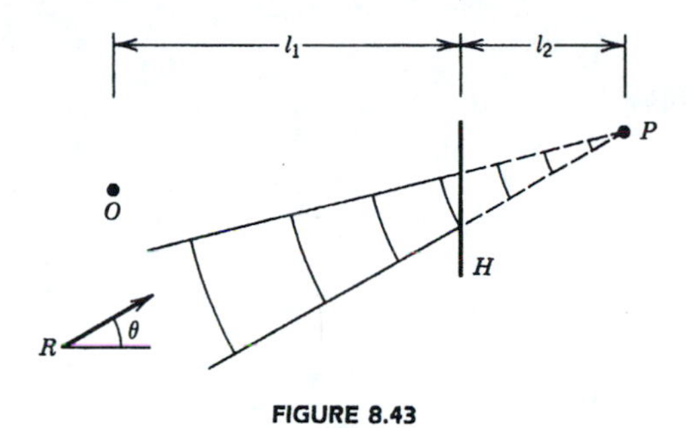

FIGURE 8.43

8.9 Assume that the off-axis hologram of Figure 8.43 is reconstructed with a conjugate divergent point source, located at point P, of the same wavelength.
(a) Draw an analog system diagram of the holographic reconstruction.
(b) Evaluate the real and virtual hologram images.

8.10 Assume that the distance parameters of the off-axis hologram in Problem 8.8 are $l_1 = 30$ cm, $l_2 = 20$ cm, and that the oblique angle θ of the convergent reference beam is 30°. If the hologram is illuminated by a conjugate white-light point source located at point P, as shown in Figure 8.43,
(a) Calculate the vertical smearing length of the real image.
(b) Identify the red and violet color images. Notice that the white-light point source has a uniform spectral density from 350 to 700 nm.

8.11 In Section 3.3 we saw that a positive lens is capable of performing a two-dimensional Fourier transformation. If a holographic construction takes place because of the Fourier transform property of a lens, as shown in Figure 8.44,
(a) Draw an analog system diagram of the holographic recording.

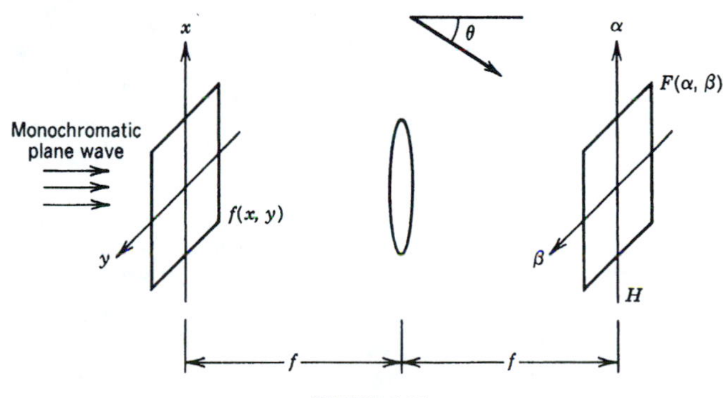

FIGURE 8.44

(b) Assuming that the holographic construction is recorded in the linear region of the T–E curve (see Section 4.5), calculate the amplitude transmittance function of the encoded hologram. (*Note:* this type of hologram is called a *Fourier hologram.*)

(c) Calculate the carrier spatial frequency of the hologram.

8.12 If the Fourier hologram of Problem 8.11 is inserted at the front focal length of a Fourier transform lens, as depicted in Figure 8.45,

(a) Draw an analog system diagram of the hologram image reconstruction.

(b) Evaluate the output of the complex light distribution.

(c) Sketch the locations of the hologram images.

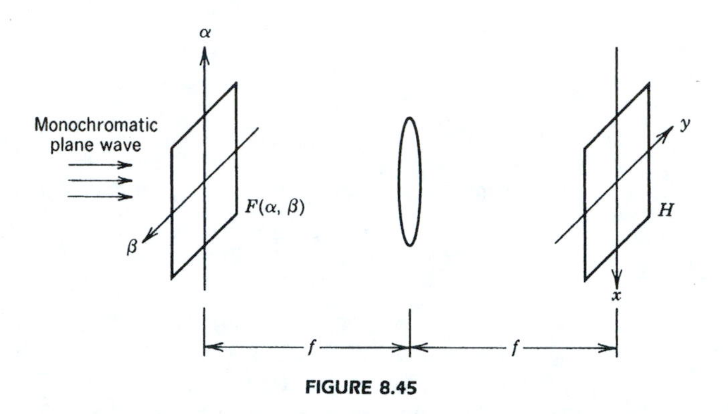

FIGURE 8.45

8.13 Assume that the wavelength of the off-axis holographic construction in Figure 8.11 is $\lambda_1 = 500$ nm, that the separation between the extended object and the holographic aperture is $l_1 = 30$ cm, and that the oblique angle θ is 45°.

(a) Calculate the carrier spatial frequency of the hologram.

(b) Assuming that the Fresnel diffraction from the object is spatial-frequency-limited (i.e., finite in spatial-frequency bandwidth), sketch the spectral distributions (i.e., the spatial-frequency contents) of the hologram along the x axis.

(c) If the spatial-frequency bandwidth of the extended object beam is about 50 lines/mm, compute the minimum oblique angle θ of the reference beam needed to separate the spectral contents of the hologram images.

8.14 The off-axis hologram of Problem 8.13 is illuminated by a conjugate divergent point source, located at $l_2 = 35$ cm, and the reconstruction wavelength is $\lambda_2 = 600$ nm.

(a) Calculate the lateral magnifications of the real and virtual hologram images.

(b) If the reconstruction wavelength is $\lambda_2 = 400$ nm, compute the corresponding lateral magnifications of hologram images.

(c) Calculate the longitudinal locations of the hologram images of parts a and b, and state the effects of the hologram images produced by the reconstruction wavelength.

8.15 Repeat Problem 8.14 for the hologram images whose reconstruction is done by a conjugate convergent wavefront that converges to a point at a distance 35 cm (i.e., $l_2 = -35$ cm) behind the hologram. Compare the results with those of Problem 8.14.

8.16 Hologram image reconstruction is also known as wavefront reconstruction.
(a) Write an expression for the off-axis wavefront construction process.
(b) If the hologram is illuminated by the same reference beam, show that an object wavefront can be generated from the hologram.
(c) On the other hand, if the hologram is illuminated by a conjugate reference beam, show that a conjugate object wave field, which represents the real-image reconstruction, can be generated.

8.17 The hologram image blurring (i.e., light dispersion) that is due to white-light illumination can be minimized by making a reflection hologram with near-field object recording, as shown in Figure 8.46.
(a) Draw an analog system diagram of the holographic construction.
(b) Evaluate the reflectance function of the hologram.
(c) Determine the spacing between the subholograms within the emulsion.

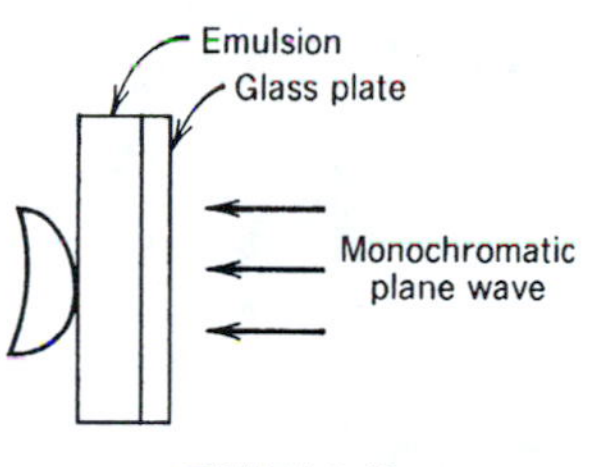

FIGURE 8.46

8.18 If the reflection hologram of Problem 8.17 is illuminated by a normally incident white light, show that color blurring (i.e., light dispersion) of the object will be minimized when the object point is closer to the recording emulsion.

8.19 Show that a true-color reflection hologram can be constructed with the optical setup of Figure 8.46, if the holographic construction is carried out with red, green, and blue coherent lights.

8.20 For the two-step rainbow holographic construction depicted in Figure 8.24, show that the hologram image is *orthoscopic* if the rainbow hologram is illuminated from behind by a conjugate plane wave.

8.21 If the oblique angle of the reference beam for the rainbow hologram of Example 8.11 is 50° (i.e., $\theta = 50°$), show that the color blurring of the hologram image observed by the unaided eye would be reduced.

8.22 Determine the location of the real-slit image if the rainbow hologram of Problem 8.21 is illuminated by a conjugate monochromatic plane wave of wavelength $\lambda_2 = 700$ nm. Notice that $l_1 = 30$ cm, $\lambda_1 = 600$ nm, $w = 2$ mm, and $\theta = 50°$.

8.23 If the slit width of the rainbow holographic construction of Example 8.11 is reduced to 1 mm instead of 2 mm, show that the color blurring of the rainbow hologram image under observation would be reduced.

8.24 If the narrow slit of the one-step rainbow holographic construction shown in Figure 8.30 is located at $f/2$ (i.e., a half of the focal length),

(a) Calculate the location and slit width of the slit image.

(b) Draw an equivalent schematic diagram representing the holographic construction.

(c) Draw an object point analog system diagram for the schematic diagram of part b.

8.25 If the rainbow hologram described in Problem 8.24 is viewed with a white-light source,

(a) Calculate the location of the white-light source for observing the rainbow hologram images. Notice that in practice the holographic emulsion is not negligibly thin.

(b) Draw a schematic diagram representing the reconstruction of the rainbow hologram images. Sketch the location of the smeared slit image.

8.26 To encode an orthoscopic rainbow hologram imaging, during the holographic construction, we place a narrow slit at $l = 1.5f$ (i.e., 1.5 times the focal length) in front of the imaging lens.

(a) Calculate the location and the slit width of the slit image produced by the imaging lens.

(b) Compare the results of part a with those of part a of Problem 8.24.

(c) Sketch an equivalent schematic diagram representing the one-step rainbow holographic construction.

8.27 (a) Draw a schematic diagram for the rainbow hologram imaging of Problem 8.26.

(b) Show that the rainbow hologram imaging is orthoscopic.

9 SIGNAL PROCESSING

The recent advances in real-time spatial-light modulators and electrooptic devices have brought optical signal processing to a new stage of development. Much attention has been focused on high-speed optical signal processing and on computing at a high rate of data processing. In this chapter we discuss the basic principles of optical signal processing under coherent, incoherent, and partially coherent illuminations.

9.1 AN OPTICAL SYSTEM UNDER COHERENT AND INCOHERENT ILLUMINATION

Let us consider the hypothetical optical system depicted in Figure 9.1. The light emitted by the source Σ is monochromatic. The complex light distribution at the input plane comes from the incremental light source $d\Sigma$. To determine the light field at the output plane, we let this complex light distribution at the input plane be $u(x, y)$. If the complex amplitude transmittance of the input plane is $f(x, y)$, the complex light field immediately behind the signal plane would be $u(x, y)f(x, y)$.

We assume that the optical system under consideration is a linear and spatially invariant with a spatial impulse response of $h(x, y)$. The complex light field at the output plane of the system, which comes from $d\Sigma$, can then be determined by the

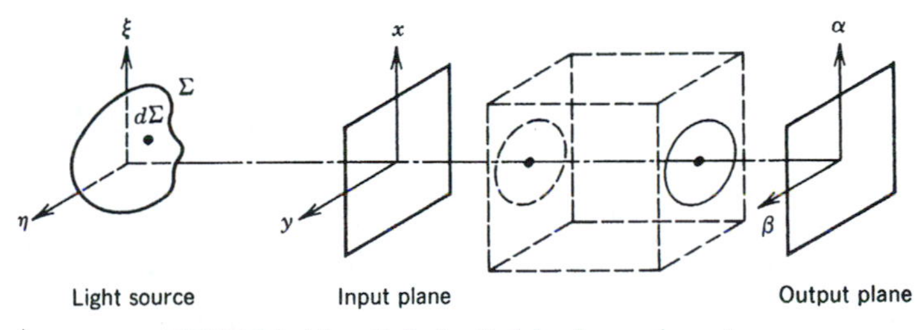

FIGURE 9.1 A hypothetical optical signal-processing system.

following convolution equation,

$$g(\alpha, \beta) = [u(x, y)f(x, y)] * h(x, y), \tag{9.1}$$

where the asterisk denotes the convolution operation.

The corresponding intensity distribution at the output plane, which is due to $d\Sigma$, is

$$dI(\alpha, \beta) = g(\alpha, \beta)g^*(\alpha, \beta)\, d\Sigma, \tag{9.2}$$

where the asterisk represents the complex conjugate. Thus, the overall intensity distribution at the output plane is

$$I(\alpha, \beta) = \iint |g(\alpha, \beta)|^2\, d\Sigma, \tag{9.3}$$

which can be written in the following convolution form,

$$I(\alpha, \beta) = \iiiint_{-\infty}^{\infty} \Gamma(x, y; x', y')h(\alpha - x, \beta - y)h^*(\alpha - x', \beta - y') \cdot f(x, y)f^*(x', y')\, dx\, dy\, dx'\, dy', \tag{9.4}$$

where

$$\Gamma(x, y; x', y') = \iint_{\Sigma} u(x, y)u^*(x', y')\, d\Sigma \tag{9.5}$$

is the *spatial coherence function*, which is also known as the *mutual intensity function.*

Let us now choose two arbitrary points Q_1 and Q_2 at the input plane. If r_1 and r_2 are the respective distances from Q_1 and Q_2 to $d\Sigma$, the complex light disturbances at Q_1 and Q_2 that come from $d\Sigma$ can be written as

$$u_1(x, y) = \frac{[I(\xi, \eta)]^{1/2}}{r_1} e^{ikr_1} \tag{9.6}$$

and

$$u_2(x', y') = \frac{[I(\xi, \eta)]^{1/2}}{r_2} e^{ikr_2} \tag{9.7}$$

where $I(\xi, \eta)$ is the intensity distribution of the light source. By substituting Eqs. 9.6 and 9.7 in Eq. 9.5, we have

$$\Gamma(x, y; x', y') = \iint_{\Sigma} \frac{I(\xi, \eta)}{r_1 r_2} \exp[ik(r_1 - r_2)]\, d\Sigma. \tag{9.8}$$

When the light rays are paraxial, $r_1 - r_2$ can be approximated by

$$r_1 - r_2 \simeq \frac{1}{r}[\xi(x - x') + \eta(y - y')], \tag{9.9}$$

where r is the distance between the source plane and the signal plane. Then Eq. 9.8 can be written as

$$\Gamma(x, y; x', y') = \frac{1}{r^2} \iint I(\xi, \eta) \exp\left\{i \frac{k}{r} [\xi(x - x') + \eta(y - y')]\right\} d\xi \, d\eta, \quad (9.10)$$

which represents the inverse Fourier transform for intensity distribution at the source plane. Equation 9.10 is also known as the Van Cittert–Zernike theorem. The normalized form of this theorem is given in Eq. 6.47.

Now let us consider two extreme situations. In one we let the light source become infinitely large, and we assume that it is uniform, that is, $I(\xi, \eta) \simeq K$. Thus, Eq. 9.10 becomes

$$\Gamma(x, y; x', y') = K_1 \delta(x - x', y - y'), \quad (9.11)$$

where K_1 is a proportionality constant. This equation describes a completely *incoherent* optical system.

On the other hand, if we let the light source be vanishingly small, then $I(\xi, \eta) \simeq K\delta(\xi, \eta)$ and Eq. 9.10 becomes

$$\Gamma(x, y; x', y') = K_2, \quad (9.12)$$

where K_2 is an arbitrary constant. This equation describes a completely *coherent* optical system. In other words, a monochromatic point source describes a strictly coherent regime, whereas an extended source describes a strictly incoherent system. Furthermore, an extended monochromatic source is also known as a *spatially incoherent* source.

For the completely incoherent optical system described in Eq. 9.11, [$\Gamma(x, y; x', y') = K_1 \delta(x - x', y - y')$, the intensity distribution at the output plane, given in Eq. 9.4, becomes

$$I(\alpha, \beta) = \iiiint_{-\infty}^{\infty} \delta(x' - x, y' - y) h(\alpha - x, \beta - y) \cdot h^*(\alpha - x', \beta - y') f(x, y) f^*(x', y') \, dx \, dy \, dx' \, dy', \quad (9.13)$$

which can be reduced to

$$I(\alpha, \beta) = \iint_{-\infty}^{\infty} |h(\alpha - x, \beta - y)|^2 |f(x, y)|^2 \, dx \, dy. \quad (9.14)$$

It is therefore apparent that for incoherent illumination, the intensity distribution at the output plane is the convolution of the input signal's intensity in relation to the intensity of the spatial impulse response. In other words, an incoherent optical system is linear in *intensity*, or

$$I(\alpha, \beta) = |h(x, y)|^2 * |f(x, y)|^2, \quad (9.15)$$

where the asterisk denotes the convolution operation. An analog system diagram of an incoherent system is shown in Figure 9.2. The output intensity response can

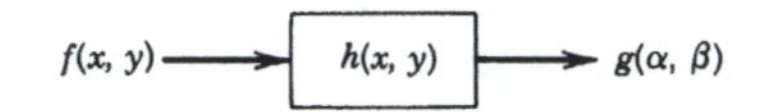

FIGURE 9.2 An analog system diagram of an incoherent system. *Note:* $h_i(x, y)$ represents the intensity of the spatial impulse response.

$f(x, y)$ → $h(x, y)$ → $g(\alpha, \beta)$

FIGURE 9.3 An analog system diagram of a coherent system. *Note:* $h(x, y)$ represents the spatial impulse response.

be determined by

$$I_o(\alpha, \beta) = \iint_{-\infty}^{\infty} I_i(x, y) h_i(\alpha - x, y - \beta)\, dx\, dy, \tag{9.16}$$

where $I_i(x, y)$ is the input intensity excitation and $h_i(x, y) = |h(x, y)|^2$ is the intensity of the spatial impulse response.

However, for the strictly coherent illumination described in Eq. 9.12, $\Gamma(x, y; x', y') = K_2$, Eq. 9.4 becomes

$$I(\alpha, \beta) = g(\alpha, \beta) g^*(\alpha, \beta) = \iint_{-\infty}^{\infty} h(\alpha - x, \beta - y) f(x, y)\, dx\, dy$$

$$\cdot \iint_{-\infty}^{\infty} h^*(\alpha - x', \beta - y') f^*(x', y')\, dx'\, dy'. \tag{9.17}$$

It is therefore apparent that the coherent optical system is linear in *complex amplitude*, or

$$g(\alpha, \beta) = \iint_{-\infty}^{\infty} h(\alpha - x, \beta - y) f(x, y)\, dx\, dy. \tag{9.18}$$

An analog system diagram of Eq. 9.18 is given in Figure 9.3.

Example 9.1

Consider the spatial impulse response of an input–output optical system that is given by

$$h(x, y) = \text{rect}\left(\frac{x}{\Delta x}\right) - \text{rect}\left(\frac{x - \Delta x}{\Delta x}\right)$$

and shown in Figure 9.4*a*.

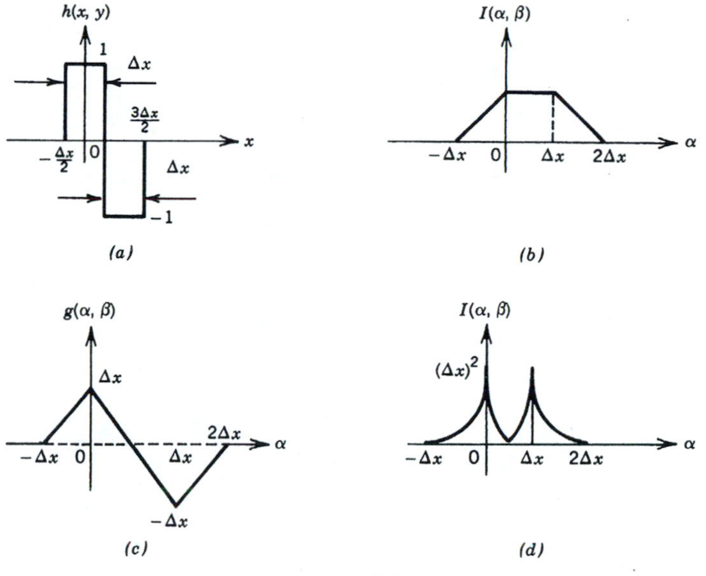

FIGURE 9.4

(a) If the optical system is illuminated by a spatially limited incoherent wavefront such as

$$f(x, y) = \operatorname{rect}\left(\frac{x}{\Delta x}\right),$$

calculate the corresponding irradiance at the output plane.

(b) If the spatially limited illumination of the optical system is a coherent wavefront, compute the corresponding complex light field at the output plane.

Answers

(a) From Eq. 9.15, under incoherent illumination we have

$$\begin{aligned} I(\alpha, \beta) &= |f(x, y)|^2 * |h(x, y)|^2 \\ &= \operatorname{rect}\left(\frac{x}{\Delta x}\right) * \operatorname{rect}\left(\frac{x - \Delta x/2}{2\Delta x}\right). \end{aligned}$$

The graphical sketch of this result is given in Figure 9.4*b*.

(b) From Eq. 9.18, under coherent illumination we have

$$\begin{aligned} g(\alpha, \beta) &= f(x, y) * h(x, y) \\ &= \operatorname{rect}\left(\frac{x}{\Delta x}\right) * \left[\operatorname{rect}\left(\frac{x}{\Delta x}\right) - \operatorname{rect}\left(\frac{x - \Delta x}{\Delta x}\right)\right]. \end{aligned}$$

A graphical sketch is given in Figure 9.4*c*, and the corresponding irradiance is shown in Figure 9.4*d*. ■

Example 9.2

The transfer function of the optical system shown in Figure 9.1 is given by

$$H(p, q) = K_1 \exp[i(\alpha_0 p + \beta_0 q)],$$

where K_1, α_0, and β_0 are arbitrary positive constants. We assume that the complex amplitude transmittance at the input plane is

$$f(x, y) = K_2 e^{i\phi(x,y)},$$

where K_2 is an arbitrary positive constant. Calculate the output responses under incoherent and coherent illuminations, respectively.

To compute the output responses, we first evaluate the spatial impulse response of the optical system by

$$h(x, y) = \mathscr{F}^{-1}[H(p, q)] = K_1 \delta(x + \alpha_0, y + \beta_0).$$

We refer to Eq. 9.15 to calculate the output response under incoherent illumination,

$$\begin{aligned} I(\alpha, \beta) &= |f(x, y)|^2 * |h(x, y)|^2 \\ &= K_2^2 * K_1^2 \delta^2(x + \alpha_0, y + \beta_0) \\ &= K_1^2 K_2^2. \end{aligned}$$

We use Eq. 9.18 to calculate the output response under coherent illumination,

$$\begin{aligned} g(\alpha, \beta) &= f(x, y) * h(x, y) \\ &= K_2 e^{i\phi(x,y)} * K_1 \delta(x + \alpha_0, y + \beta_0) \\ &= K_1 K_2 \exp[i\phi(x + \alpha_0, y + \beta_0)]. \end{aligned}$$

Notice that the phase distribution is preserved under coherent illumination. ■

9.2 COHERENT OPTICAL SIGNAL PROCESSING

Referring to the phase transform properties of lenses discussed in Section 3.3, we see that transform lenses can be put together to construct a coherent optical signal processor, as depicted in Figure 9.5, where P_1, P_2, and P_3 represent the input, the Fourier, and the output planes, respectively, and a monochromatic point source S is located at the front focal length of a collimating lens. If an object transparency of amplitude transmittance $f(x, y)$ is inserted at the input plane P_1, the complex light field distributed at P_2 would be the Fourier transform of $f(x, y)$, or

$$F(p, q) = \mathscr{F}[f(x, y)], \tag{9.19}$$

where $p = (2\pi/f\lambda)x$ and $q = (2\pi/f\lambda)y$ are the angular spatial-frequency coordinates. A positive lens will perform a direct Fourier transformation, in which the transform kernel, $e^{-i(px+qy)}$, is negative. But a positive transform kernel, $e^{i(px+qy)}$, is needed

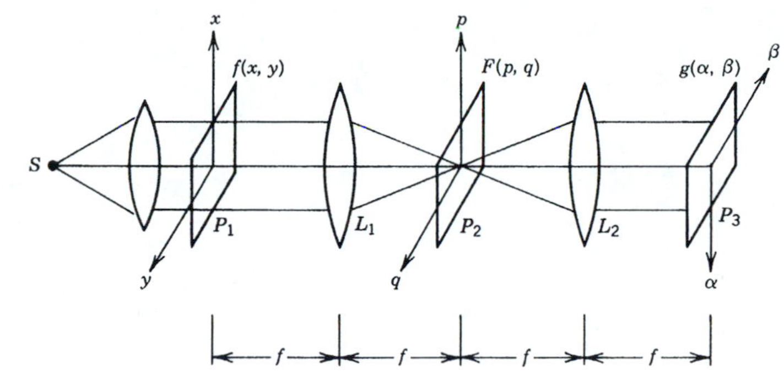

FIGURE 9.5 A coherent optical signal processor.

for an inverse Fourier transformation. It is therefore apparent that a positive transform kernel can be introduced, with the help of a second positive lens, by simply inverting the output coordinate system (α, β), as shown in the output plane P_3 of Figure 9.5. Thus, the complex light distributions at the output plane P_3 can be shown to be

$$f(\alpha, \beta) = \mathscr{F}^{-1}[F(p, q)]. \tag{9.20}$$

An analog system diagram of Figure 9.5 is given in Figure 9.6.

Let us now assume that a spatial filter of complex amplitude transmittance $H(p, q)$ is inserted in the Fourier plane P_2. This complex light field P_2 immediately behind the spatial filter is then

$$E(p, q) = KF(p, q)H(p, q), \tag{9.21}$$

where K is a proportionality constant.

Since the second lens L_2 performs an inverse Fourier transformation of the complex light field $E(p, q)$ to the output plane P_3, the complex-amplitude light distribution at P_3 can be shown by

$$g(\alpha, \beta) = K \iint_S F(p, q)H(p, q)e^{i(p\alpha + q\beta)}\, dp\, dq, \tag{9.22}$$

where the surface integration is taken over the spatial-frequency domain P_2.

Alternatively, according to the Fourier multiplication property, Eq. 9.22 can be written as

$$g(\alpha, \beta) = K \iint_S f(x, y)h(\alpha - x, \beta - y)\, dx\, dy = Kf(x, y) * h(x, y), \tag{9.23}$$

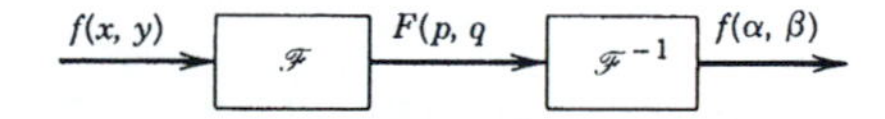

FIGURE 9.6 An analog system diagram of the signal processor in Figure 9.5.

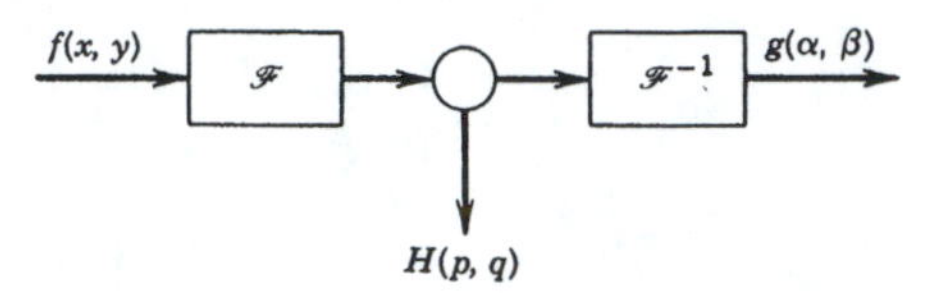

FIGURE 9.7 An analog system diagram of the signal processor in Figure 9.5, with a spatial filter insertion.

where the integral is taken at the input spatial domain, and $h(x, y)$ is the spatial impulse response of the filter:

$$h(x, y) = \mathscr{F}^{-1}[H(p, q)]. \tag{9.24}$$

An analog system diagram of the optical signal processor is shown in Figure 9.7.

It is important to stress that the spatial filter $H(p, q)$ can consist of apertures or slits of any shape. Depending on the arrangement of the apertures, it can act as a low-pass, high-pass, or band-pass spatial filter. Clearly any opaque portion in the filter represents a rejection of the spatial-frequency band. In addition, inclusion of a phase plate with the filter would produce a phase delay. Since we are able to construct amplitude filters and phase filters separately, in principle we are able to construct any complex spatial filter. However, Anthony Vander Lugt developed an interferometric technique for constructing a complex spatial filter, as we show in the next section. Vander Lugt constructed the filter using a Fourier hologram.

Example 9.3

Assume that the amplitude transmittance of an object function is given by

$$f(x, y)[1 + \cos(p_0 x)],$$

where p_0 is an arbitrary angular carrier spatial frequency, and $f(x, y)$ is assumed to be spatial-frequency-limited. If this object transparency is inserted at the input plane of the coherent optical signal processor in Figure 9.5,

(a) Determine the corresponding spectral distribution at the Fourier plane P_2.
(b) Design a stop band filter for which the light distribution at the output plane P_3 will be $f(x, y)$.

Answers

(a) Since lens L_1 will perform a direct Fourier transform, the complex light distribution at P_2 can be shown to be

$$\mathscr{F}\{f(x, y)[1 + \cos(p_0 x)]\} = F(p, q) + \tfrac{1}{2}F(p - p_0, q) + \tfrac{1}{2}F(p + p_0, q).$$

This distribution is sketched in Figure 9.8*a*.

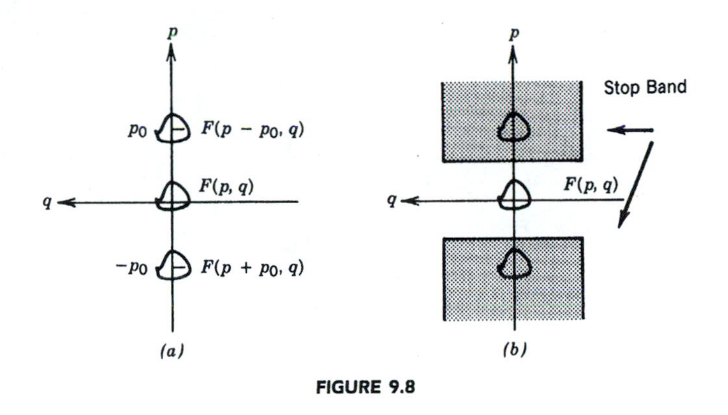

FIGURE 9.8

(b) A stop band filter for the spectral distribution shown in Figure 9.8*a* is sketched in Figure 9.8*b*. This is essentially a low-pass filter which allows $F(p, q)$ to pass through. Since L_2 will perform an inverse Fourier transformation, the complex light field at P_3 is

$$\mathscr{F}^{-1}[F(p, q)] = f(x, y).$$

■

■ Example 9.4

Consider the coherent optical signal processor shown in Figure 9.9*a*. The spatial filter is a one-dimensional sinusoidal grating,

$$H(p) = \tfrac{1}{2}[1 + \sin(\alpha_0 p)],$$

where α_0 is an arbitrary constant that is equal to the separation of the input object functions $f_1(x, y)$ and $f_2(x, y)$. Compute the complex light field at the output plane P_3.

By applying the Fourier translation property, we can show that the complex light field at P_2 is

$$F_1(p, q)e^{-i\alpha_0 p} + F_2(p, q)e^{i\alpha_0 p}.$$

The complex light distribution immediately behind the filter is

$$\begin{aligned} E(p, q) &= [F_1(p, q)e^{-i\alpha_0 p} + F_2(p, q)e^{i\alpha_0 p}]H(p) \\ &= \tfrac{1}{2}F_1(p, q)e^{-i\alpha_0 p} + \tfrac{1}{2}F_2(p, q)e^{i\alpha_0 p} \\ &\quad + \frac{1}{4i} F_1(p, q) - \frac{1}{4i} F_1(p, q)e^{-i2\alpha_0 p} - \frac{1}{4i} F_2(p, q) \\ &\quad + \frac{1}{4i} F_2(p, q)e^{i2\alpha_0 p}. \end{aligned}$$

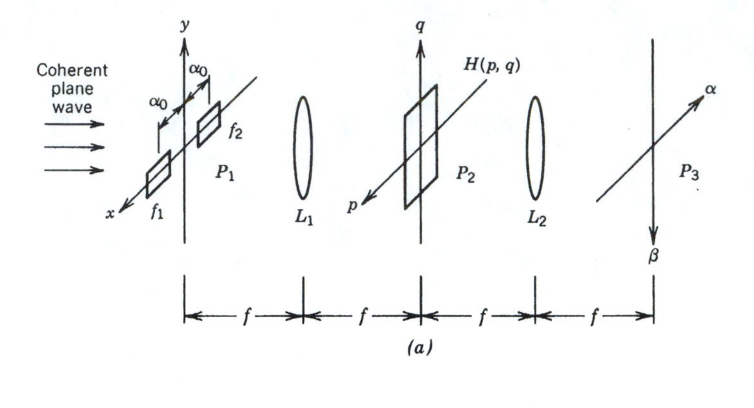

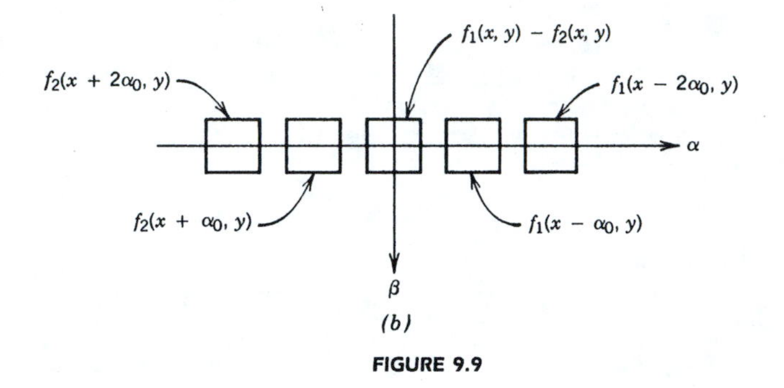

FIGURE 9.9

The light distribution at the output plane is therefore

$$
\begin{aligned}
g(\alpha, \beta) &= \mathscr{F}^{-1}[E(p, q)] \\
&= \tfrac{1}{2} f_1(x - \alpha_0, y) + \tfrac{1}{2} f_2(x + \alpha_0, y) \\
&\quad + \frac{1}{4i} [f_1(x, y) - f_2(x, y)] \\
&\quad - \frac{1}{4i} f_1(x - 2\alpha_0, y) + \frac{1}{4i} f_2(x + 2\alpha_0, y).
\end{aligned}
$$

A sketch of $g(\alpha, \beta)$ is given in Figure 9.9*b*. Note that the coherent optical signal processor is capable of performing image subtraction, that is, $f_1(x, y) - f_2(x, y)$, which is diffracted at the origin of the output plane. ■

9.3 SYNTHESIS OF A COMPLEX SPATIAL FILTER

In general, a spatial filter can be described by a complex amplitude transmittance distribution:

$$H(p, q) = |H(p, q)| e^{i\phi(p,q)}. \tag{9.25}$$

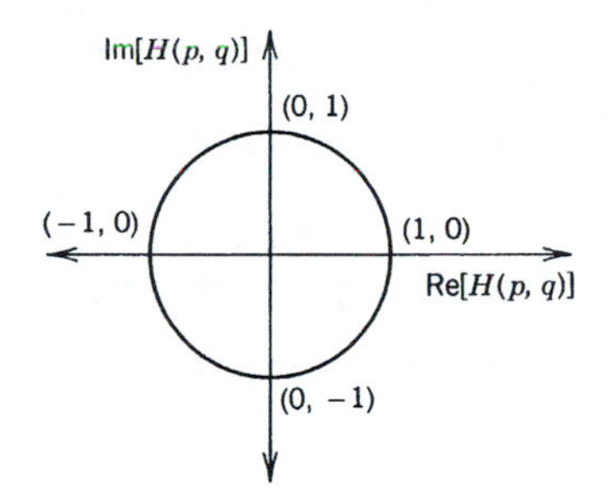

FIGURE 9.10 A complex amplitude transmittance.

In practice, optical spatial filters are generally of the passive type. The physically realizable conditions of optical spatial filters are

$$|H(p,q)| \leq 1 \tag{9.26}$$

and

$$0 \leq \phi(p,q) < 2\pi. \tag{9.27}$$

We note that such a transmittance function can be represented by a set of points within or on a unit circle in the complex plane, as shown in Figure 9.10. The amplitude transmission of the filter changes with the optical density, and the phase delay varies with the thickness. Thus, a complex spatial filter may be constructed by combining an amplitude filter and a phase delay filter.

Let us now discuss the technique developed by Vander Lugt for constructing a complex spatial filter using an interferometric method, as shown in Figure 9.11. The complex light field over the spatial-frequency plane is

$$E(p,q) = F(p,q) + e^{i\alpha_0 p}, \tag{9.28}$$

where $\alpha_0 = f \sin\theta$, f is the focal length of the transform lens, and $F(p,q) = |F(p,q)|e^{-i\phi(p,q)}$.

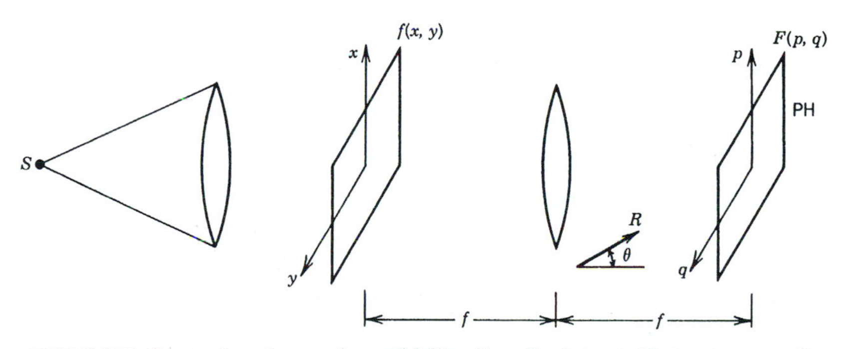

FIGURE 9.11 Construction of a complex spatial filter. *Note:* $f(x, y)$, input object transparency; R, reference plane wave; PH, photographic plate, the recording medium.

The corresponding intensity distribution over the recording medium is

$$I(p,q) = 1 + |F(p,q)|^2 + 2|F(p,q)|\cos[\alpha_0 p + \phi(p,q)]. \tag{9.29}$$

We assume that if the recording is linear in amplitude transmittance, the corresponding amplitude transmittance function of the spatial filter is

$$H(p,q) = K\{1 + |F(p,q)|^2 + 2|F(p,q)|\cos[\alpha_0 p + \phi(p,q)]\}, \tag{9.30}$$

which is, in fact, a *real positive function.*

If this complex spatial filter is inserted in the Fourier plane of a coherent optical signal processor, as shown in Figure 9.5, the complex light immediately behind the spatial filter would be

$$\begin{aligned} E(p,q) &= F(p,q)H(p,q) \\ &= K[F(p,q) + F(p,q)|F(p,q)|^2 + F(p,q)F(p,q)\cdot e^{ip\alpha_0} \\ &\quad + F(p,q)F^*(p,q)e^{-ip\alpha_0}], \end{aligned} \tag{9.31}$$

where the asterisk denotes the complex conjugate. The complex light field at the output plane can be determined by

$$\begin{aligned} g(\alpha,\beta) &= \iint E(p,q)e^{-i(\alpha p + \beta q)}\,dp\,dq \\ &= K[f(x,y) + f(x,y) * f(x,y) * f^*(-x,-y) \\ &\quad + f(x,y) * f(x+\alpha_0, y) + f(x,y) * f^*(-x+\alpha_0, -y)], \end{aligned} \tag{9.32}$$

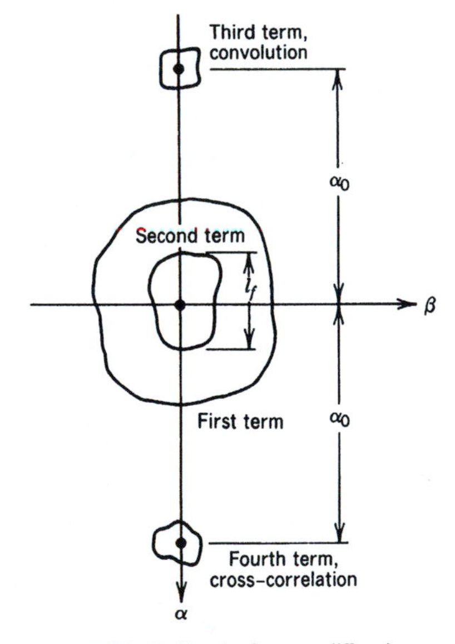

FIGURE 9.12 Sketch of output diffractions.

where the first and second terms represent the zero-order diffraction, which appears at the origin of the output plane, and the third and fourth terms are the convolution and cross-correlation terms, which are diffracted in the neighborhood of $\alpha = \alpha_0$ and $\alpha = -\alpha_0$, respectively, as sketched in Figure 9.12.

Example 9.5

Assume that a noisy object transparency, such as $f(x, y) + n(x, y)$, is inserted in the input plane of the coherent optical signal processor of Figure 9.5. If a matched spatial filter $H(p, q)$ is inserted in the Fourier plane of the optical processor,

$$H(p,q) = K_1 + 2K_2|F(p,q)|\cos[\alpha_0 p + \phi(p,q)],$$

where the K's are proportionality constants and $F(p,q) = \mathscr{F}[f(x,y)]$, calculate the light distribution at the output plane.

The light distribution at the output plane can be obtained by

$$g(\alpha, \beta) = [f(x,y) + n(x,y)] * h(x,y),$$

where

$$\begin{aligned} h(x,y) &= \mathscr{F}^{-1}[H(p,q)] \\ &= K_1\delta(x,y) + K_2 f(x+\alpha_0, y) + K_2 f^*(-x+\alpha_0, -y). \end{aligned}$$

Thus, we have

$$\begin{aligned} g(\alpha,\beta) = {} & K_1[f(x,y) + n(x,y)] * \delta(x,y) \\ & + K_2[f(x,y) + n(x,y)] * f(x+\alpha_0, y) \\ & + K_2[f(x,y) + n(x,y)] * f^*(-x+\alpha_0, -y). \end{aligned}$$

If we assume that the additive noise is white and Gaussian-distributed [notice that $n(x,y) * f(x,y) = 0$], the light distribution at the output plane reduces to

$$\begin{aligned} g(\alpha,\beta) = {} & K_1[f(x,y) + n(x,y)] + K_2 f(x,y) * f(x+\alpha_0, y) \\ & + K_2 f(x,y) \circledast f^*(x-\alpha_0, y), \end{aligned}$$

where $\circledast$ denotes the correlation operation. ■

Example 9.6

If the input object function is translated to a new location, that is, $f(x - x_0, y - y_0)$, show that the output correlation peak is also translated to the same location.

Using the last term of Eq. 9.32, we show that

$$\iint f(x - x_0, y - y_0) f^*(\alpha + x - \alpha_0, \beta + y)\, dx\, dy = R_{11}(\alpha - \alpha_0 - x_0, \beta - y_0).$$

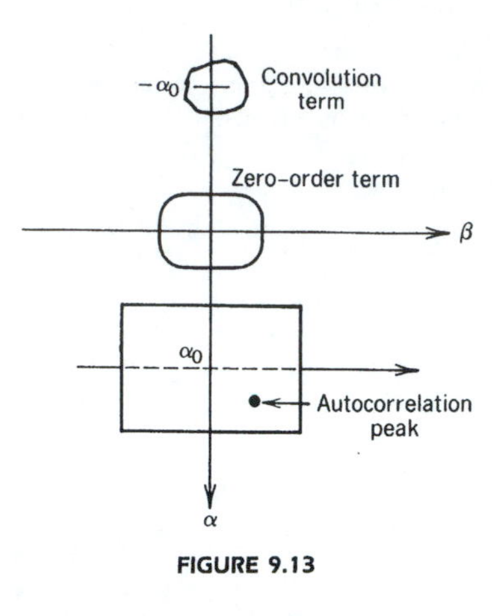

FIGURE 9.13

Thus, we see that the autocorrelation function R_{11}, the correlation peak, moves to the location to which the input object function moved. A sketch of the output diffraction is shown in Figure 9.13. ■

9.4 THE JOINT TRANSFORM CORRELATOR

Complex spatial filtering can also be performed with an optical joint Fourier transform processor, as shown in Figure 9.14. An analog system diagram of the optical architecture appears in Figure 9.15. Since the input objects are illuminated by a coherent plane wave, the complex light distribution arriving at the square-law detector in the Fourier plane P_2 will be

$$E(p,q) = F_1(p,q)e^{-i\alpha_0 p} + F_2(p,q)e^{i\alpha_0 p}, \tag{9.33}$$

where $F_1(p,q)$ and $F_2(p,q)$ are the Fourier spectra of input objects $f_1(x,y)$ and $f_2(x,y)$, respectively. The corresponding irradiance at the output end of the square-

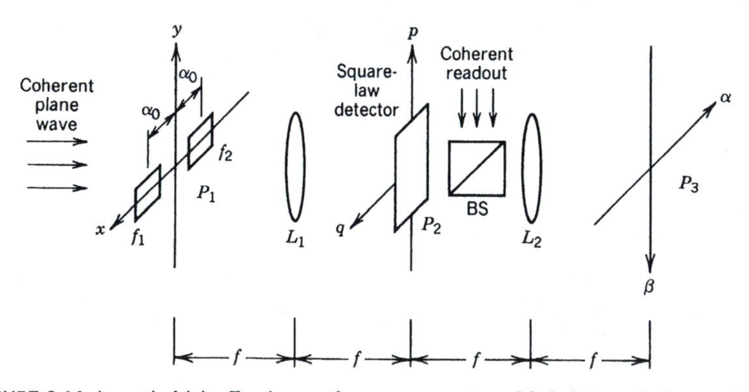

FIGURE 9.14 An optical joint Fourier transform processor. *Note:* BS designates the beam splitter.

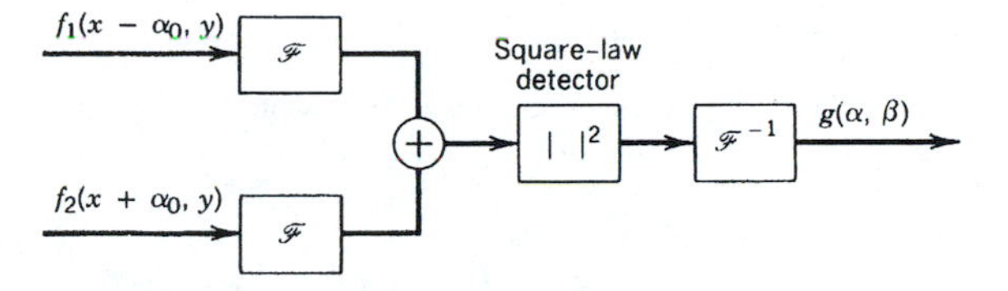

FIGURE 9.15 An analog system diagram of the optical architecture of the joint transform processor in Figure 9.14.

law detector is

$$\begin{aligned} I(p,q) &= |E(p,q)|^2 \\ &= |F_1(p,q)|^2 + |F_2(p,q)|^2 + F_1(p,q)F_2^*(p,q)e^{-i2\alpha_0 p} \\ &\quad + F_1^*(p,q)F_2(p,q)e^{i2\alpha_0 p}, \end{aligned} \tag{9.34}$$

which can be written as

$$\begin{aligned} I(p,q) = |F_1(p,q)|^2 + |F_2(p,q)|^2 + 2|F_1(p,q)|\,|F_2(p,q)| \\ \cdot \cos[2\alpha_0 p - \phi_1(p,q) + \phi_2(p,q)], \end{aligned} \tag{9.35}$$

where

$$F_1(p,q) = |F_1(p,q)|e^{i\phi_1(p,q)}. \tag{9.36}$$

and

$$F_2(p,q) = |F_2(p,q)|e^{i\phi_2(p,q)}. \tag{9.37}$$

If the irradiance of Eq. 9.35 is read out by a coherent plane wave, the complex light distribution at the output plane P_3 will be

$$\begin{aligned} g(\alpha,\beta) &= f_1(x,y) \circledast f_1^*(x,y) + f_2(x,y) \circledast f_2^*(x,y) \\ &\quad + f_1(x,y) \circledast f_2^*(x - 2\alpha_0, y) + f_1^*(x,y) \circledast f_2(x + 2\alpha_0, y), \end{aligned} \tag{9.38}$$

where $\circledast$ denotes the correlation operation. The first two terms represent overlapping correlation functions, $f_1(x, y)$ and $f_2(x, y)$, which are diffracted at the origin of the output plane. The last two terms are the two cross-correlation terms, which are diffracted around $\alpha = 2\alpha_0$ and $\alpha = -2\alpha_0$, respectively. Notice that a square-law converter, such as a photographic plate, a liquid crystal light valve, or a charge-coupled camera, can be used.

Example 9.7

In the optical joint Fourier transform processor of Figure 9.14, we assume that $f_1(x - \alpha_0, y)$ is imbedded in additive white, Gaussian noise, that is, $f_1(x - \alpha_0, y) + n(x - \alpha_0, y)$, and $f_2(x + \alpha_0, y)$ is replaced by $f_1(x + \alpha_0, y)$.

(a) Draw an analog system diagram of the optical joint Fourier transform correlator.

(b) Evaluate the complex light distribution at the output plane.

Answers

(a) An analog system diagram of this problem appears in Figure 9.16.

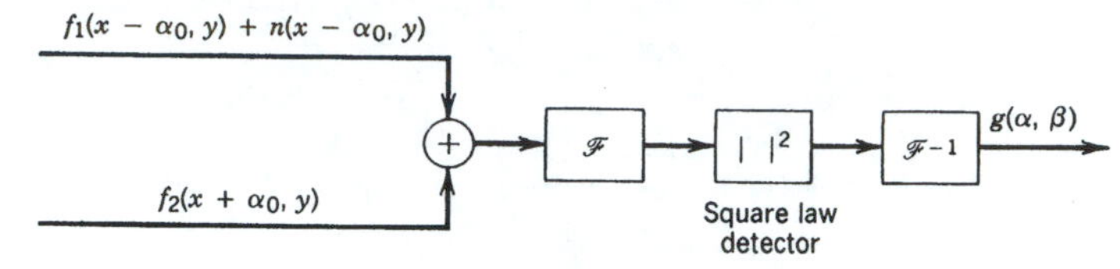

FIGURE 9.16

(b) The complex light field arriving at the input end of the square-law detector is

$$E(p, q) = F_1 e^{-i\alpha_0 p} + N e^{-i\alpha_0 p} + F_1 e^{i\alpha_0 p}.$$

The corresponding irradiance can be shown as

$$I(p, q) = 2|F_1|^2 + |N|^2 + F_1 N^* + N F_1^* + (F_1 F_1^* + N F_1^*)e^{-i2\alpha_0 p} + (F_1 F_1^* + F_1 N^*)e^{i2\alpha_0 p}.$$

Since the noise is assumed to be additive and Gaussian-distributed, we note that

$$\iint f_1(x, y) n(\alpha + x, \beta + y)\, dx\, dy = 0.$$

Thus, the complex light field at the output plane would be

$$g(\alpha, \beta) = 2 f_1(x, y) \circledast f_1^*(x, y) + n(x, y) \circledast n^*(x, y) + f_1(x, y) \circledast f_1^*(x - 2\alpha_0, y) + f_1(x, y) \circledast f_1^*(x + 2\alpha_0, y).$$

The first two terms will be diffracted around the origin of the output plane, and the two autocorrelation peaks, the third and fourth terms, will be diffracted at $\alpha = 2\alpha_0$ and $\alpha = -2\alpha_0$, respectively. ■

■ Example 9.8

Using the optical joint Fourier transform processor of Figure 9.14, show that the spatial frequency and orientation of the interference fringes within a joint power spectrum determine the spatial content of the object. And sketch the locations of the autocorrelation peaks in the output plane.

Let the reference function be $f_2(x, y) = f_1(x, y)$, which is centered at $x = -\alpha_0$. If the input object function is $f_1(x - \alpha_0, y - y_0)$, as shown in Figure 9.17*a*, the complex light distribution at the input end of the square-law detector is

$$E(p, q) = F_1 \exp[-i(\alpha_0 p + y_0 q)] + F_1 e^{i(\alpha_0 p)}.$$

The corresponding power spectral distribution at the output plane can be shown as

$$I(p, q) = 2|F_1|^2 + F_1 F_1^* \exp[-i(2\alpha_0 p + y_0 q)] + F_1 F_1^* \exp[i(2\alpha_0 p + y_0 q)],$$

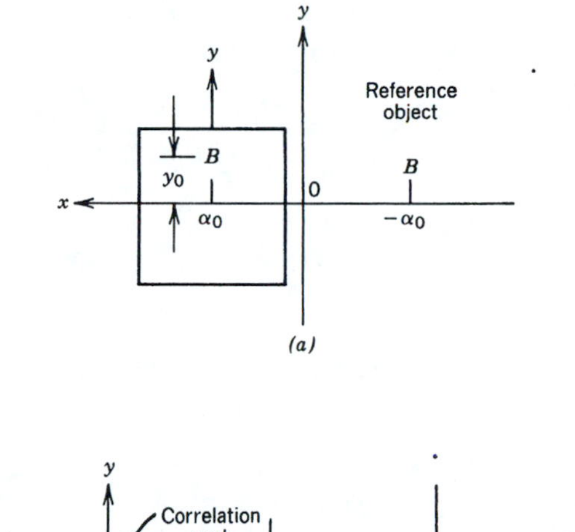

(a)

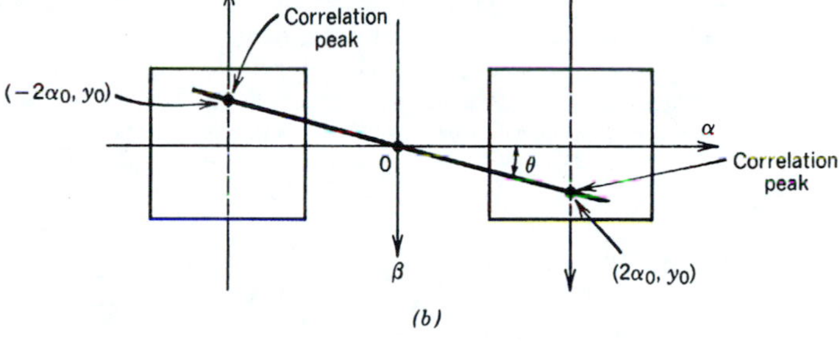

(b)

FIGURE 9.17

which can also be written as

$$I(p, q) = 2|F_1|^2[1 + \cos(2\alpha_0 p + y_0 q)].$$

From this result we see that there is a linear-phase distribution that can be expressed in terms of the fringe direction. Thus, by orienting the spatial-frequency coordinate of p to p', we have

$$I(p', q') = 2|F_1(p', q')|^2[1 + \cos(\sqrt{(2\alpha_0)^2 + (y_0)^2}p')],$$

where the angular orientation of the fringes is

$$\theta = \tan^{-1}\left(\frac{y_0}{2\alpha_0}\right).$$

The positions of the autocorrelation peaks in the output plane, which are located at $(r, -\theta)$ and $(r, \pi - \theta)$, respectively, can therefore be determined by

$$\gamma = \sqrt{(2\alpha_0)^2 + (y_0)^2}$$

and

$$\theta = \tan^{-1}\left(\frac{y_0}{2\alpha_0}\right).$$

By taking the inverse Fourier transform of $I(p, q)$, we have the following complex light distribution at the output plane,

$$g(\alpha, \beta) = 2f_1(x, y) \circledast f_1^*(x, y) + f_1(x, y) \circledast f_1^*(x - 2\alpha_0, y - y_0)$$
$$+ f_1^*(x, y) \circledast f_1(x + 2\alpha_0, y + y_0),$$

which is sketched in Figure 9.17*b*. Thus, we see that the spatial frequency and the orientation of the fringes determine the spatial content of the object. ■

9.5 WHITE-LIGHT OPTICAL SIGNAL PROCESSING

Although coherent optical signal processors can perform a variety of complex signal operations, coherent processing systems are usually plagued by coherent artifact noise. This difficulty has prompted optical engineers to look for an alternative, for example, using a partially coherent source for optical signal processing. The basic advantages of partially coherent processing are that (1) it can suppress the coherent artifact noise, (2) partially coherent sources are inexpensive, (3) the processing environment is very relaxed, (4) partially coherent processors are relatively easy and economical to operate, and (5) they are suitable for color image processing.

We now discuss an achromatic, partially coherent processing technique that can be carried out by a white-light source, as shown in Figure 9.18. This partially coherent processing system is similar to a coherent processing system, except that it uses an extended white-light source, a source-encoding mask, a signal-sampling grating, multiple spectral-band filters, and achromatic transform lenses. For example, if we place an input object transparency $s(x, y)$ in contact with an image-sampling phase grating, for every wavelength λ the complex wave field at the Fourier plane P_2 would be, assuming a white-light point source,

$$E(p, q; \lambda) = \iint s(x, y) e^{ip_0 x} e^{-i(px + qy)}\, dx\, dy$$
$$= S(p - p_0, q). \tag{9.39}$$

Here the integral is over the spatial domain of the input plane P_1, (p, q) denotes the angular spatial-frequency coordinate system, p_0 is the angular spatial frequency

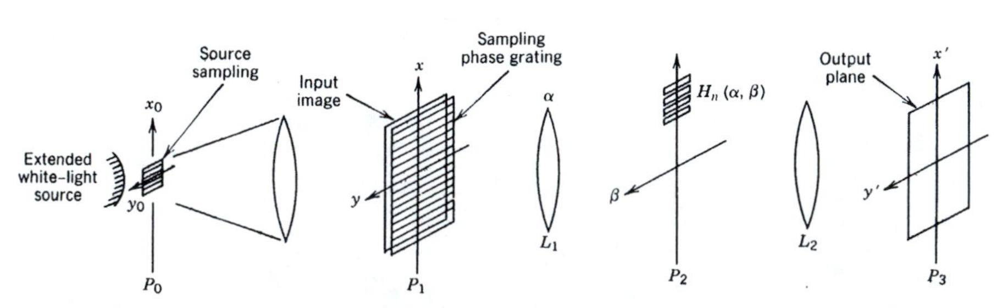

FIGURE 9.18 A white-light optical signal processor.

of the sampling phase grating, and $S(p, q)$ is the Fourier spectrum of $s(x, y)$. The preceding equation can be written in (α, β) spatial variables,

$$E(\alpha, \beta; \lambda) = S\left(\alpha - \frac{\lambda f}{2\pi} p_0, \beta\right), \tag{9.40}$$

where $p = (2\pi/\lambda f)\alpha$, $q = (2\pi/\lambda f)\beta$, and f is the focal length of the achromatic transform lens. Thus, we see that the Fourier spectra disperse into rainbow colors along the α axis, and that each Fourier spectrum for a given wavelength λ is centered at $\alpha = (\lambda f/2\pi)p_0$.

In complex signal filtering, a set of complex spatial filters with a narrow spectral bandwidth is provided. We assume that the input object is spatial-frequency-limited; and that the spatial bandwidth of each spectral-band filter $H(p_n, q_n)$ is given by

$$H(p_n, q_n) = \begin{cases} H(p_n, q_n), & \alpha_1 < \alpha < \alpha_2, \\ 0, & \text{otherwise,} \end{cases} \tag{9.41}$$

where $p_n = (2\pi/\lambda_n f)\alpha$, $q_n = (2\pi/\lambda_n f)\beta$; λ_n is the main wavelength of the filter; $\alpha_1 = (\lambda_n f/2\pi)(p_0 + \Delta p)$ and $\alpha_2 = (\lambda_n f/2\pi)(p_0 - \Delta p)$ are the upper and lower spatial limits of $H(p_n, q_n)$, respectively; and Δp is the spatial bandwidth of the input object $s(x, y)$.

Since the limiting wavelengths of each $H(p_n, q_n)$ can be shown to be

$$\lambda_h = \lambda_n \frac{p_0 + \Delta p}{p_0 - \Delta p} \quad \text{and} \quad \lambda_l = \lambda_n \frac{p_0 - \Delta p}{p_0 + \Delta p}, \tag{9.42}$$

its spectral bandwidth can be approximated by

$$\Delta\lambda_n = \lambda_n \frac{4p_0 \Delta p}{p_0^2 - (\Delta p)^2} \simeq \frac{4\Delta p}{p_0} \lambda_n. \tag{9.43}$$

If we place this set of spectral-band filters side-by-side and position them properly over the smeared Fourier spectra, the complex light field at the output plane, which comes from λ_n, would be

$$g(x, y) = s(x, y; \lambda_n) * h(x, y; \lambda_n), \tag{9.44}$$

where the asterisk represents the convolution operation, and $h(x, y; \lambda_n)$ is the spatial impulse response of $H_n(p_n, q_n)$. Since the spectral lines of the white-light source are *mutually incoherent*, the intensity distribution at the output plane is

$$I(x, y) = \sum_{n=1}^{N} \Delta\lambda_n |s(x, y; \lambda_n) * h(x, y; \lambda_n)|^2. \tag{9.45}$$

Thus, the partially coherent processor is capable of processing the signal in a complex wave field. Since the output intensity is the sum of the mutually incoherent narrow-band spectral irradiances, the coherent artifact noise can be eliminated. It is also apparent that the white-light source emits all the visible wavelengths, and that it is suitable for color image processing.

Example 9.9

Draw an analog system diagram of the processing operation carried out by the partially coherent optical signal processor of Figure 9.18. Since we assume a white-light point source, the object transparency inserted at the input plane is illuminated by a *spatially coherent* white-light plane wave. The complex wave field immediately behind the object transparency is

$$\sum_n s(x, y; \lambda_n).$$

An analog system diagram of the white-light process is shown in Figure 9.19. ■

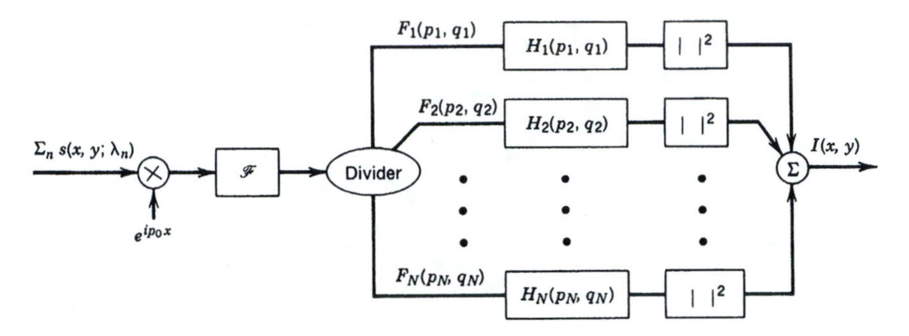

FIGURE 9.19 An analog system diagram of the signal processor in Figure 9.18. We assume a white-light point source.

Example 9.10

The amplitude transmittance function of a multiplexed transparency, with a positive and a negative image, is given by

$$t(x, y) = t_1(x, y)(1 + \cos p_0 x) + t_2(x, y)(1 + \cos q_0 y),$$

where t_1 and t_2 are the positive and negative image functions, and p_0 and q_0 are the spatial sampling frequencies along the x and y axes, respectively.

This encoded transparency is inserted at the input plane of the white-light processor shown in Figure 9.20*a*.

(a) Evaluate the smeared spectra at the Fourier plane.

(b) If the focal length of the transform lens is $f = 300$ mm, and the sampling frequencies are $p_0 = 80\pi$ and $q_0 = 60\pi$ radians/mm, compute the smearing length of the Fourier spectra. Assume that the spectral lines of the white-light source vary in length from 350 to 750 nm.

(c) Design a set of transparent color filters by which the density or gray levels of the image can be encoded in pseudocolors at the output plane.

(d) Compute the irradiance of the pseudocolor image.

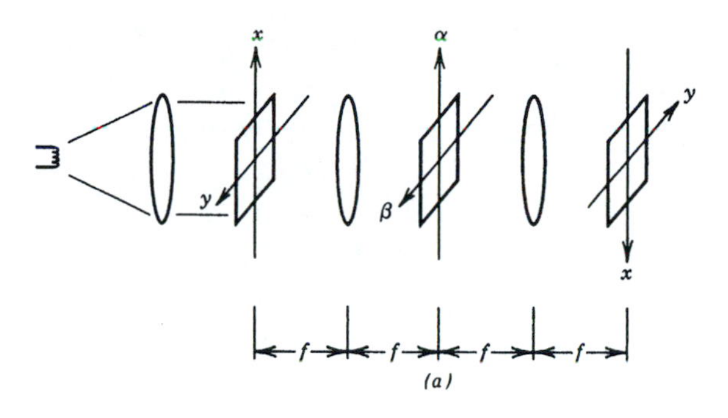

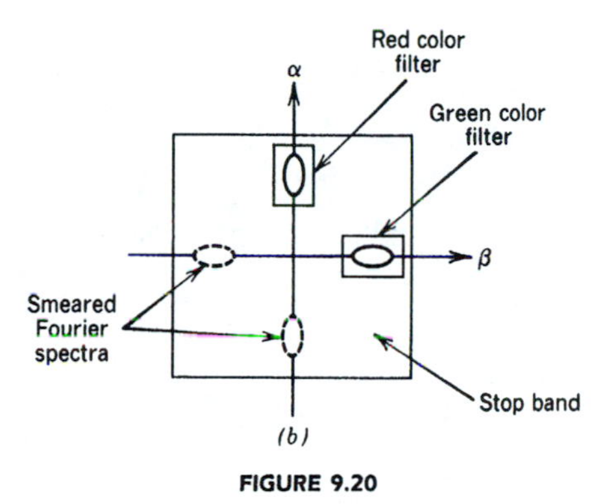

FIGURE 9.20

Answers

(a) The smeared Fourier spectra can be evaluated as follows,

$$\mathscr{F}[t(x, y)] = T_1(p, q) + 2T_1(p - p_0, q) + 2T_1(p + p_0, q) + T_2(p, q) + 2T_2(p, q - q_0) + 2T_2(p, q + q_0),$$

where T_1 and T_2 are the Fourier spectra of t_1 and t_2, respectively. Using Eq. 9.40, we get

$$T_1(p \mp p_0, q) = T_1\left(\alpha \mp \frac{\lambda f}{2\pi} p_0, \beta\right)$$

and

$$T_2(p, q \mp q_0) = T_2\left(\alpha, \beta \mp \frac{\lambda f}{2\pi} q_0\right).$$

Thus, we see that the positive and negative image spectra smear into rainbow colors along the α and the β axes, respectively.

(b) The smearing length along the α axis can be calculated as follows:

$$\alpha_1 - \alpha_2 = \frac{f p_0}{2\pi}(\lambda_1 - \lambda_2).$$

Substituting $f = 300$ mm, $p_0 = 80\pi$, $\lambda_1 = 750$ nm, and $\lambda_2 = 350$ nm, we get

$$\alpha_1 - \alpha_2 = \frac{(300)(80\pi)}{2\pi}(750 - 350)(10^{-6})$$
$$= 4.8 \text{ mm},$$

as shown in Figure 9.20*b*. The smearing length along the β axis can be similarly computed:

$$\beta_1 - \beta_2 = \frac{(300)(60\pi)}{2\pi}(750 - 350) \times 10^{-6}$$
$$= 3.6 \text{ mm}.$$

(c) A set of transparent color filters for encoding the density of the image is shown in Figure 9.20*b*. Notice that the positive image is encoded in red and the negative image in green.

(d) The irradiance of the pseudocolor image at the output plane can be evaluated as

$$I(x, y) = \left|\mathscr{F}^{-1}[2T_1(p - p_0, q)]\right|^2_{\lambda = \lambda_r} + \left|\mathscr{F}^{-1}[2T_2(p, q - q_0)]\right|^2_{\lambda = \lambda_g},$$

which can be reduced to

$$I(x, y) = K[T_{1r}^2(x, y) + T_{2g}^2(x, y)],$$

where λ_r and λ_g represent the red and green wavelengths, respectively. The output irradiance is essentially a combination of a red positive image and a green negative image. Thus, an image whose density levels are pseudocolor-encoded can be viewed at the output plane. ■

REFERENCES

1. J. W. Goodman, *Introduction to Fourier Optics*, McGraw-Hill, New York, 1968.
2. F. T. S. Yu, *Optical Information Processing*, Wiley-Interscience, New York, 1983.
3. F. T. S. Yu, *White-Light Optical Signal Processing*, Wiley-Interscience, New York, 1985.

PROBLEMS

9.1 Given an input–output linear and spatially invariant optical system under strictly incoherent illumination.

(a) Derive the incoherent transfer function in terms of the spatial impulse response of the system.

(b) State some basic properties of the incoherent transfer function.

9.2 Derive the cutoff frequency of an optical system under coherent and incoherent illuminations.

9.3 Given an optical imaging system that is capable of combining two Fraunhofer diffractions resulting from input objects $f_1(x, y)$ and $f_2(x, y)$, as shown in Figure 9.21. Compute the irradiance at the output plane under coherent and incoherent illuminations.

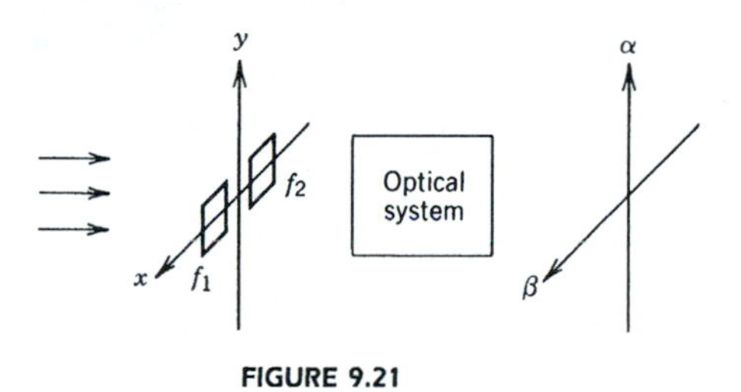

FIGURE 9.21

9.4 Assume an *all-pass* optical system in which the spatial impulse response can be represented by a delta function,

$$h(x, y) = \delta(x, y).$$

If the binary-phase grating of Figure 9.22 is inserted at the input plane of the optical system, compute the image irradiance at the output plane under coherent and incoherent illuminations. Give a conclusive distinction of the observed images.

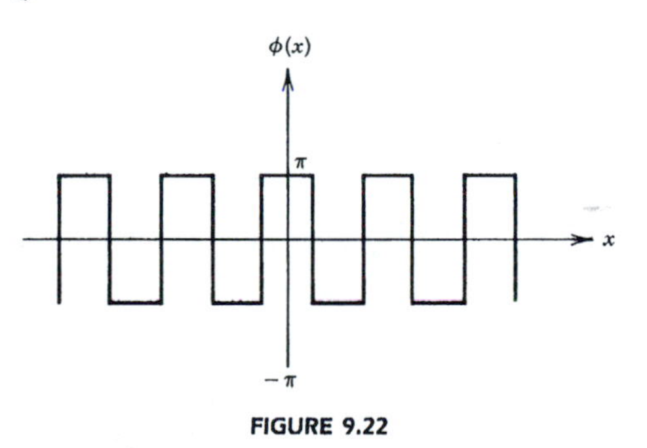

FIGURE 9.22

9.5 We assume that the input object function for the coherent optical signal processor of Figure 9.5 is given as

$$f(x, y) = \tfrac{1}{2}[1 + \cos(60\pi x)],$$

and that the focal length of the processor is $f = 300$ mm.

(a) Evaluate the complex light field at the Fourier plane when the processor is illuminated by a monochromatic plane wave of $\lambda = 400$ nm.

(b) Repeat part a for $\lambda = 600$ nm. What scale changes does the shift in wavelength cause?

9.6 If the zero-order spectral distribution for the coherent optical signal processor described in Problem 9.5 is blocked by a high-pass spatial filter,

(a) Calculate the intensity distribution $I(\alpha, \beta)$ at the output plane.

(b) Compute the basic spatial frequency of the intensity distribution calculated in part a.

9.7 We assume that the coherent optical signal processor in Figure 9.5 has a monochromatic point source S of $\lambda = 600$ nm, and that the focal length of its transform lens is $f = 1000$ mm. If the input object $f(x, y)$ is an open square aperture of size $s = 5$ mm,

(a) Calculate the size of its Fourier spectrum (i.e., the first lobe) at the spatial-frequency plane P_2.

(b) Repeat part a for focal length $f = 100$ mm.

9.8 Assume that an object function embedded in an additive white, Gaussian noise, $f(x, y) + n(x, y)$, is inserted in the input plane of the coherent optical signal processor of Figure 9.5. A matched spatial filter, $H(p, q) = KF^*(p, q)$, is inserted in the Fourier plane.

(a) Compute the complex light distribution at the output plane. Assume that $f(x, y) * n(x, y) = 0$.

(b) If $f(x, y)$ is a complicated object function, such as a tank, sketch the shape of irradiance at the output plane.

9.9 If the object function referred to in Problem 9.8 is moved to a new location, that is, $f(x - x_0, y - y_0)$,

(a) Show that the corresponding spectrum is shift-invariant.

(b) Evaluate the complex light distribution at the output plane.

9.10 (a) Draw an analog system diagram of the coherent optical signal processor in Figure 9.23.

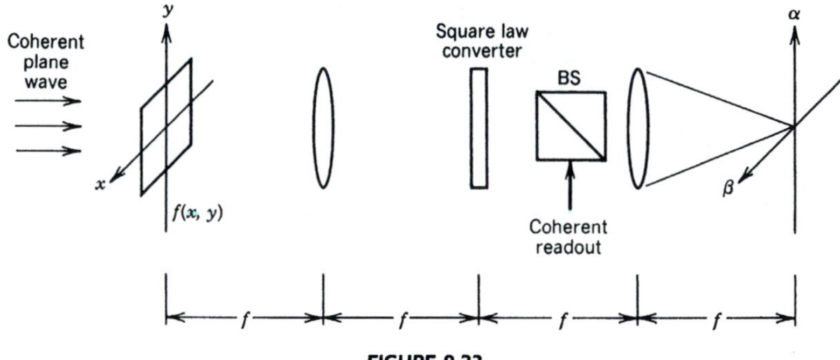

FIGURE 9.23

(b) Evaluate the complex light distribution at its output plane.

9.11 A spatial filter, $H(p, q) = i(p + q)$, is inserted at the Fourier plane of the coherent optical signal processor in Figure 9.5.

(a) Calculate the complex light distribution at the output plane. *Hint:*

$$\mathscr{F}\left[\frac{\partial^2 f(x, y)}{\partial x \, \partial y}\right] \simeq [i(p + q)][F(p, q)].$$

(b) If the input object is an open rectangular aperture, what would the image irradiance at the output plane be?

9.12 Suppose that the input object function of the coherent optical signal processor in Figure 9.5 is a rectangular grating of spatial frequency p_0, as shown in Figure 9.24.

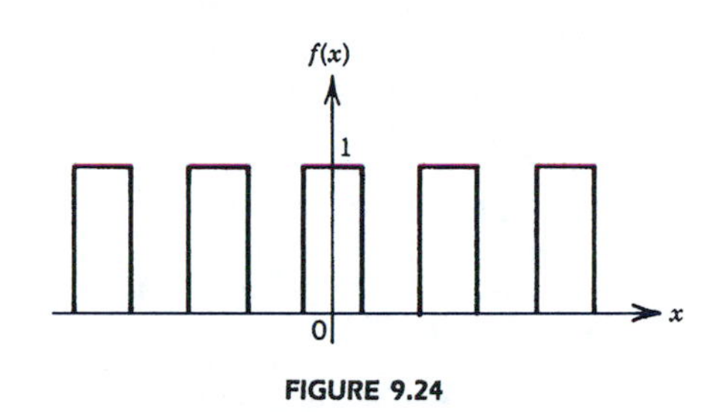

FIGURE 9.24

(a) Evaluate and sketch the spectral content of the object at the Fourier plane.

(b) If we insert a π-phase filter,

$$H(r) = \begin{cases} e^{i\pi}, & r \leq r_0, \\ 0, & \text{otherwise,} \end{cases}$$

at the origin of the spatial-frequency plane, sketch the light distribution at the output plane and state your conclusion.

9.13 For construction of the matching filter in Figure 9.11, assume that the focal length of the transform lens is $f = 300$ mm, that the wavelength is $\lambda = 6500$ Å, and that the oblique angle of the reference beam is $\theta = 30°$.

(a) Compute the carrier spatial frequency of the matching filter.

(b) What resolution must the recording medium have (e.g., film)?

9.14 Given an image transparency which has been distorted by linear motion and whose image points are each described by a small line segment. The corresponding transmittance of a smeared object point is represented by a rectangular function,

$$f(x) = \text{rect}\left(\frac{x}{\Delta x}\right),$$

where Δx is the smeared length. If we wish to restore the image with the coherent optical signal processor of Figure 9.5, show that the inverse filter $H(p)$ is, strictly speaking, physically unrealizable.

9.15 The matching filter of Eq. 9.30 is placed at the Fourier plane of the coherent optical signal processor of Figure 9.5, and no object transparency is inserted in the input plane.

(a) Compute the complex light distribution at the output plane.

(b) Sketch your observation at the output plane.

9.16 The separation of the input objects for the optical joint Fourier transform processor of Figure 9.14, is $2\alpha_0 = 2$ cm, the wavelength of the coherent source is $\lambda = 6000$ Å, and the focal length of the transform lenses is $f = 500$ mm.

(a) Calculate the spatial frequency of the fringes at the Fourier plane.

(b) What resolution must the square-law converter have?

9.17 The reference object f_2 of a joint transform correlator is located at the origin, $f_2(x, y)$, and the input object is located at $x = 10$ mm, $y = 10$ mm, that is, $f_1(x - 10, y - 10)$.

(a) Calculate the intensity distribution at the input end of the square-law converter, if $\lambda = 6000$ Å and $f = 500$ mm.

(b) Using the result of part a, determine the locations in the output plane of the cross-correlation functions with readout illuminations of $\lambda = 6000$ Å and $\lambda = 4000$ Å, respectively.

9.18 Given an input transparency with an object size of 5 mm that is to be processed by the white-light processor of Figure 9.18. We assume that the focal length of the transform lenses is $f = 500$ mm.

(a) Calculate the size of the source required.

(b) If the spatial frequency of the input object is assumed to be two lines per millimeter, estimate the required size of the spatial filter H_n and the required spatial frequency of the phase grating.

9.19 We wish to perform spatial-frequency pseudocolor encoding of an input object's transparency with a white light.

(a) Sketch a white-light processor that can perform this task.

(b) We wish to encode low spatial frequency in red and high spatial frequency in blue. Sketch a spatial filter that can perform this operation.

9.20 It is known that a color image can be encoded in a black-and-white transparency in such a way that the encoded transmittance is represented by the following equation,

$$t(x, y) = K + t_r(x, y)\cos(p_0 x) + tg(x, y)\cos(p_0 y) + t_b(x, y)\cos(2p_0 y),$$

where K is an arbitrary constant and t_r, t_g, and t_b represent the red, green, and blue color images. If the encoded transparency is inserted in the white-light processor of Figure 9.20*a*,

(a) Evaluate the corresponding smeared Fourier spectra at the spatial-frequency plane.

(b) Design a set of transparent color filters for retrieving color images.

(c) Compute the irradiance of the color image at the output plane.

10 FIBER OPTICS: AN INTRODUCTION

In Chapter 7 we briefly mentioned that an optical fiber may be used to send a modulated laser signal from the transmitter downstream to the demodulation and reception system. Figure 10.1 is a schematic of a light beam propagating in a typical fiber. The basic principle is rather simple. If the incident cone of light strikes the interface between the two materials of the fiber at an angle ϕ greater than the critical angle ϕ_c ($\sin \phi_c = \eta_1/\eta_2$: see Section 2.5), it will be totally reflected and thus propagate along the inner core. Over the past few decades, this simple way of guiding light and its applications in communication and other fields relying on signal processing have seen rapid refinement and advancement. These advances are due in part to continued improvement in semiconductor lasers, which are capable of modulation at a rate in tens or hundreds of gigabits per second, and in part to advancements in the fabrication of low-loss, high-quality optical fibers.

In this chapter we discuss some of the basic principles underlying the fabrication of optical fibers, their operational characteristics, and the propagation of light in optical fibers. We will see that fiber optics have several potential advantages over conventional electronics, especially in applications requiring a broad bandwidth, good electrical insulation, and freedom from electromagnetic interference and intrusion.

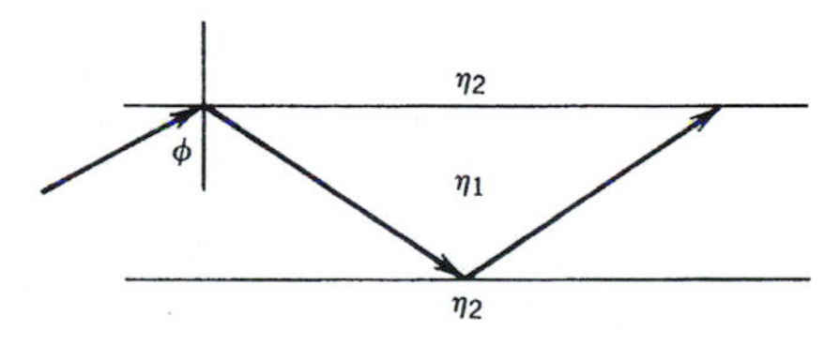

FIGURE 10.1 Total internal reflection.

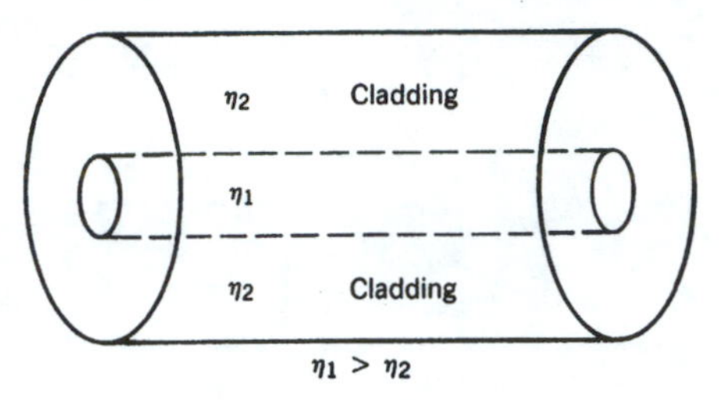

FIGURE 10.2 Typical optical fiber with a cladding.

10.1 FIBER CONSTRUCTION

In practice, an optical fiber (see Figure 10.2) consists of a central cylindrical core with refractive index η_1, surrounded by a layer of material, called the *cladding*, with a lower refractive index η_2. The core transmits the light waves, and the cladding keeps the light waves within the core and strengthens the fiber. In actual systems an outer jacket protects the fiber from moisture and abrasion (see Figure 10.3).

The core and the cladding are made of either glass or plastic. Three major combinations of these two materials are used to make the fibers: plastic core with plastic cladding, glass core with plastic cladding, and glass core with glass cladding. When plastics are used, the core can be made of either polystyrene or polymethyl methacrylate, and the cladding is generally made of silicone or Teflon.

For optical fibers the silica must be extremely pure; however, very small amounts of dopants such as boron, germanium, or phosphorus may be added to change the refractive indices of the fiber. Boron oxide is added to silica to form borosilicate glass, which is used in some claddings.

In comparison with glass, plastic fibers are flexible and inexpensive, are easy to install and connect, can withstand greater stress than glass fibers, and weigh 40 percent less. However, they do not transmit light as efficiently. Because of the

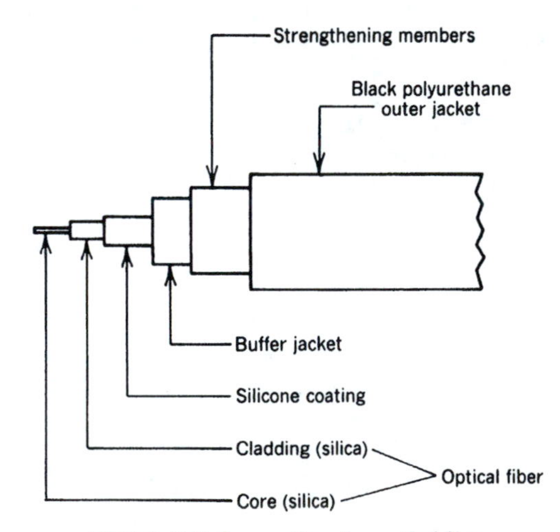

FIGURE 10.3 Cross section of an optical fiber.

considerable losses from plastic fibers, they are used only for short runs, such as within buildings. Since glass core fibers are so much more widely used than plastic, subsequent references to fibers in this chapter are to glass fibers. In comparison to copper wire, optical fibers are much lighter. For example, a 40-km length of fiber core weighs only 1 kg, compared to the 30-kg weight of a 40-km length of copper wire with a 0.32-mm outer diameter.

Information is transmitted through the core of the fiber by means of electromagnetic (EM) wave propagation. Therefore, the EM wave must be kept within the core, and not be allowed to leak into the cladding. As we shall see, the basic principle of fiber optics communication is that the refractive index of the core, η_1, must be greater than that of the cladding, η_2.

The diameters of the core and the cladding determine many of the optical characteristics of the fiber (see Section 10.3), as well as some of its physical characteristics. For example, the fiber must be large enough to allow splicing and the attachment of connectors, but if it is too large, it will be too stiff to bend and will take up too much material and space.

Core diameters range from 5 to 500 μm, and cladding diameters vary from 100 to 700 μm. To keep the light within the core, the cladding must have a minimum thickness of one to two wavelengths of the light transmitted. The protective jacket may add as much as 100 μm to the fiber's total diameter. Typical fiber dimensions are given in Figures 10.4*a*, *b*, and *c*.

Although fibers with the minimum core diameter of 5 μm have a much broader bandwidth than those with larger diameters, their small size causes severe handling and connection problems. Moreover, within any given fiber, tight tolerances are necessary, for even slight variations in the dimensions can cause significant changes in optical characteristics. Because of this, the core diameter may be specified as ± 2 μm.

In contrast to ordinary glass, which is brittle, optical glass fibers have great tensile strength and are able to withstand hard pulling and stretching. The toughest fibers are as strong as stainless steel wires of the same diameter. Indeed, in comparison with copper wire, optical fiber has the tensile strength of copper wire twice its thickness. One-kilometer lengths of these fibers have withstood pulling forces of more than 500,000 pounds per square inch before breaking, and a fiber 10 m long can be stretched by 50 cm and still spring back to its original shape. Yet a fiber with a 400-μm diameter can be bent into a circle with a radius as small as 2 mm.

To produce fibers of this tenacity, manufacturers try to keep the glass core and cladding free from microscopic cracks on the surface and flaws in the interior. When a fiber is under stress, it can break at either of these flaws. Flaws can develop during or after manufacture. Even a tiny particle of dust or a soft piece of Teflon can give the fiber's surface a fatal scratch.

To prevent such abrasions, manufacturers coat the fiber immediately after it is made with a protective jacket of plastic. Plastics are polymers, usually organic. In addition, the plastic jacket protects the fiber surface from moisture, and it cushions the fiber when it is pressed against irregular surfaces. This cushioning reduces the effect of small random bends and microbends, which would otherwise cause transmission losses. Finally, the jacket compensates for some of the contractions and expansions caused by temperature variations.

Although single-fiber cables are used, a number of fibers are usually placed together in one cable. These cables are designed so that there is little or no stress on the fibers themselves.

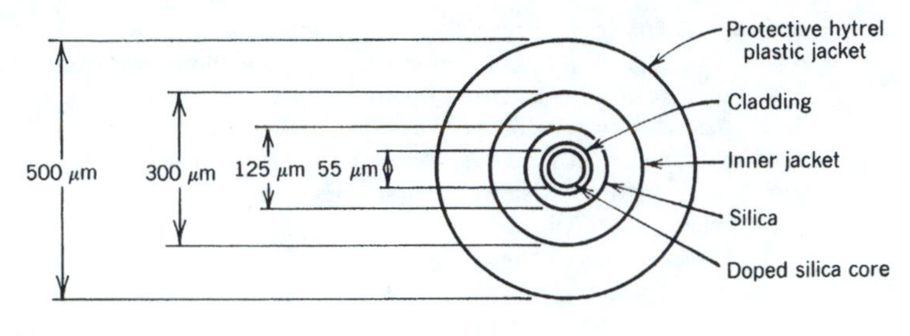

(a)

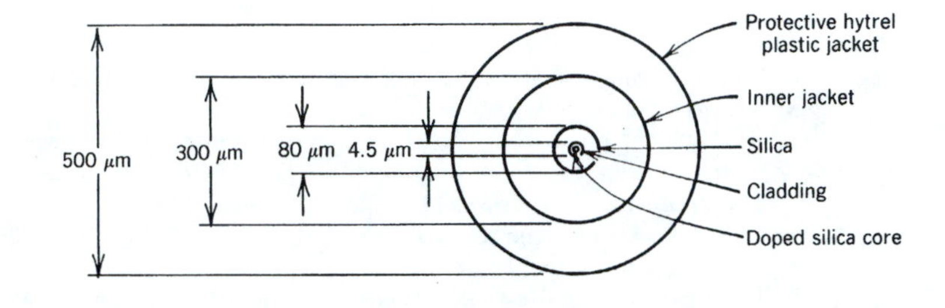

(b)

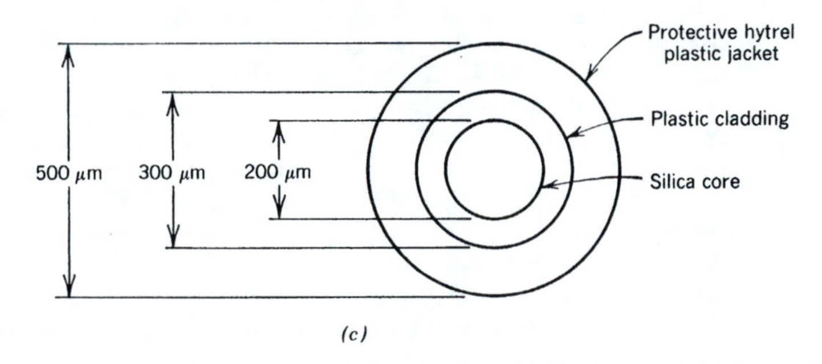

(c)

FIGURE 10.4 Nominal dimensions of typical optical fibers. (*a*) Wide-band graded-index multimode fiber. (*b*) Step-index single-mode fiber. (*c*) Large-core plastic-clad optical fiber.

10.2 DISPERSION AND ATTENUATION IN FIBERS

Figure 10.5 is a schematic that shows how a light ray is coupled into the optical fiber. If the incident light ray is within the acceptance cone angle θ_a given by (see Example 10.1)

$$\theta_a = \sin^{-1}\left[\left(\frac{\eta_1^2 - \eta_2^2}{\eta_0}\right)^{1/2}\right], \tag{10.1}$$

it will be guided along the core.

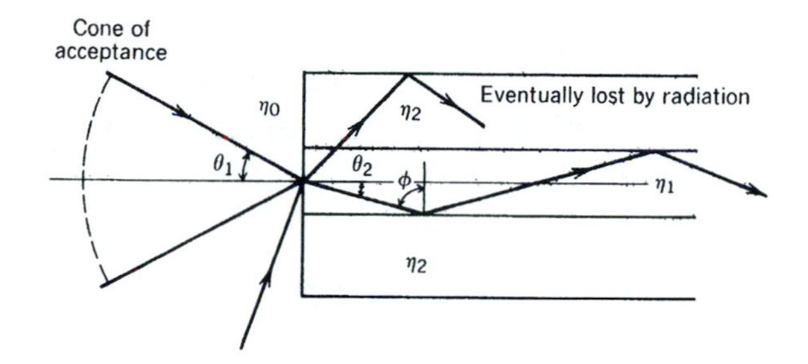

FIGURE 10.5 The acceptance angle of the fiber. *Note:* ϕ, incident angle at the core–cladding boundary.

Rays at greater incident angles will be guided along the cladding, as is also shown in Figure 10.5. The cladding material allows greater loss than the core, so that after a short distance the light being guided along it will diminish through scattering, absorptions, and so on.

In practice, coupling of the light into the fiber can be accomplished very simply with the help of a lens (see Figure 10.6), or by putting the fiber in close proximity to the light source and linking them with index-matching liquid to reduce the losses (see Figure 10.7). After the light has been properly launched into the fiber, two physical mechanisms take place during the transmission, namely *attenuation* and *dispersion*.

Attenuation of transmission, the reduction of its intensity over an interval, is mainly due to the chemical impurities in the core. If the absorption resonance of these impurities is within the optical frequencies being transmitted, there will be considerable attenuation through absorption. A tremendous research and development effort has been focused on lowering the amounts of these impurities. Optical fibers with attenuation of less than 1 dB/km have been made, and attenuation of 6 dB/km is commonplace.

There are, however, two other sources of attenuation, the effects of bending and of microbending, as illustrated in Figures 10.8 and 10.9, respectively. These diagrams depict what happens to a light ray that is just within the critical angle

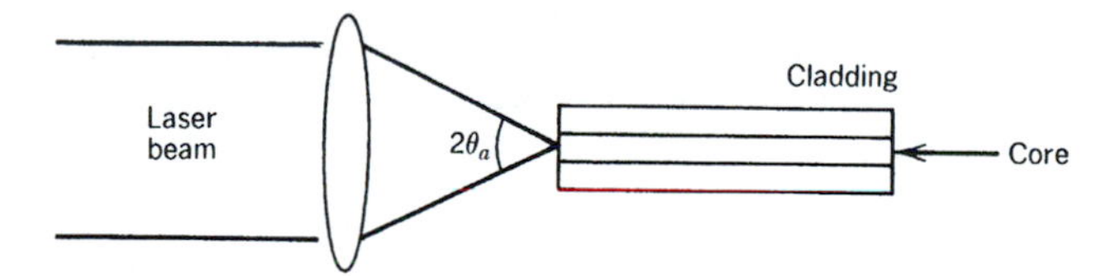

FIGURE 10.6 Generation of the correct angular spread of light for transmission.

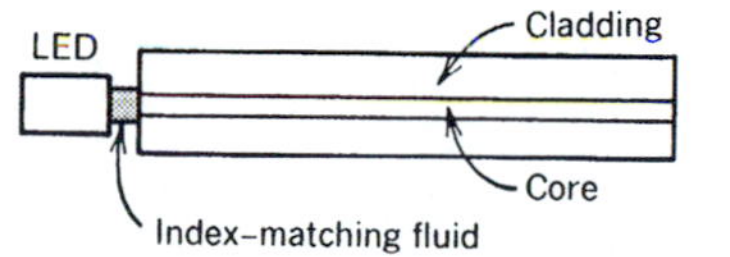

FIGURE 10.7 Light transmission through an optical fiber with a light-emitting diode (LED).

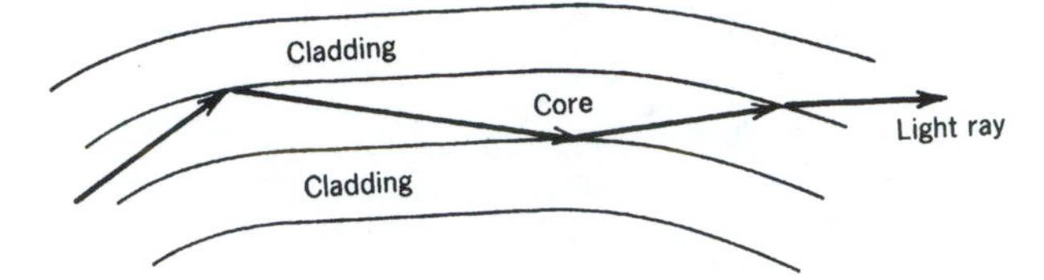

FIGURE 10.8 Attenuation caused by bending.

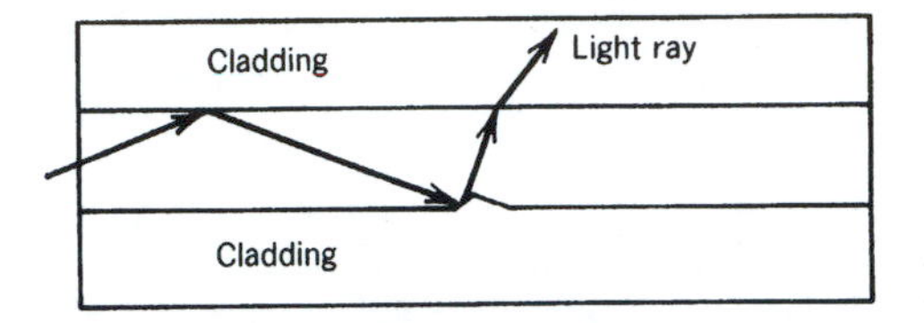

FIGURE 10.9 Attenuation caused by microbending.

before a bend or microbend occurs. A bend is the actual bending of the whole optical fiber, and a microbend is a slight kink at the boundary between the core and the cladding. Notice that a considerable number of light rays will leak when the bend in the fiber is sharp. In this context, a sharp bend is one with a radius of about 50 mm or less, which corresponds to a loss of a few decibels per kilometer.

Dispersion is the separation of a complex of electromagnetic waves into its various frequency components. Dispersion of light transmission in the optical fiber has two general causes. One, *material dispersion*, occurs because the velocity of light propagation within the core depends on the wavelength. This phenomenon is due primarily to dependence of the wavelength on the refractive index, $\eta(\lambda)$, of the material, a circumstance that is similar to the chromatic aberration of a lens.

In the second type of dispersion, known as *intermode dispersion*, the light rays launched within the acceptance angle of the core take two separate optical paths, straight and reflective, even though they are from a purely monochromatic source. Intermode dispersion is easily seen in the diagram in Figure 10.10. The light ray labeled 1 travels down the middle, whereas the ray labeled 2 is reflected up and down at the critical angle. It is apparent that ray 1 and ray 2 cover different distances along the core. The difference in the optical paths between the points where ray 2 is reflected can be calculated by

$$\Delta l = \left(1 - \frac{\eta_2}{\eta_1}\right) l, \tag{10.2}$$

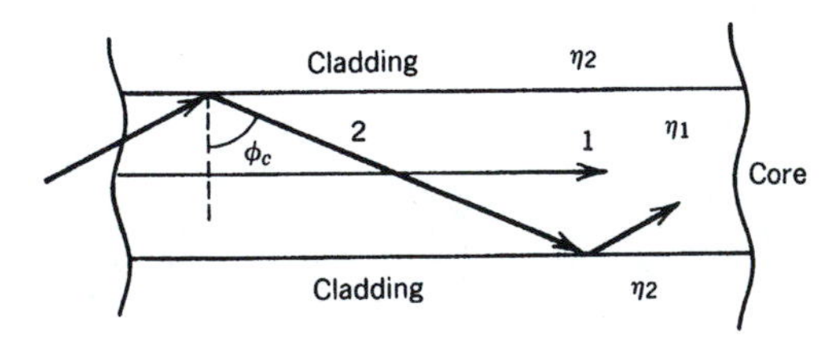

FIGURE 10.10 Dispersion when light transmission takes two different optical paths.

where l is the length between the reflections. Thus, the fractional difference would be

$$\frac{\Delta l}{l} \simeq \frac{\eta_1 - \eta_2}{\eta_1}. \tag{10.3}$$

Intermode dispersion causes a fractional difference in the periods of time it takes the two light rays to traverse the fiber, given by

$$\Delta T \simeq \frac{\eta_1 - \eta_2}{\eta_1} \frac{L}{c} \text{ sec},$$

where L is the length of the optical fiber and c is the velocity of light.

Example 10.1

Use Snell's law of refraction to evaluate the optical fiber of Figure 10.5.

(a) Calculate the numerical aperture of the fiber.
(b) Calculate the acceptance angle of the fiber.

Answers

(a) Using Snell's law, we have

$$\eta_0 \sin \theta_1 = \eta_1 \sin \theta_2,$$

where θ_1 and θ_2 are the incident and refraction angles, and η_0 and η_1 are the refractive indices of the air and the core, respectively. We also note that the incident angle at the core–cladding interface is

$$\phi = \frac{\pi}{2} - \theta_2,$$

where ϕ is greater than the critical angle of reflection at the core–cladding boundary.

Substituting in Snell's law, and using $\sin^2 \phi + \cos^2 \phi = 1$, we have

$$\eta_0 \sin \theta_1 = \eta_1 (1 - \sin^2 \phi)^{1/2}.$$

When ϕ equals ϕ_c, the critical angle of the core–cladding interface, then

$$\sin \phi_c = \frac{\eta_0}{\eta_1}$$

and

$$\eta_0 \sin \theta_a = (\eta_1^2 - \eta_2^2)^{1/2},$$

where θ_a denotes the acceptance angle of the fiber. Thus, the numerical aperture (NA) of the fiber is

$$\text{NA} = \eta_0 \sin \theta_a = (\eta_1^2 - \eta_2^2)^{1/2}.$$

(b) The angle of acceptance of the fiber can be written as

$$\theta_a = \sin^{-1}\left[\left(\frac{\eta_1^2 - \eta_2^2}{\eta_0}\right)^{1/2}\right],$$

which is identical to Eq. 10.1. ■

■ Example 10.2

The core and the cladding of a silica optical fiber have refractive indices of $\eta_1 = 1.5$ and $\eta_2 = 1.4$, respectively.

(a) Calculate the critical angle of reflection for the core–cladding boundary.
(b) Calculate the acceptance angle of the fiber.

Answers

(a) The core–cladding critical angle of reflection can be determined by

$$\phi_c = \sin^{-1}\left(\frac{\eta_2}{\eta_1}\right)$$
$$= \sin^{-1}\left(\frac{1.4}{1.5}\right) = 68.96°.$$

(b) By applying Eq. 10.1, we can find the acceptance angle of the fiber:

$$\theta_a = \sin^{-1}[(\eta_1^2 - \eta_2^2)^{1/2}]$$
$$= \sin^{-1}[(1.5^2 - 1.4^2)^{1/2}] = 32.58°.$$ ■

10.3 MODAL DESCRIPTION

A modal description of an optical fiber is a more exact way of treating the propagation of light rays within the core of the fiber. In such a description, which indicates the manner in which a process takes effect, the electromagnetic nature of light is accounted for, and its propagation along the fiber is expressed in terms of waveguide theories. For simplicity, we assume that the optical fiber behaves as a waveguide with perfect conducting walls. In this instance the electric field of the transverse electric mode of order 2 (TE_2) at the boundary is zero, as shown in Figure 10.11.

The electric field *across* the waveguide would consist of interference patterns between the incident and reflected beams. To satisfy the boundary condition (i.e., $E = 0$) of the waveguide, the diameter of the guide must intersect an integral number of half-wavelengths of the incident beam. This can be expressed mathematically as

$$\sin\theta = \frac{n\lambda}{2d}, \qquad n = 1, 2, \ldots, \tag{10.5}$$

where θ is the angle of the incident beam, d is the separation of the conducting walls, and n is the modal number of propagation. Thus, we see that only certain

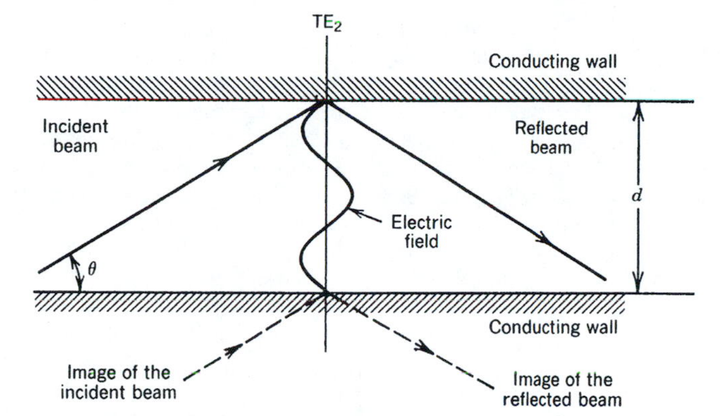

FIGURE 10.11 A conducting waveguide. *Note:* TE_2, transverse electric mode of order 2.

values of the angle of incidence are allowed; otherwise the boundary condition would be violated.

An exact solution of the modal description of a fiber optic should begin with a set of Maxwell's equations (Chapter 5) that are subject to the cylindrical boundary conditions of the fiber. This solution has been carried out in extensive detail in a number of texts (e.g., reference 1), but such a treatment is beyond the scope of this chapter. However, a similar solution for the conducting waveguide applies to the core–cladding (i.e., $\eta_1 > \eta_2$) boundary in an optical fiber, although the exact required boundary conditions are slightly different. The consequence is that only specific angular directions of the incident beams are allowed within the core. And each of the allowed beam directions corresponds to a different mode of wave propagation in the waveguide.

For another aspect of the modal description of optical fibers, we would examine qualitatively the appearance of the modal field in a planar dielectric-slab waveguide, as shown in Figure 10.12. This guide is composed of a dielectric core (or slab) of refractive index η_1 sandwiched between dielectric cladding of refractive

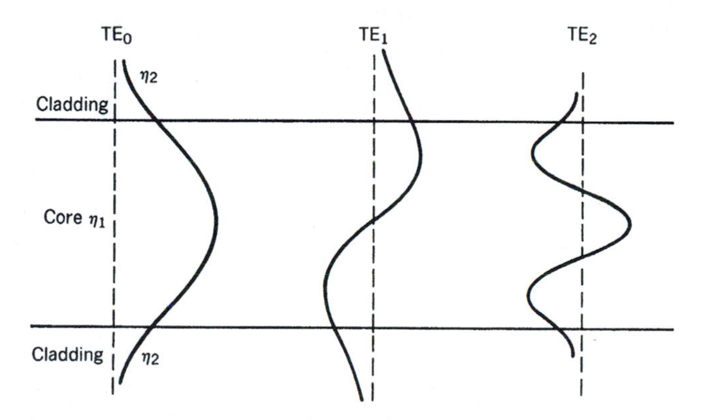

FIGURE 10.12 Electric field distribution of the lower-order guided modes in a slab waveguide. *Note:* $\eta_1 > \eta_2$.

index η_2, in which we assume $\eta_2 < \eta_1$. This is a simple example of an optical waveguide that will also serve to describe wave propagation in optical fibers. The figure shows a set of lower-order transverse electric (TE) guided modes, which were obtained from the solutions of Maxwell's equations. The order of mode is equal to the number of transverse field maxima across the guide, and it is also related to the angle of the incident beam, which may be approximated by Eq. 10.5. In other words, the steeper the angle of the incident beam, the higher the order of the guided mode. We note that the electric fields of the guided modes are not completely confined within the central core, but go beyond the core–cladding boundary and decay very rapidly in the cladding materials.

For lower-order guided modes, the electric fields are concentrated near the center of the core. For higher-order modes, however, the electric field distribution spreads toward the core–cladding interface and penetrates deeper into the cladding material.

Example 10.3

Consider an optical fiber designed to guide monochromatic light of $\lambda = 400$ nm. If the diameter of the core is about 10 μm, calculate the allowable angles of the incident beam.

Answers

The allowable angles can be calculated with the help of Eq. 10.5,

$$\theta = \sin^{-1}\left(\frac{n\lambda}{2d}\right)$$
$$= \sin^{-1}\left(\frac{0.4n}{2 \times 10}\right) = \sin^{-1}(0.02n),$$

where n is the order of the guided mode. Thus, we have

$$\theta = 1.15°, \quad n = 1$$
$$\theta = 2.29°, \quad n = 2$$
$$\theta = 3.94°, \quad n = 3$$
$$\theta = 4.58°, \quad n = 4, \ldots$$

■

Example 10.4

If the refractive indices of the core and cladding of the fiber described in Example 10.3 are provided, $\eta_1 = 1.5$ and $\eta_2 = 1.35$,

(a) Calculate the largest allowable angle of the incident beam.
(b) Calculate the actual acceptance angle of the fiber.
(c) Calculate the numerical aperture of the fiber.

Answers

(a) The critical angle of reflection at the core–cladding interface can be shown as

$$\phi_c = \sin^{-1}\left(\frac{\eta_2}{\eta_1}\right) = \sin^{-1}\left(\frac{1.35}{1.5}\right) = 64.1°.$$

Thus, the permitting angle of the incident beam would be

$$\theta < \frac{\pi}{2} - 64.1° = 25.9°.$$

According to Eq. 10.5, the allowable angles of the incident beam are determined as

$$\theta = 24.8°, \qquad n = 1$$

and

$$\theta = 26.1°, \qquad n = 22.$$

Thus, the largest allowable angle of the incident beam is $\theta = 24.8°$.

(b) The actual acceptance angle θ_1 of the fiber for $\theta = 24.8°$ is given by

$$\eta_0 \sin\theta_1 = \eta_1 \cos\left(\frac{\pi}{2} - \theta\right)$$

$$\theta_1 = \sin^{-1}[\eta_1 \cos(65.2°)] = \sin^{-1}[(1.5^2)(0.42)] = 39°.$$

Notice that the computed acceptance angle is, according to Eq. 10.1,

$$\theta_a = \sin^{-1}[(\eta_1^2 - \eta_2^2)^{1/2}] = \sin^{-1}[(1.5^2 - 1.35^2)^{1/2}] = 40.8°,$$

which is larger than the actual acceptance angle θ_1.

(c) The corresponding numerical aperture of the fiber would be

$$\mathrm{NA} = \eta_0 \sin\theta_1 = 0.63.$$

■

10.4 TYPES OF OPTICAL FIBERS

In general, optical fibers can be divided into two broad categories, *single mode* and *multimode.*

In single-mode fibers, only one mode is allowed to propagate in the fiber. The angle of the incident beam is determined by the transverse field in *one* wavelength across the fiber, and the incident angle at the core–cladding interface is large enough that it exceeds the critical angle of reflection. The expression for a single-mode fiber is, in terms of Eq. (10.5), for $n = 1$,

$$\sin\theta \simeq \theta = \frac{\lambda}{d}, \tag{10.6}$$

where d is diameter of the core.

Since the incident angle at the core–cladding boundary is larger than the critical angle ϕ_c, the angle of the incident beam is given by

$$\theta = \frac{\lambda}{d} > \frac{\pi}{2} - \phi_c. \tag{10.7}$$

In practice, this implies that the diameter of the core is on the order of 2 to 8 μm, which is very small.

The dispersion in a single-mode fiber is due primarily to absorption by the material of the core. Thus, a single-mode fiber is the most suitable for long-distance communication and is capable of handling very broad-band signals. For example, figures of 10 Gbits/sec over 10 km have been reported. Since the core of a single-mode fiber is very small, as can be seen from Eq. 10.7, launching light into the fiber is a difficult task, but keeping the insertion of light constant over a long period of time is even more difficult. Nevertheless, these basic problems have been solved, and single-mode fibers are expected to be in common use for high-capacity communication links in the near future.

In a *multimode fiber*, from two to thousands of modes can be propagated simultaneously. There are two principal types of multimode fibers, *step index* and *graded index*.

A *step-index* fiber is in fact the type of fiber that we have been discussing. In this type of fiber, the refractive index is constant within the core, but it changes abruptly across the fiber, as shown in Figure 10.13. Ray diagrams for the two different modes of propagation, by axial rays and reflected rays, are shown in the figure. The step-index fiber is highly dispersive; one percent in the different optical path lengths of rays in the two modes of propagation is common. This implies an order of 40-nsec difference in the time periods of propagation for axial and reflective rays in a 1-km length of fiber (see Example 10.6). Thus, transmission of a pulse

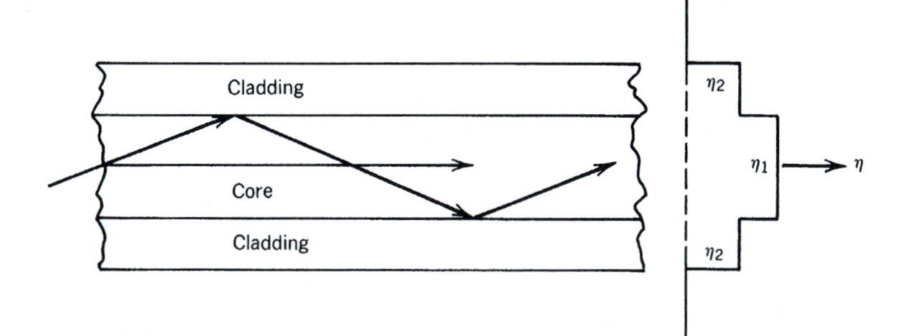

FIGURE 10.13 Step-index fiber.

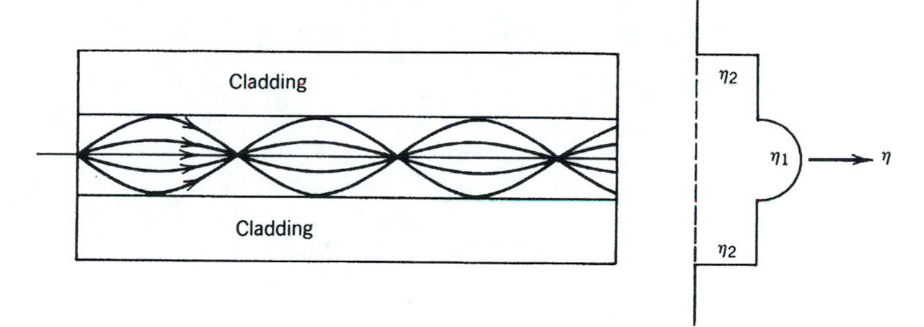

FIGURE 10.14 Graded-index fiber.

rate of about 20 Mbits/sec over a kilometer communication link made with step-index fiber is not feasible.

In a *graded-index fiber*, on the other hand, the profile of its refractive index, $\eta(r)$, varies radially. The refractive index varies from a maximum value of η_1 at the center of the core to a constant value of η_2 at and beyond the core's radius in the cladding region, as shown in Figure 10.14. The profile of the fiber's refractive index can generally be described by the following expression (from reference 2)

$$\eta(r) = \begin{cases} \eta_1\left(1 - 2\Delta \dfrac{r}{a}\right)^{1/2}, & r < a, \\ \eta_2 = \eta_1(1 - 2\Delta)^{1/2}, & r \geq a, \end{cases} \tag{10.8}$$

where a is the radius of the core, r is the radial distance, and

$$\Delta \simeq \frac{\eta_1 - \eta_2}{\eta_1} \tag{10.9}$$

is known as the relative refractive index difference, meaning the difference in the refractive indices of the core and cladding. The profile parameter for refractive indices is α and is illustrated in Figure 10.15. The refractive index profile varies

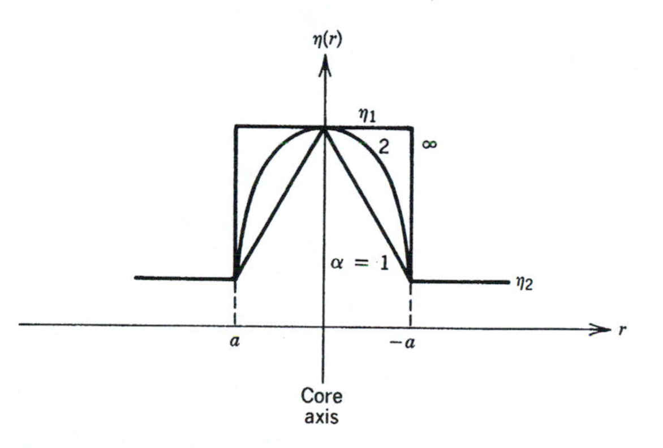

FIGURE 10.15 Possible profiles of the refractive indices of a graded fiber.

with the profile parameter α; that is, when $\alpha = \infty$, it represents a step profile; $\alpha = 2$ describes a parabolic profile; and $\alpha = 1$ illustrates a triangular profile.

The ray paths in the graded core are curved. Therefore, since the inner rays travel through a higher index, the difference in the time periods of propagation for the outer rays and the inner rays can be zero. The time difference, that is, the time dispersion ΔT, can be reduced to a value on the order of one nanosecond per kilometer for a core diameter of about 50 μm. Since the core is fairly straight, it is quite permissible to use a butt joint coupling with a semiconductor light source, for example, a light-emitting diode (LED) (see Figure 10.7).

Example 10.5

The refractive indices of a single-mode fiber core and its cladding are $\eta_1 = 1.5$ and $\eta_2 = 1.47$, and the wavelength of the launching light is $\lambda = 600$ nm. Calculate the core diameter required.

Answers

From Eq. 10.7, the diameter of the core can be written as

$$d < \frac{\lambda}{0.5\pi - \phi_c},$$

where ϕ_c is the critical angle of reflection, which can be calculated by

$$\phi_c = \sin^{-1}\left(\frac{\eta_1}{\eta_2}\right) = \sin^{-1}\left(\frac{1.47}{1.5}\right)$$
$$= 78.5° = 1.37 \text{ radians.}$$

The diameter of the core is therefore

$$d < \frac{600 \times 10^{-3}}{0.5\pi - 1.37} = 3\ \mu\text{m}.$$

Thus, the core diameter should be smaller than 3 μm. ■

Example 10.6

Given that the refractive index of the step-index fiber's core is $\eta_1 = 1.5$, and that the critical angle of reflection at the core–cladding interface is 80°, calculate the difference in the time periods of propagation for the two kinds of rays, the axial rays and those reflected at the critical angle, for a 1-km-long fiber.

Answers

From the rectangular profile step-index shown in Figure 10.13, the difference in the optical paths of the two kinds of rays can be calculated by

$$\frac{1}{\cos(20°)} - 1 = 1.064 - 1 = 0.064 \text{ km.}$$

Thus, the difference in time periods is

$$\frac{0.064}{v} = \frac{\eta_1}{c}(0.064)$$

$$= \frac{(1.5)(0.064)}{3 \times 10^5}$$

$$= 0.032 \times 15^5 \text{ sec} = 320 \text{ nsec}$$

■

10.5 ELEMENTS FOR FIBER OPTIC COMMUNICATION AND DISTRIBUTION SYSTEMS

The function of optical fiber in channeling light from one end of a fiber to the other is analogous to the function of a metallic cable channeling electromagnetic waves. Just as these electromagnetic waves are modulated at the launching end, they will be detected as signals at the receiving end. Thus, myriad distribution elements are required to fashion a system capable of maintaining high-capacity multidirectional communication links over long distances.

Figure 10.16 is a schematic of some of the basic elements needed for a typical system. At one end are various light sources, such as LEDs and semiconductor lasers, which are modulated, mostly electronically, by signals and messages from some electronic circuit. These light sources are also modulated by other means, such as with acoustooptical switches and by controlling the injection current of the semiconductor lasers (see Chapter 7). At the other end, the light signals are detected by various kinds of detectors (see Chapter 4) according to a direct heterodyne scheme.

In between these two endpoints are various switches, multiplexers, repeaters, and directional couplers for distributing various signals and handling tasks. In

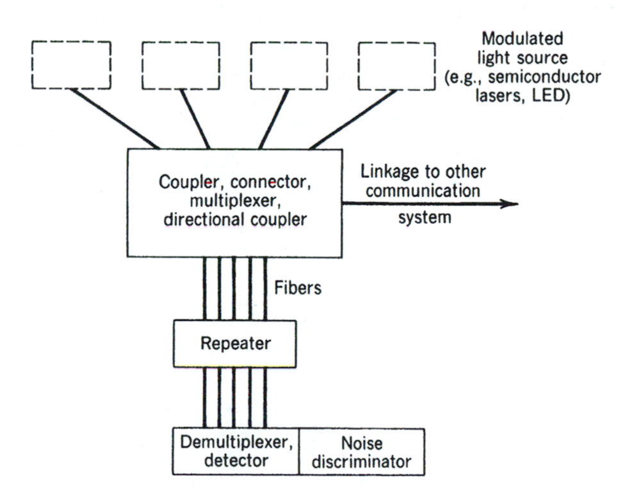

FIGURE 10.16 Schematic diagram of a fiber optic communication system.

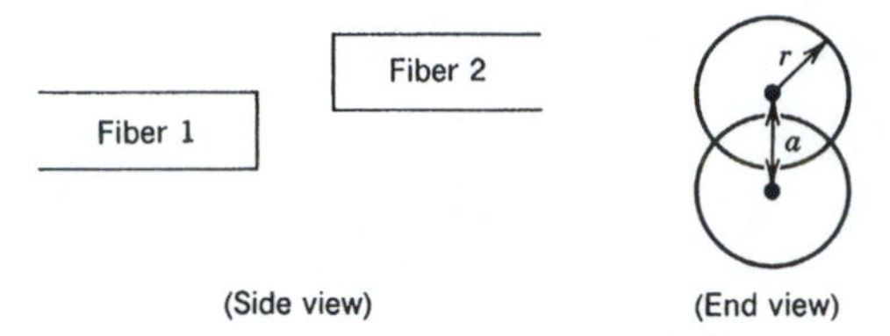

FIGURE 10.17 Lateral misalignment in a fiber connection.

traditional electronic communication system, these elements are quite well established, with internationally adapted standards for both digital and analog systems. With photonic communication system becoming a reality, the optical counterparts of these elements have also become commonplace.

This section is therefore devoted to a discussion of some of the basic optical elements involved in this vast distribution network in which light waves are the signal carrier.

10.5.1 Connectors and Couplers

Just as electronic networks are necessary, so in fiber optic distribution systems do we need fiber connections for extending distances, for synthesizing operations, and for repair. The principal problem with the simple procedure of joining fibers is loss through several mechanisms such as misalignment (see Figure 10.17). Depending on the distribution of the optical power within the fiber, this incomplete overlap is clearly a source of loss. The *air gap* between the two fibers creates two fiber–air boundaries, which cause reflection loss. The ends of the fibers need to be properly fused with an index-matching substance, one having the same index of refraction as that of the material of the fibers. Two other common ways in which fibers do not connect are by angular misalignment—the ends that should meet are positioned angularly to each other (Figure 10.18*a*); and by having surface ends that are jagged and not flat (Figure 10.18*b*).

Couplers are used to direct light from the source to the fiber, and they can be very inefficient for several reasons. Light coming from typical sources can be

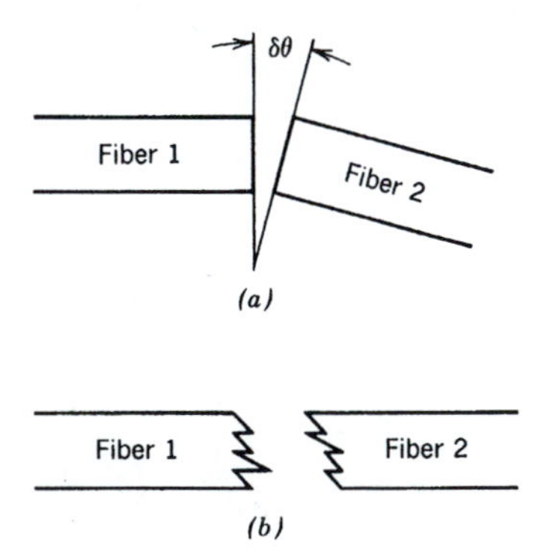

FIGURE 10.18 (*a*) Angular misalignment, (*b*) nonflat surfaces. *Note:* $\delta\theta$, displacement angle.

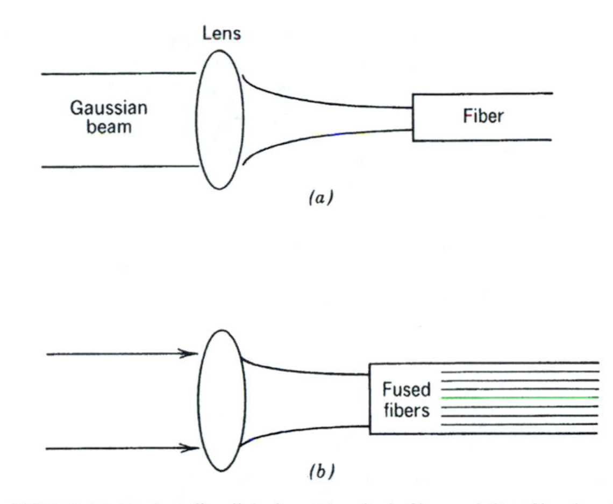

FIGURE 10.19 Coupling light into (*a*) a single fiber and (*b*) a fiber bundle.

coupled into optical fiber in a variety of ways, such as coupling light into one (see Figure 10.19*a*) or several fibers (see Figure 10.19*b*) with a focusing lens. With the fiber end placed at the focal plane, there will be reflection loss at the entry plane unless the end surface of the fiber has a proper antireflection coating. In addition, there can be loss through model mismatch if the input mode of the laser (Gaussian beam, non-Gaussian beam, etc.) is not the mode that can be launched into the fiber.

10.5.2 Switches, Repeaters, and Multiplexers

Optical switches, which are needed to perform on–off, directional gating, and trunking tasks, are made up of a variety of electronic, electrooptical, and all-optical materials (see Figures 10.20 and 10.21). Their operational principles, speeds, limitations, and so on, depend on the kind of material used. One of the most distinguishing

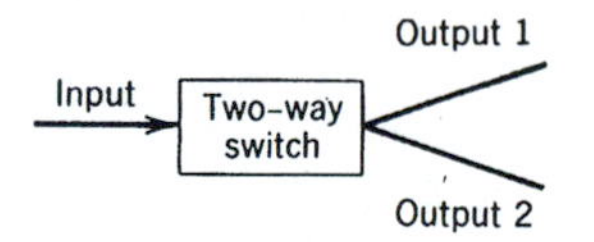

FIGURE 10.20 A two-way fiber switch.

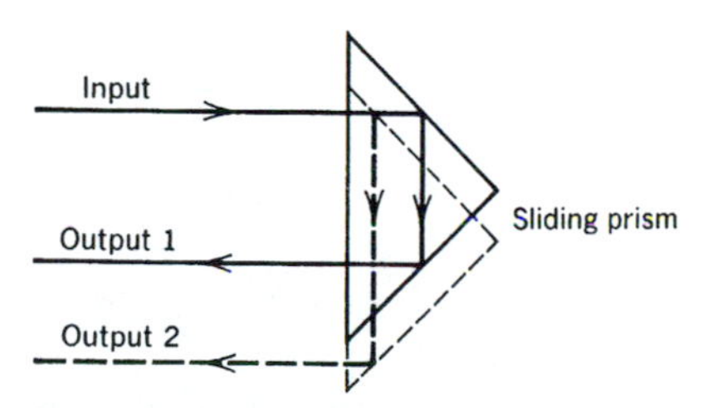

FIGURE 10.21 An example of a two-way fiber optic switch.

and useful characteristics of optical communication, however, is the noninterference between two light waves of different wavelengths as they propagate along the fiber. This means that several channels of information, with their different carrier wavelengths (frequencies), can be included in a single fiber. This is called *wavelength division multiplexing.* Since optical frequencies are in the 10^{14} range—radio and microwave frequencies are a few orders of magnitude smaller—optical communication may provide broader bandwidths and considerable capacity for handling information. A *wavelength multiplexer* channels light of different frequencies into a fiber (see Figure 10.22*a*) by using such dispersive elements as prisms (Figure 10.22*b*) and gratings (Figure 10.22*c*). A *wavelength demultiplexer* works in exactly the reverse fashion, directing light of various wavelengths from a fiber into their respective channels.

At some point in the communication channel, losses caused by the various couplings, by switching and wavelength multiplexing, and by absorption in the fiber will be so severe that the optical signals will require a repeater. The purpose of the repeater is to amplify the signal (but not the background noise) and send it downstream. A schematic of the elements necessary for an electronic–photonic repeater is shown in Figure 10.23. Moreover, other methods for amplifying the light signals propagating along the fibers are currently being researched.

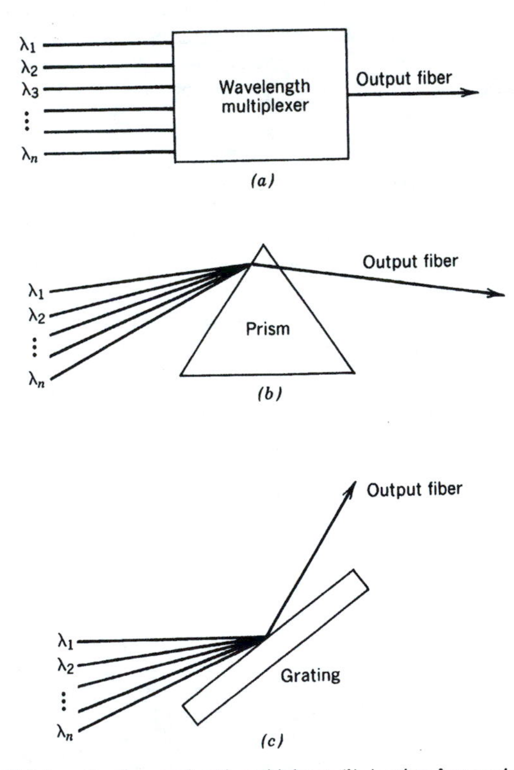

FIGURE 10.22 (*a*) Schematic of a wavelength multiplexer. (*b*) A prism for wavelength multiplexing. (*c*) A grating for wavelength multiplexing.

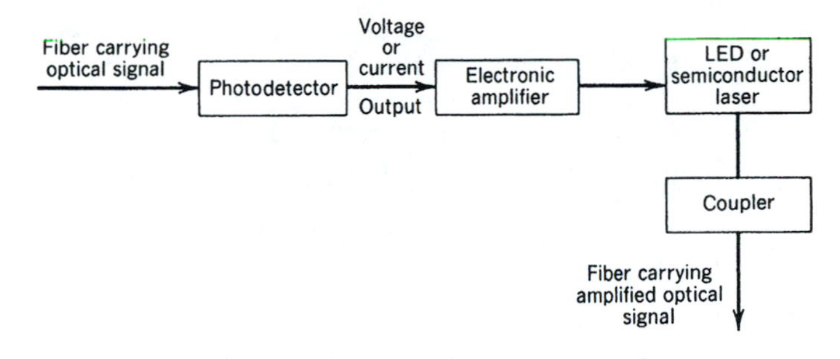

FIGURE 10.23 Elements making up a repeater station.

Fiber optic communication is a rapidly developing field, with an ever-increasing assortment of physical and optical elements reported daily. Aspects of design and system considerations, standards, and linkages with other communication systems—microwave, satellite, conventional electronics—are undergoing scrutiny and improvements. Unlike the basic principles of refraction, reflection, diffraction, and lasers and the electromagnetic nature of light, which were the subjects of earlier chapters, and all of which will remain almost invariant for a long time to come, many of the facets of optical communication technology will undoubtedly be modified and revised. This chapter was intended as an introduction to some of the basics of fiber optics, but the interested reader will find additional and more up-to-date information in the literature.

REFERENCES

1. G. Keiser, *Optical Fiber Communications*, McGraw-Hill, New York, 1983.
2. J. Senoir, *Optical Fiber Communications: Principles and Practice*, Prentice-Hall, Englewood Cliffs, N.J., 1985.
3. J. C. Palais, *Optical Fiber Communications*, second edition, Prentice-Hall, Englewood Cliffs, N.J., 1988.
4. G. Galliano and F. Tosco, *Optical Fiber Cables in Optical Fiber Communications*, McGraw-Hill, New York, 1980.
5. D. L. Lee, *Electromagnetic Principles of Integrated Optics*, Wiley, New York, 1986.

PROBLEMS

10.1 Assume that a glass fiber has a refractive index of $\eta_1 = 1.48$, and that it is surrounded by air.

(a) Calculate the critical angle at the core–air boundary.

(b) Calculate the acceptance angle of the fiber.

(c) Find the numerical aperture of the fiber.

10.2 Assume that an optical fiber with a core diameter large enough to be considered by ray theory analysis has a core refractive index of 1.46 and a cladding refractive index of 1.43.

(a) Determine the critical angle of reflection at the core–cladding boundary.

(b) Calculate the acceptance angle if the fiber is submerged in water whose refractive index is 1.33.

(c) Determine the numerical aperture of the fiber in water.

10.3 If the refractive indices of the core and cladding of an optical fiber are $\eta_1 = 1.46$ and $\eta_2 = 1.45$, respectively,

(a) Determine the acceptance angle of the fiber.

(b) Calculate the diameter of the core.

10.4 To launch a parallel laser beam into the optical fiber of Problem 10.3 requires a condenser lens as shown in Figure 10.6. Determine the required diameter and focal length of the condenser.

10.5 The typical difference in the refractive indices of the core and cladding of an optical fiber for long-distance transmissions is about 1 percent. If the refractive index of the core is 1.47,

(a) Estimate the numerical aperture of the fiber.

(b) Calculate the critical angle of reflection at the core–cladding interface.

10.6 The ratio of the cladding and core refractive indices of a 2-km length of optical fiber is $\eta_2/\eta_1 = 0.98$. Calculate the dispersion of the fiber.

10.7 Estimate the diameter of the core of a single-mode optical fiber that is suitable for transmitting a monochromatic light beam of $\lambda = 650$ nm. Assume that the refractive index of the core is $\eta_1 = 1.48$ and that of the cladding $\eta_2 = 1.46$.

10.8 If the cladding refractive index in Problem 10.7 is reduced to $\eta_2 = 1.42$, estimate the new core diameter.

10.9 The relative refractive index difference for a fiber is defined as

$$\Delta = \frac{\eta_1^2 - \eta_2^2}{2\eta_1^2}.$$

(a) Show that Δ can be approximated by

$$\Delta \simeq \frac{\eta_1 - \eta_2}{\eta_1}$$

for $\Delta \ll 1$.

(b) Show that the numerical aperture of the fiber can be approximated by

$$\text{NA} \simeq \eta_1(2\Delta)^{1/2}.$$

10.10 A step-index multimode fiber with a core diameter of 100 μm and a relative refractive index difference of 1 percent is used to transmit a light wave of 0.80 μm. Assume that the refractive index of the core is 1.46.

(a) Estimate the normalized frequency of the fiber. *Hint:* The normalized frequency can be defined as (references 1, 2)

$$J = \frac{2\pi}{\lambda} a(\eta_1^2 - \eta_2^2)^{1/2},$$

where a is the radius of the fiber's core.

(b) Estimate the number of guided modes.

10.11 The core of a graded-index fiber has a diameter of 80 μm and a refractive index with a parabolic profile. If the numerical aperture of the fiber is 0.2, and the fiber is operating at a wavelength of $\lambda = 0.7$ μm,

(a) Estimate the normalized frequency of the fiber.

(b) Estimate the number of guided modes. *Hint:* The number of guided modes for a graded fiber can be shown as (references 1, 2)

$$M_g = \frac{\alpha}{\alpha + 2} (\eta_1 ka)^2 \Delta \simeq \frac{\alpha}{\alpha + 2} \frac{J^2}{2},$$

where $k = 2\pi/\lambda$, α is the profile parameter of the reflective index, Δ is the relative refractive index difference, and J is the normalized frequency of the filter.

10.12 A graded-index fiber whose core has a refractive index with a parabolic profile can be used for single-mode transmission, although there is no apparent advantage to this as there is for a step-index fiber. The refractive index of the core is $\eta_1 = 1.5$, and the relative refractive index difference Δ is 0.01 percent. Assuming that the normalized frequency for single-mode transmission is 3.4, and that the fiber is operating at a wavelength of $\lambda = 0.6$ μm, estimate the maximum diameter of the core for single-mode transmission.

10.13 A step-index multimode fiber has a relative refractive index difference of 1.6 percent and a wavelength of $\lambda = 600$ nm. If the number of modes of propagation in the fiber is 1000, calculate the diameter of the core.

10.14 The core of a step-index single-mode fiber has a diameter of 5 μm, a refractive index of $\eta_1 = 1.5$, and the cladding has a refractive index of $\eta_2 = 1.45$.

(a) Calculate the relative refractive index difference of the fiber.

(b) What is the shortest wavelength allowed in the single-mode transmission?

10.15 If we double the diameter of the core of the fiber in Problem 10.14 but keep its refractive index the same and maintain the same single-mode transmission, calculate what the refractive index of the cladding must be.

10.16 A graded-index multimode optical fiber has a core with a refractive index of $\eta_1 = 1.52$ and an acceptance angle in air of 7.5°. Determine the required refractive index of the cladding.

10.17 If a graded-index fiber has a core with a diameter of 60 μm and the axis has a refractive index of 1.5 with a profile parameter α of 1.8, calculate the number of guided modes of the fiber when it is operating at a wavelength of $\lambda = 0.7$ μm.

10.18 Using Figure 10.17, calculate the overlap area between two fibers with radius r, if the axis is displaced a distance a.

10.19 In Figure 10.19a the end surface of the fiber is in the focal plane of the beam. If the beam waist ω_0 in this focal plane is 5 μm, what is the focal length of the lens that is located 1 cm from the fiber? What is the size of the beam at the lens?

INDEX